纳 米 光 子 学 丛 书

Nanoscale Photonics and Spectroscopy II
纳米光子学与光谱

（第2版）

孙萌涛　程宇清　崔　琳　著
Sun Mengtao　Cheng Yuqing　Cui Lin

清华大学出版社
北京

内容简介

本书系统介绍了纳米尺度下光子学和光谱学的基本物理原理,以及它们的相互作用机制。还介绍了金属表面等离激元的激发、探测和对分子光谱的增强机制,其中包括表面等离基元增强的吸收、荧光、拉曼、光学成像等。此外,还揭示了表面等离基元和激子耦合相互作用的物理机制,进一步阐释了这种相互作用在表面催化反应中的应用。

本书结构完整,物理原理阐述清晰,列举的应用例子是最新的科学前沿研究。本书适合作为研究生和高年级本科生的光学教材,也可作为相关领域科研工作者的基础参考书。

版权所有,侵权必究。举报:010-62782989,beiqinquan@tup.tsinghua.edu.cn。

图书在版编目(CIP)数据

纳米光子学与光谱=Nanoscale Photonics and Spectroscopy.Ⅱ:英文/孙萌涛,程宇清,崔琳著.—2版.—北京:清华大学出版社,2023.11
(纳米光子学丛书)
ISBN 978-7-302-64762-1

Ⅰ.①纳… Ⅱ.①孙…②程…③崔… Ⅲ.①纳米技术-光子-英文 Ⅳ.①TB383②O572.31

中国国家版本馆 CIP 数据核字(2023)第 193945 号

责任编辑:	鲁永芳
封面设计:	常雪影
责任校对:	王淑云
责任印制:	刘海龙

出版发行:清华大学出版社
网　　址:http://www.tup.com.cn,http://www.wqbook.com
地　　址:北京清华大学学研大厦 A 座　　邮　编:100084
社 总 机:010-83470000　　邮　购:010-62786544
投稿与读者服务:010-62776969,c-service@tup.tsinghua.edu.cn
质量反馈:010-62772015,zhiliang@tup.tsinghua.edu.cn
印 装 者:涿州汇美亿浓印刷有限公司
经　销:全国新华书店
开　本:170mm×240mm　　印　张:29.25　　字　数:489 千字
版　次:2019 年 10 月第 1 版　2023 年 11 月第 2 版　印　次:2023 年 11 月第 1 次印刷
定　价:129.00 元

产品编号:103032-01

《纳米光子学丛书》编委会

主编：孙萌涛　北京科技大学

编委：梁文杰　中国科学院物理研究所
　　　陈佳宁　中国科学院物理研究所
　　　杨志林　厦门大学
　　　肖湘衡　武汉大学
　　　徐　平　哈尔滨工业大学
　　　刘立伟　中国科学院苏州纳米技术研究所
　　　石　英　吉林大学
　　　孙树清　清华大学深圳研究生院
　　　方蔚瑞　大连理工大学
　　　黄映洲　重庆大学
　　　张正龙　陕西师范大学
　　　董　军　西安邮电大学
　　　李　敬　中国科学院理化技术研究所

CONTENTS

CHAPTER 1 Introduction ·· 1

    1.1 Concept of Spectroscopy ··· 1

    1.2 Concept of Photonics and Plasmonics ·· 4

    1.3 Concept of Plasmon-Enhanced Spectroscopy ······································· 10

        1.3.1 Plasmon-enhanced fluorescence ··· 10

        1.3.2 Plasmon-enhanced Resonance fluorescence energy transfer ·· 12

        1.3.3 Surface-enhanced Raman scattering ·· 13

        1.3.4 The remote-excitation of SERS ··· 16

        1.3.5 Tip-enhanced Raman scattering spectroscopy ························ 18

        1.3.6 Remote excitation-TERS microscopy ·· 18

        1.3.7 Plasmon-enhanced coherence anti-Stokes Raman scattering images ··· 21

    References ··· 23

CHAPTER 2 Molecular Spectroscopy ·· 25

    2.1 Jablonski Diagram ·· 25

    2.2 Electronic State Transition ·· 27

        2.2.1 Ultraviolet-visible-near IR absorption spectroscopy ············· 27

        2.2.2 Two-photon absorption spectroscopy ······································· 28

        2.2.3　Fluorescence spectroscopy ………………………… 31
        2.2.4　Fluorescence resonance energy transfer ……………… 36
    2.3　Vibration spectroscopy …………………………………………… 39
        2.3.1　Raman spectroscopy ………………………………… 40
        2.3.2　Infrared spectroscopy ……………………………… 42
        2.3.3　Modes of molecular vibration ……………………… 45
        2.3.4　The difference between Raman and spectra ………… 46
    2.4　Rotational State …………………………………………………… 47
    2.5　Electronic and Vibrational Spectroscopy by Circularly
         Polarized Light …………………………………………………… 49
        2.5.1　Electronic circular dichroism ……………………… 49
        2.5.2　Raman optical activity ……………………………… 51
    References …………………………………………………………………… 52

CHAPTER 3　Photonics and Plasmonics ……………………………………… 57

    3.1　Introduction ………………………………………………………… 57
    3.2　Exciton ……………………………………………………………… 57
        3.2.1　Brief introduction of excitons ……………………… 57
        3.2.2　Exciton classification ……………………………… 59
    3.3　Polariton ……………………………………………………………… 69
        3.3.1　Brief introduction of polariton …………………… 69
        3.3.2　Polariton types ……………………………………… 69
    3.4　Plasmon and surface plasmons …………………………………… 76
        3.4.1　Plasmons ……………………………………………… 76
        3.4.2　Surface plasmons …………………………………… 78
        3.4.3　Surface plasmon polaritons ………………………… 79
    3.5　Plasmon-Exciton Coupling: Plexciton ………………………… 81
    References …………………………………………………………………… 81

CHAPTER 4 2D Borophene excitons ········ 85

4.1 Introduction ········ 85
4.2 Monolayer borophene ········ 89
    4.2.1 Monolayer borophene on Ag(111) ········ 89
    4.2.2 Monolayer borophene on Al(111) ········ 94
    4.2.3 Monolayer borophene on Ir(111) ········ 95
    4.2.4 Monolayer borophene on Au(111) ········ 97
    4.2.5 Monolayer borophene on Cu(111) ········ 99
4.3 Bilayer borophene ········ 100
    4.3.1 Bilayer borophene on Ag(111) ········ 100
    4.3.2 Bilayer borophene synthesis on Cu(111) ········ 111
4.4 Borophene heterostructure ········ 113
    4.4.1 Borophene-PTCDA lateral heterostructure ········ 113
    4.4.2 Borophene-Black phosphorus heterostructure ········ 116
    4.4.3 2D/1D borophene-graphene nanoribbons heterostructure ········ 120
    4.4.4 Borophene-graphene heterostructure ········ 123
References ········ 125

CHAPTER 5 Surface Plasmons ········ 130

5.1 Brief Introduction of SPs ········ 130
5.2 Physical Mechanism of SPs ········ 132
    5.2.1 Drude model ········ 132
    5.2.2 Relationship between Refractive Index and Dielectric Constant ········ 133
    5.2.3 Dispersion relations ········ 135
5.3 Localized SPs ········ 138
    5.3.1 LSPs in metallic nanosphere ········ 138
    5.3.2 LSPs in coupled metallic NPs: parallel-polarized

      excitation ··· 141

   5.3.3 LSPs in coupled metallic NPs: vertical-polarized

      excitation ··· 155

   5.3.4 Plexciton model: coupling between plasmon and

      exciton ·· 170

   5.3.5 Fano Resonant Propagating Plexcitons and

      Rabi-splitting Local Plexcitons ·· 190

   5.3.6 Plexciton revealed in experiment ······································ 209

   5.3.7 LSPs in coupled metallic NPs: many-body ························· 219

 5.4 Plasmonic Waveguide ·· 234

   5.4.1 The EM theory for calculating nanowires ························· 234

   5.4.2 The decay rate in the plasmon mode ······························· 235

   5.4.3 The spontaneous emission near the nanotip ····················· 236

   5.4.4 SPP modes of Ag NW by One-End Excitation ····················· 237

   5.4.5 Optical non-reciprocity with multiple modes

      based on a hybrid metallic NW ······································· 238

   5.4.6 Strongly enhanced propagation and non-reciprocal

      properties of CdSe NW ·· 247

 5.5 Unified treatments for LSPs and PSPs ·· 259

 5.6 Plexciton in TERS and in PSPs ·· 272

 References ··· 283

CHAPTER 6 Plasmon-Enhanced Fluorescence Spectroscopy ················ 289

 6.1 The principle of plasmon-enhanced fluorescence ······················· 289

 6.2 Plasmon-Enhanced Upconversion Luminescence ······················· 292

   6.2.1 Brief introduction ·· 292

   6.2.2 Physical principle and mechanism ··································· 294

 6.3 Principle of Plasmon-Enhanced FRET ·· 303

 References ··· 307

CHAPTER 7 Plasmon-Enhanced Raman Scattering Spectra 310

- 7.1 Surface-Enhanced Raman Scattering Spectroscopy 310
  - 7.1.1 Brief history of SERS spectroscopy 310
  - 7.1.2 Physical mechanism of SERS spectroscopy 313
- 7.2 Tip-Enhanced Raman Scattering Spectroscopy 317
  - 7.2.1 Brief introduction of TERS spectroscopy 317
  - 7.2.2 Physical mechanism of TERS spectroscopy 318
  - 7.2.3 Setup of TERS 320
- 7.3 Remote-Excitation SERS 324
- References 327

CHAPTER 8 High-Vacuum Tip-Enhanced Raman Scattering Spectroscopy 329

- 8.1 Brief Introduction 329
  - 8.1.1 Brief description of setup of HV-TERS 331
  - 8.1.2 Detailed description of setup of HV-TERS 333
- 8.2 The Application of HV-TERS Spectroscopy in *in situ* Plasmon-Driven Chemical Reactions 337
- 8.3 Plasmonic Gradient Effect 342
- 8.4 Plasmonic Nanoscissors 346
- References 354

CHAPTER 9 Physical Mechanism of Plasmon-Exciton Coupling Interaction ... 358

- 9.1 Brief Introduction of Plexcitons 358
- 9.2 Plasmon-Exciton Coupling Interaction 360
  - 9.2.1 Strong plasmon-exciton coupling interaction 360
  - 9.2.2 Application of strong plasmon-exciton coupling interaction 364
  - 9.2.3 Weak plasmon-exciton coupling interaction 371
  - 9.2.4 Application of weak plasmon-exciton coupling interaction 374

9.2.5 Plexcitons ········ 377
9.3 Application ········ 377
 9.3.1 Plasmonic electrons-enhanced resonance Raman scattering and electrons-enhanced fluorescence spectra ········ 377
 9.3.2 Tip-enhanced photoluminescence spectroscopy ········ 383
 9.3.3 Femtosecond pump-probe transient absorption spectroscopy ········ 390
References ········ 398

CHAPTER 10 Plasmon-Exciton-Co-Driven Surface Catalysis Reactions ········ 404

10.1 Plasmon-Exciton-Co-Driven Surface Oxidation Catalysis Reactions ········ 404
10.2 Plasmon-Exciton-Co-Driven Surface Reduction Catalysis Reactions ········ 412
10.3 Unified Treatment for Plasmon-Exciton-Co-Driven Oxidation and Reduction Reactions ········ 420
References ········ 424

CHAPTER 11 Nonlinear Optical Microscopies of CARS, TPEF, SHG, SFG and SRS ········ 428

11.1 Principles of Nonlinear Optical Microscopies ········ 429
11.2 Applications of Nonlinear Optical Microscopies ········ 434
 11.2.1 Optical characterizations of 2D materials ········ 434
 11.2.2 Highly efficient photocatalysis of g-C_3N_4 ········ 437
 11.2.3 Optical characterizations of 3D materials ········ 440
 11.2.4 Advances of biophotonics ········ 446
 11.2.5 MSPR-enhanced nonlinear optical microscopy ········ 447
References ········ 451

致谢 ········ 456

CHAPTER 1

Introduction

All the color figures please scan the QR code.

1.1 Concept of Spectroscopy

Spectroscopy is the study of the interaction between matter and electromagnetic radiation(Spectrum + copy). In 1672, Isaac Newton described an experiment in which he permitted sunlight to pass through a small hole and then through a prism(Fig. 1-1). Newton found that sunlight, which looks white to us, is actually made up of a mixture of all the colors of the rainbow.

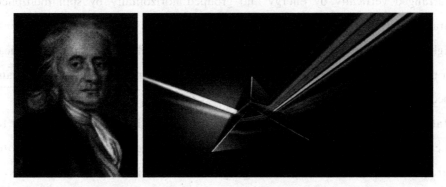

Fig. 1-1 Isaac Newton and his experiment
(For colored figure please scan the QR code on page 1)

In 1802, William Hyde Wollaston built an improved spectrometer that included a lens to focus the sun's spectrum on a screen(Fig. 1-2). Upon use, Wollaston realized that the colors were not spread uniformly, but instead had

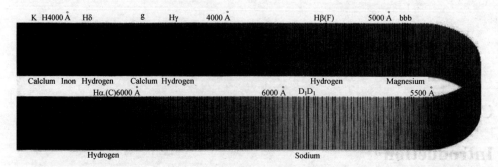

Fig. 1-2　Lines in the solar spectrum. The range illustrated is from 3900 Å to 6900 Å
(For colored figure please scan the QR code on page 1)

missing patches of colors, which appeared as dark bands in the spectrum. In 1815, Joseph Fraunhofer also examined the solar spectrum, and found about 600 such dark lines(missing colors), which are now known as Fraunhofer lines, or absorption lines.

Spectroscopy is the study of the interaction between matter and electromagnetic radiation. The entire electromagnetic spectrum is shown in Fig. 1-3.

In molecular spectroscopy, a Jablonski diagram(Fig. 1-4) is used to illustrate the electronic states of a molecule and the transitions between them. The states are arranged vertically by energy and grouped horizontally by spin multiplicity. Nonradiative transitions are indicated by squiggly arrows and radiative transitions by straight arrows. The vibrational ground states of each electronic state are indicated with thick lines, the higher vibrational states with thinner lines[1]. The diagram is named after Aleksander Jabłoński[2].

Radiative transitions involve the absorption, if the transition occurs to a higher energy level, or the emission, in the reverse case, of a photon. Nonradiative transitions arise through several different mechanisms, all differently labeled in the diagram. Relaxation of the excited state to its lowest vibrational level is called vibrational relaxation. This process involves the dissipation of energy from the molecule to its surroundings, and thus it cannot occur for isolated molecules. A second type of nonradiative transition is internal conversion(IC), which occurs when a vibrational state of an electronically

CHAPTER 1　Introduction

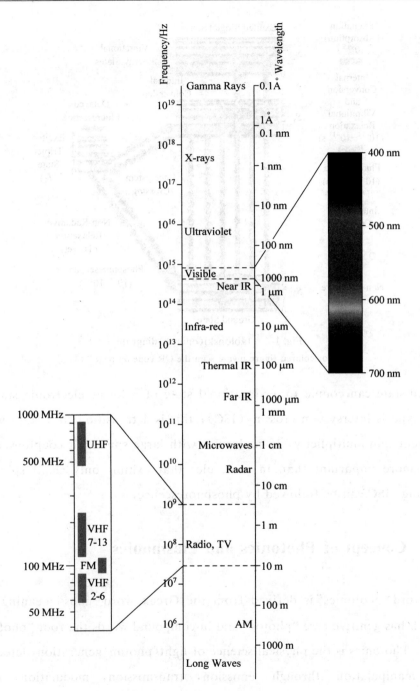

Fig. 1-3　The entire electromagnetic spectrum
(For colored figure please scan the QR code on page 1)

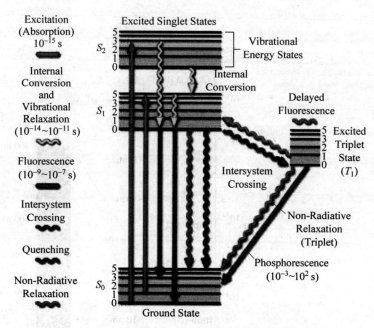

Fig. 1-4 Jablonski(energy)diagram
(For colored figure please scan the QR code on page 1)

excited state can couple to a vibrational state of a lower electronic state. A third type is intersystem crossing(ISC); this is a transition to a state with a different spin multiplicity. In molecules with large spin-orbit coupling, ISC is much more important than in molecules that exhibit only small spin-orbit coupling. ISC can be followed by phosphorescence.

1.2 Concept of Photonics and Plasmonics

The word "photonics" is derived from the Greek word "phos" meaning light (which has genitive case "photos" and in compound words the root "photo-" is used). Photonics is the physical science of light(photon) generation, detection, and manipulation through emission, transmission, modulation, signal processing, switching, amplification, and detection/sensing. Photonics is closely related to optics. Classical optics long preceded the discovery that light is quantized, when Albert Einstein famously explained the photoelectric effect in 1905. Optics tools include the refracting lens, the reflecting mirror, and various optical components and instruments developed throughout the 15th to 19th centuries. Key tenets of classical optics, such as Huygens principle, developed in the 17th century, Maxwell's equations and the wave equations, developed in the 19th century, do not depend on quantum properties of light. Photonics is related to quantum optics, optomechanics, electro-optics, optoelectronics, and quantum electronics. However, each area has slightly different connotations by scientific and government communities and in the marketplace. Quantum optics often connotes fundamental research, whereas photonics is used to connote applied research and development. Photonics also relates to the emerging science of quantum information and quantum optics. The science of photonics includes investigation of the emission, transmission, amplification, detection, and modulation of light.

In photonics, metals are not usually thought of being very useful, perhaps except mirrors. In most cases, metals are strong absorbers of light, a consequence of their large free-electron density. However, in the miniaturization of photonic circuits, it is now being realized that metallic structures can provide unique ways of manipulating light at length scales smaller than the wavelength. Maxwell's equations tell us that an interface between a dielectric(e. g. silica glass)and a metal(e. g. Ag or Au)can support an surface plasmon(SP). An SP is a coherent electron oscillation that propagates along the interface together with an electromagnetic wave. These unique interface waves result from the special dispersion characteristics (dependence of dielectric constant on frequency)of metals. Similar to photonics, by means of SPs, suchresearch is referred to as plasmonics, which can be considered as optics at the nanoscale[3].

What distinguishes SPs from "regular" photons is that they have a much smaller wavelength at the same frequency. For example, a HeNe laser, whose free-space emission wavelength is 633 nm, can excite an SP at a Si/Ag interface with a wavelength of only 70 nm. When the laser frequency is tuned very close to the SP resonance, SP wavelengths in the nanometer range can be achieved. The short-wavelength SPs enable the fabrication of nanoscale optical integrated circuits, in which light can be guided, split, filtered, and even amplified using plasmonic integrated circuits that are smaller than the optical wavelength.

SPs are easily accessible excitations in metals and semiconductors and involve a collective motion of the conduction electrons. These excitations can be exploited to manipulate electromagnetic waves at optical frequencies("light") in new ways that are unthinkable in conventional dielectric structures. The field of surface plasmon nanophotonics is rapidly developing and impacting a wide range of areas including: electronics, photonics, chemistry, biology, and medicine. SPs are coherent delocalized electron oscillations that exist at the interface between any two materials where the real part of the dielectric function changes sign across the interface (e. g. a metal-dielectric interface, such as a metal sheet in air). The existence of surface plasmons was first predicted in 1957 by Rufus Ritchie[4]. Schematic representation of an electron density wave propagating along a metal-dielectric interface can be seen from Fig. 1-5. The charge motion in a surface plasmon always creates electromagnetic fields outside (as well as inside) the metal. The charge density oscillations and associated electromagnetic fields are called SPP waves. The exponential dependence of the electromagnetic field intensity on the distance away from the interface is shown on the right in Fig. 1-5[5]. These waves can be excited very efficiently with light in the visible range of the electromagnetic spectrum.

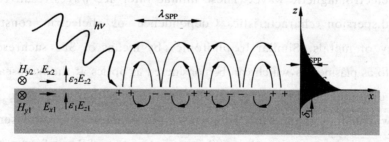

Fig. 1-5 Schematic representation of an electron density wave propagating along a metal-dielectric interface

(For colored figure please scan the QR code on page 1)

Surface plasmon polaritons (SPPs) are infrared (IR) or visible-frequency electromagnetic waves that travel along a metal-dielectric or metal-air interface. The term "SPP" explains that the wave involves both charge motion in the metal ("surface plasmon") and electromagnetic waves in the air or dielectric ("polariton")[6]. SPPs can be excited by electrons or photons. Excitation by electrons is created by firing electrons into the bulk of a metal. As the electrons scatter, energy is transferred into the bulk plasma. The component of the scattering vector parallel to the surface results in the formation of a SPP[7]. For a photon to excite an SPP, both must have the same frequency and momentum. However, for a given frequency, a free-space photon has less momentum than an SPP because the two have different dispersion relations. This momentum mismatch is the reason that a free-space photon from air cannot couple directly to an SPP. For the same reason, an SPP on a smooth metal surface cannot emit energy as a free-space photon into the dielectric(if the dielectric is uniform). The coupling of photons into SPPs can be achieved using a coupling medium such as a prism or grating or a defect on the metal surface to match the photon and SPP wave vectors(and thus match their momenta). The SPP is a non-radiative electromagnetic surface wave that propagates in a direction parallel to the negative permittivity/dielectric material interface.

Dispersion curve for SPPs can be seen from Fig. 1-6. At low frequency, an SPP approaches a Sommerfeld-Zenneck wave, where the dispersion relation (between frequency and wavevector)is the same as in free space. At a higher frequency,the dispersion relation bends over and reaches an asymptotic limit called the "surface plasma frequency"[8-9]. SPPs are usually shorter in wavelength than the incident light(photons). Hence, SPPs can have tighter spatial confinement and higher local field intensity. Perpendicular to the interface,they have subwavelength-scale confinement. An SPP will propagate along the interface until its energy is lost either to absorption in the metal or

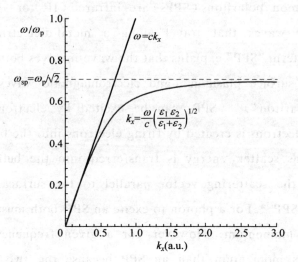

Fig. 1-6 Dispersion curve for SPPs. At low k, the surface plasmon curve (red) approaches the photon curve (blue)

(For colored figure please scan the QR code on page 1)

scattering into other directions (such as into free space). The k is a complex wave vector, $k_x = k'_x + ik''_x$, and in the case of a lossless SPP, it turns out that the x components are real and the z components are imaginary—the wave oscillates along the x direction and exponentially decays along the z direction.

As an SPP propagates along the surface (Fig. 1-5), it loses energy to the metal due to absorption. It can also lose energy due to scattering into free space or into other directions. The intensity of the surface plasmon decays with the square of the electric field, so at a distance x, the intensity has decreased by a factor of $\exp(2k''_x x)$. The propagation length is defined as the distance for the SPP intensity to decay by a factor of $1/e$. This condition is satisfied at a length, $L = \dfrac{1}{2k''_x}$[10]. The electric field falls off evanescently perpendicular to the metal surface. At low frequencies, the SPP penetration depth into the metal is commonly approximated using the skin depth formula. In the dielectric, the field will fall off far more slowly. The decay lengths in the metal and dielectric medium can be expressed as, $Z_i = \frac{\lambda}{2\pi}\left(\frac{|\varepsilon_1'| + \varepsilon_2}{\varepsilon_i^2}\right)^{1/2}$, where λ is the wavelength, ε_1' is the real part of the dielectric function of a metal, ε_2 is the permittivity of the dielectric, and i indicates the medium of propagation[10]. SPPs are very sensitive to slight perturbations within the skin depth and, therefore, are often used to probe inhomogeneities of a surface.

The charge motion in a surface plasmon always creates electromagnetic fields outside(as well as inside) the metal. The total excitation, including both the charge motion and associated electromagnetic field, is called as either a SPP at a planar interface, or a localized surface plasmon(LSP) for the closed surface of a small particle. A LSP is the result of the confinement of a surface plasmon in a nanoparticle of size comparable to or smaller than the wavelength of light used to excite the plasmon, see Fig. 1-7 (a). The LSP has two important effects: electric fields near the particle's surface are greatly enhanced and the particle's optical absorption has a maximum at the plasmon resonant frequency. The enhancement falls off quickly with distance from the surface and, for noble metal nanoparticles, the resonance occurs at visible wavelengths[10]. Surface plasmon resonance(SPR) is the resonant oscillation of conduction electrons at the interface between the negative and the positive permittivity material stimulated by incident light. LSPRs(localized SPRs) are collective electron charge oscillations in metallic nanoparticles that are excited by light. They exhibit enhanced near-field amplitude at the resonance wavelength. This field is highly localized at the nanoparticle and decays rapidly away from the nanoparticle/dielectric interface into the dielectric background, though the far-field scattering by the particle is also enhanced by the resonance. Light intensity enhancement is a very important aspect of LSPRs and localization means the LSPR has very high spatial resolution (subwavelength), limited only by the size of nanoparticles. Peaks of LSPRs can be well manipulated by the nanostructures, see Fig. 1-7(b)[11].

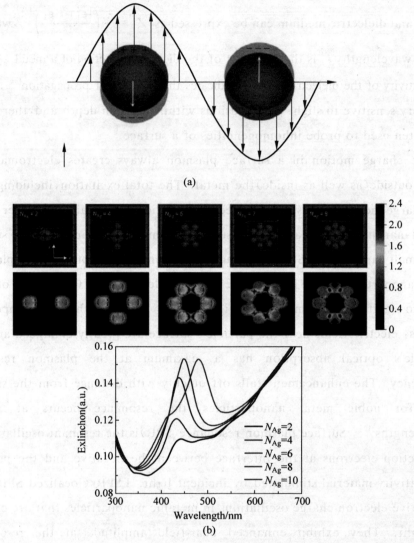

Fig. 1-7 (a) Local surface plasmon, and (b) local surface plasmon resonance by nanoparticles
(For colored figure please scan the QR code on page 1)

1.3 Concept of Plasmon-Enhanced Spectroscopy

1.3.1 Plasmon-enhanced fluorescence

The plasmon-enhanced nonlinear optical microscopy of g-C_3N_4 was studied by using the multiple surface plasmon resonances (MSPRs) of the Au @ Ag nanorods, see Fig. 1-8 in subfigure (a)[12]. The scanning election microscapy (SEM) images of the plasmonic Au@Ag nanorods; and (b)~(d) demonstrate that the Au@Ag nanorods are successfully synthesized, and a silver shell is coated onto each Au nanorod with a thickness of around 10 nm. Compared with the Au nanorods, the resonance peaks at 910 nm and 520 nm indicate a blue shift to 800 nm and 500 nm for the Au@Ag nanorods, and a new resonance peak at 400 nm can be obtained. The plasmon-enhanced two-photon-excited fluorescence (TPEF) of g-C_3N_4 covered with the Au@Ag nanorods was first studied (Fig. 1-9)[12]. According to the absorption and

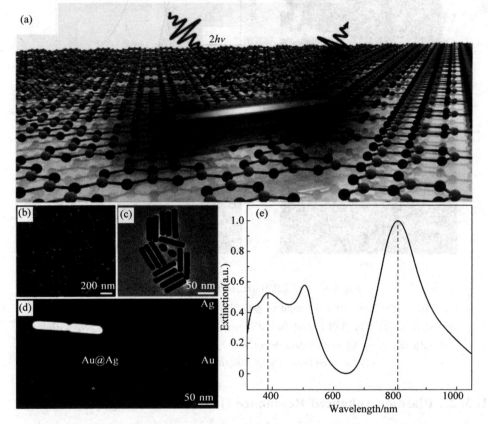

Fig. 1-8 Characterization and optical properties of Au@Ag nanorods
(a) Plasmon-enhanced nonlinear optical microscopy of g-C_3N_4 by the Au@Ag nanorods; (b) and (c) The SEM and TEM images of the Au@Ag nanorods, and (d) The elemental mapping images of the two Au@Ag nanorods; (e) UV-visible spectra of the chemically synthesized Au@Ag nanorods
(For colored figure please scan the QR code on page 1)

photoluminescence (PL) spectra of g-C_3N_4 shown in Fig. 1-9 (b), a strong extinction peak can be observed around 400 nm, which matches the fundamental frequency at 800 nm of the incident light. The TPEF images of g-C_3N_4 without (Fig. 1-9(c)) and with (Fig. 1-9(d)) the Au@Ag nanorods were obtained by 800 nm excitation. By using strong MSPRs at 800 nm and 400 nm, the TPEF can be strongly enhanced via the Au@Ag nanorods covering on the g-C_3N_4 surface.

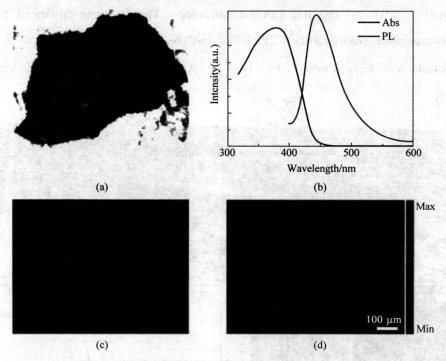

Fig. 1-9 Optical characterization of g-C_3N_4
(a) The bright field optical image of g-C_3N_4; (b) The absorption and PL spectra of g-C_3N_4; (c) The TPEF and (d) Plasmon-enhanced TPEF of g-C_3N_4 without and with the Au@Ag nanorods, respectively
(For colored figure please scan the QR code on page 1)

1.3.2 Plasmon-enhanced Resonance fluorescence energy transfer

Ozel firstly reports the mechanism of the independent control about plasmon coupling to donor or acceptor. They select Au MNPs and CdTe QDs which were assembled by layer-by-layer assembly technique[13]. They made experiments about the plasmon only coupling with donor or acceptor QDs (Fig. 1-10(a) ~ (c)). Note that there is a strong spectral overlap (Fig. 1-10(d)) when the FRET condition is satisfied[14].

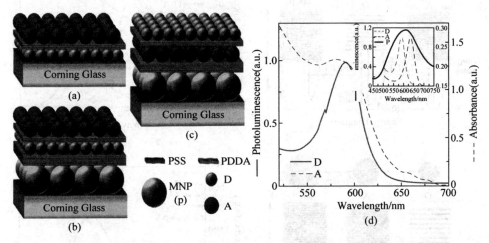

Fig. 1-10 (a) ~ (c) The conventional FRET, the plasmon-mediated FRET with plasmon coupling only to the donor QDs, and the plasmon-mediated FRET with plasmon coupling only to the acceptor QDs, respectively; (d) The PL spectrum
(For colored figure please scan the QR code on page 1)

In a concerned donor-acceptor system, a comparison was made between the PL spectra of steady states about the plasmon-enhanced donor (pD-A) (see Fig. 1-11(a) and (b)) and the plasmon-enhanced acceptor (D-pA) (see Fig. 1-11(c) and (d)). It is concluded that FRET is occurred during D-A in two cases. For the D-pA, there is not FRET, but an fluorescence enhancement for A, as well as pA; for the pD-A, it has FRET and enhanced fluorescence of D almost all transfer to A[13].

1.3.3 Surface-enhanced Raman scattering

Since the discovery of SERS[15], it has been extensively studied experimentally and theoretically because of its extremely high surface sensitivity and powerful application on fingerprint vibrational spectroscopy in qualitative and quantitative analysis. The pyridine molecule is an important probe molecule in

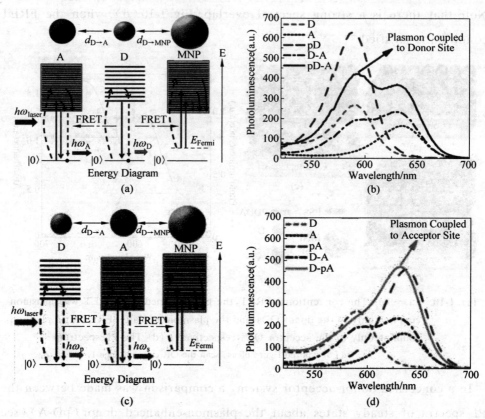

Fig. 1-11 Plasmon-Acceptor'Exciton coupling[13]
(For colored figure please scan the QR code on page 1)

SERS field, and the enhancement effect is strongly dependent on the SERS substrates[16]. In Fig. 1-12, (a) is the absorption spectroscopy of pyridine adsorbed on Au @ Pd substrate, it can be found that most of electrons and holes are distributed on the first layer of the substrate. (b) demonstrate that chemical enhancement can enhance Raman spectra of pyridine about 10 times. (c) reveals that electromagnetic enhancement can reach up to 10^3, where the distance of two nanoparticles is 2 nm; and the (d) reveals that electromagnetic enhancement is strongly dependent on the distance of nanoparticles. The total enhancements, including chemical and electromagnetic enhancements, are up to 10^4, which is consistent with the experimental reports[17].

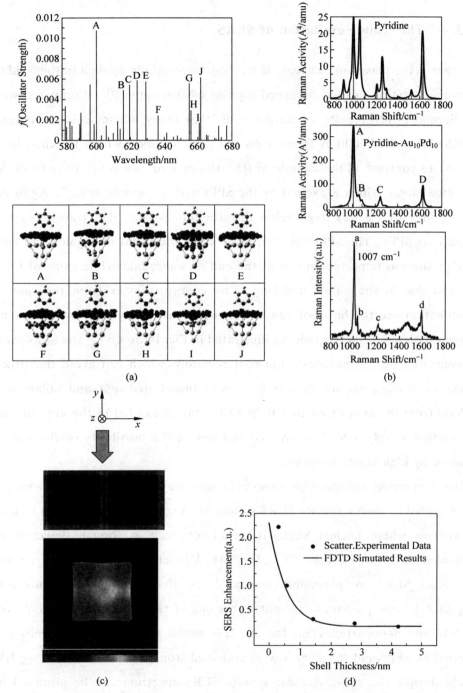

Fig. 1-12 The chemical and electromagnetic enhancements on pyridine adsorbed on the Au@Pd substrate

(For colored figure please scan the QR code on page 1)

1.3.4 The remote-excitation of SERS

Currently, the plasmonic waveguide has been successfully applied in the field of remote-excitation surface-enhanced Raman spectroscopy[18-21]. Compared with traditional SERS, remote excitation SERS has many advantages. In normal SERS, the excited light is focused on the detected spot, which is called local SERS. In contrast, in the "remote SERS", the excited spot is far away from the detected target, which is excited by the SPPs with a "remote mode". Ag or Au waveguides offer a way to go below the size limit because they transfer optical signals via SPPs. The SPPs can propagate along the metal waveguide and emit free photons at imperfections or at the end of waveguide, while some of them are lost due to the ohmic damping. The energy transfer from the spot of illumination and the hot-spot can be supported mainly by the remote systems based on plasmonic waveguide, as illustrated in Fig. 1-13(A)[19]. The SPPs could of course excite the nanoscale hot-spot remotely, which can avoid the strong background noise because of a large area of the excited spot, and isolate heat effects from the area of excited light to the target excited in the area of sub-diffraction wavelength. Moreover, it can reduce the possibility of the sample damage by high laser excitation.

Based on surface-enhanced spectroscopy, a novel way of sensing measurements can be provided by such a remote SERS system, see Fig. 1-13(A), and it can be used in systems where normal SERS is unsuitable, such as Raman detection of biomolecules in living cells[18-22]. In 2014, Ujii et al. reported the remote excitation SERS by plasmonic waveguide in living cell[22]. As shown in Fig. 1-13(B), the plasmonic hotspot(at the gap of two Ag nanowires(NWs), or of NW and nanoparticles) on the Ag NW probe inserted into the cells was excited by SPPs, and the signals were collected from the other side the Ag NW probe outside the cells. A clear remote SERS spectrum can be observed by using an Ag NW probe hotspot inserted into a living HeLa cell[22].

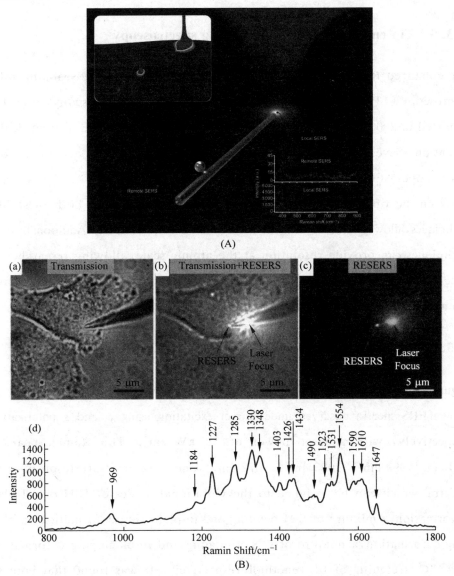

Fig. 1-13 (A) A sketch of the local SERS and the remote SERS in an Ag NP-NW system(Raman molecule adsorbed on the Ag NW). The Raman signals obtained at the illuminating terminal are local SERS, while the signals collected at the junction whose energy come from the propagating SPPs are remote SERS. (B) Scheme of the positions chosen for remote-excitation SERS spectroscopy inside a live HeLa cell during Ag NW probe endoscopy
(a)～(c) Remote excitation SERS endoscopy of a live HeLa cell. (a) Optical transmission; (b) Combination of optical transmission and remote excitation SERS; (c) Remote excitation SERS only images of an Ag NW probe in a live HeLa cell; (d) A remote excitation SERS spectrum from the nucleus of the live HeLa cell
(For colored figure please scan the QR code on page 1)

1.3.5 Tip-enhanced Raman scattering spectroscopy

Tip-enhanced Raman scattering(TERS), the combination of scanning probe microscopy (SPM) with surface-enhanced Raman spectroscopy, was first reported in 2000[23]. This technique has subsequently attracted a great deal of attention. Several groups have built TERS systems based on an atomic force microscope(AFM)or scanning tunneling microscope(STM)[24-26]. The distance between the tip and the film can be precisely controlled in both these SPM techniques, allowing fine adjustment during enhancement studies. Additionally, both techniques can provide resolution at the atomic scale, allowing researchers to "see clearly" target molecules. Figure 1-14 demonstrates the plasmon-driven chemical reaction in high-vacuum tip-enhanced TERS(HV-TERS)[26].

1.3.6 Remote excitation-TERS microscopy

Figure 1-15 demonstrates the Remote excitation-TERS[27]. Subfigures (a) and (b) show TERS spectra observed under direct excitation using p and s polarization, respectively, with a laser power of 50 kW/cm^2. The Raman peaks at 700 cm^{-1}, 999 cm^{-1}, 1023 cm^{-1}, 1072 cm^{-1}, and 1572 cm^{-1} are resolved in the spectra, which can be assigned to the vibrational modes of C-H out-of-plane deformation, a mixture of S-H bending and in-plain ring deformation, in-plane ring deformation, a mixture of C-S stretching and in-plane ring deformation, and C-C stretching of benzenethiol, respectively. It was found that both the Raman and background intensities were higher with p-polarization than with s-polarization, resulting from the excitation of gap-mode plasmons between the apex of the Ag NW and the Au(111) substrate. Conversely, with remote excitation this polarization dependence was reversed. (c) and (d) show TERS spectra observed under remote excitation with p-and s-polarization, respectively, using the same laser power(50 kW/cm^2). The spectra in (c) and

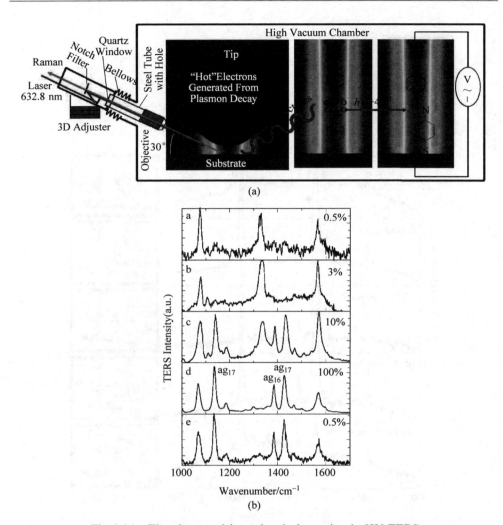

Fig. 1-14 The plasmon-driven chemical reaction in HV-TERS
(a) The scheme of set up and experimental details, and (b) The processes of the plasmon-driven chemical reaction
(For colored figure please scan the QR code on page 1)

(d) are multiplied by 10 times for easier comparison. Note that, with remote excitation, when the laser power is above 100 kW/cm^2, the remote-TERS signal tends to fluctuate and indeed sometimes the NW was cut at the focused spot. This could be due to a temperature rise at the NP/NW junction upon laser irradiation. Here, both the Raman and background intensities were higher with s-polarization than with p-polarization. This is in line with expectation,

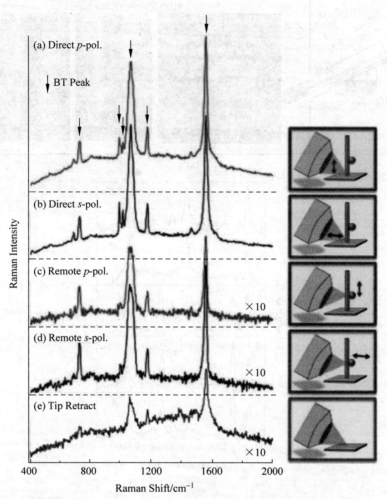

Fig. 1-15　Raman spectrum from benzenethiol-modified Au(111) substrate in the presence(a)～(d), and the absence of the tip(e). (a) and (b) TERS spectra measured when HeNe laser light (632.8 nm, 50 kW/cm^2) is focused at the end of a Ag NW with p-and s-polarization (direct excitation), respectively. (c) and (d) TERS spectra measured when the laser light is focused at the Au NP with p-and s-polarization (remote excitation), respectively. The tip is approached to the substrate under shear-force AFM feedback. Spectra shown in(c)～(e) are multiplied 10 times for easier comparison. Raman peaks that are assigned to Raman scattering of benzenethiol are indicated by an arrow in(a)

(For colored figure please scan the QR code on page 1)

giving the fact that light polarized perpendicular to the long axis of a NW can more efficiently excite propagating SP when a laser is irradiated at a coupling point. This stronger coupling therefore leads to a stronger excitation of propagating SP travelling along the NW, which can thus explain the above observation.

1.3.7 Plasmon-enhanced coherence anti-Stokes Raman scattering images

Raman scattering is a widely used method to identify molecules by means of their vibrational fingerprint. However, spontaneous Raman is incredibly weak with a cross-section of $\sim 10^{-30}$ cm^{2}[28], and therefore various techniques have been applied to increase its sensitivity. CARS[29] utilizes a nonlinear four-wave mixing process resulting in much higher sensitivity than spontaneous Raman and has been employed in vibrational bioimaging since the late nineties[30]. CARS is a third-order nonlinear effect where three photons ω_1, ω_2, and ω_3 of two different frequencies ω_s and ω_p interact coherently through the third-order susceptibility ($\chi^{(3)}$) of the material producing a spectrally separated, blue-shifted photon at the anti-Stokes frequency ω_{as} (Fig.1-16(A)). Although CARS is more sensitive than conventional Raman due to its higher-order dependence on incident power, its sensitivity is still not enough for monitoring or imaging molecules present in extremely low concentrations. Moreover, reduced photodamage and faster real-time acquisition is highly desirable in bioimaging applications and hence is crucial to achieve higher sensitivities. One way to achieve higher sensitivities is to employ SP generated on nanostructures to enhance CARS signals. This is illustrated schematically in Fig.1-16, which shows the transitions and field dependence on signals in various Raman processes.

To demonstrate the molecular imaging abilities of SECARS we used the commercially available plasmonic substrate Klarite. This is based on a different (pyramidal) void architecture and is uniformly reproducible over large area

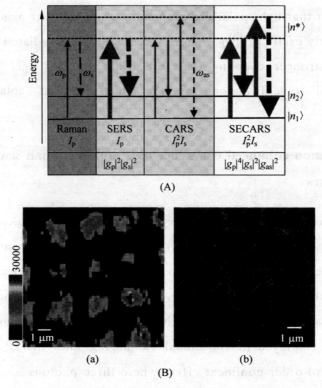

Fig. 1-16 (A) Schematic band energy diagram showing transitions in different Raman processes and their dependence on the pump and Stokes intensity (I_{pump} and I_{Stokes}) and local electric field enhancement (g). The Dirac notation $|n^*\rangle$ denotes virtual states of the molecules while $|n_1\rangle$ and $|n_2\rangle$ symbolize its vibrational ground and first excited state. In (B), (a) SECARS image of benzenethiol monolayer on the commercially available plasmonic surface Klarite. The color of each pixel tracks the peak intensity at 1070 cm^{-1} extracted from each SECARS spectrum taken with integration time of 10 ms every 300 nm using a 100× objective. (b) SEM image of the same area showing the pyramidal pits of Klarite

(For colored figure please scan the QR code on page 1)

though of slightly weaker enhancement. An SEM image of a Klarite surface consisting of micropyramid pits with a lattice pitch of 2.5 μm is shown in Fig.1-16(B). Klarite supports strongly localized plasmons and has been successfully applied to SERS. Due to the adiabatic taper into the pits, their plasmon absorption is also very broad, with strong absorption from 600 nm to 900 nm, which makes them an ideal substrate for SECARS.

References

[1] ATKINS P, DE PAULA J. Atkins' physical chemistry[M]. New York: Oxford University Press,2002.

[2] JABŁOŃSKI A. Efficiency of anti-Stokes fluorescence in dyes[J]. Nature,1933,131: 839-840.

[3] ALBERT P, ATWATER H A. Plasmonics: optics at the nanoscale[J]. Materials Today,2005,8: 56.

[4] RITCHIE R H. Plasma losses by fast electrons in thin films[J]. Physical Review,1957, 106: 874-881.

[5] ZENG S, DOMINIQUE B, HO H, et al. Nanomaterials enhanced surface plasmon resonance for biological and chemical sensing applications[J]. Chemical Society Reviews,2014,43: 3426-3452.

[6] ZENG S,YU X,LAW W C,et al. Size dependence of Au NP-enhanced surface plasmon resonance based on differential phase measurement[J]. Sensors and Actuators B: Chemical.,2013,176: 1128-1133.

[7] RAETHER H. Surface plasmons on smooth and rough surfaces and on gratings[M]. New York: Springer-Verlag,1988.

[8] COTTAM M G. Introduction to surface and superlattice excitations[M]. New York: Cambridge University Press. 1989.

[9] HOMOLA J. Surface plasmon resonance based sensors[M]. Berlin: Springer-Verlag,2006.

[10] RYCENGA M,COBLEY C M,ZENG J,et al. Controlling the synthesis and assembly of silver nanostructures for plasmonic applications[J]. Chem. Rev.,2011,111(6): 3669-3712.

[11] CHOI H,KO S J,CHOI Y,et al. Versatile surface plasmon resonance of carbon-dot-supported silver nanoparticles in polymer optoelectronic devices[J]. Nature Photonics, 2013,7: 732-738.

[12] MIA X,WANG Y, LIA R, et al. Multiple surface plasmon resonances enhanced nonlinear optical microscopy[J]. Nanophotonics,2019,8: 487.

[13] OZEL T, HERNANDEZ-MARTINEZ P L, MUTLUGUN E, et al. Observation of selective plasmon-exciton coupling in nonradiative energy transfer: donor-selective versus acceptor-selective plexcitons[J]. Nano Lett.,2013,13: 3065-3072.

[14] ZONG H,WANG J,WANG X,et al. Plasmon-enhanced fluorescence resonance energy transfer[J]. Chem. Rec., 2019,19: 818.

[15] FLEISCHMANN M, HENDRA P J, MCQUILLAN A J. Raman spectra of pyridine adsorbed at a silver electrode[J]. Chem. Phys. Lett.,1974,26: 163.

[16] YANG Z,LI Y,LI Z,et al. Surface enhanced Raman scattering of pyridine adsorbed on Au@Pd core/shell nanoparticles[J]. J. Chem. Phys.,2009,130: 234705.

[17] SHENG J J,LI J F,YIN B S,et al. A preliminary study on surface-enhanced Raman scattering from au and Au@ Pd nanocubes for electrochemical applications[J]. Can. J. Anal. Sci. Spectrosc.,2007,52: 178.

[18] HUANG Y, FANG Y, SUN M. Remote excitation of surface-enhanced Raman scattering on single Au nanowire with quasi-spherical termini[J]. J. Phys. Chem. C, 2011,115: 3558-3561.

[19] HUANG Y, FANG Y R, ZHANG Z L, et al. Nanowire-supported plasmonic waveguide for remote excitation of surface-enhanced Raman scattering[J]. Light: Sci. Appl.,2014,3: e199.

[20] FANG Y R, WEI H, HAO F, et al. Remote-excitation surface-enhanced Raman scattering using propagating Ag nanowire plasmons [J]. Nano Lett., 2009, 9: 2049-2053.

[21] SUN M T,ZHANG Z,WANG P,et al. Remotely excited Raman optical activity using chiral plasmon propagation in Ag nanowires[J]. Light: Sci. Appl.,2013,2: e112.

[22] LU G,KEERSMAECKER H D,SU L,et al. Live-cell SERS endoscopy using plasmonic nanowire waveguides[J]. Adv. Mater.,2014,26: 5124-5129.

[23] AYARS E J,HALLEN H D,JAHNCKE C L. Electric field gradient effects in Raman spectroscopy[J]. Phys. Rev. Lett.,2000,85: 4180-4183.

[24] PETTINGER B,REN B,PICARDI G,et al. Nanoscale probing of adsorbed species by tip-enhanced Raman spectroscopy[J]. Phys. Rev. Lett.,2004,92: 096101.

[25] ZHANG R,ZHANG Y,DONG Z C,et al. Chemical mapping of a single molecule by plasmon-enhanced Raman scattering[J]. Nature,2013,498: 82-86.

[26] SUN M T,ZHANG Z L,ZHENG H R,et al. *In situ* plasmondriven chemical reactions revealed by high-vacuum tip-enhanced Raman spectroscopy[J]. Sci. Rep., 2012, 2: 647.

[27] YASUHIKO F, PETER W, STEVEN D F, et al. Remote excitation-tip-enhanced Raman scattering microscopy using silver nanowire[J]. Jap. J. Appl. Phys., 2016, 55: 851.

[28] MCREERY R. Raman spectroscopy for chemical analysis[M]. New York: John Wiley & Sons,2000.

[29] MINCK R W, TERHUNE R W, RADO W G. Laser-stimulated Raman effect and resonant four-photon interactions in gases H_2, D_2, and CH_4 [J]. Appl. Phys. Lett., 1963,3: 181-184.

[30] ZUMBUSCH A,HOLTOM G R,XIE X S. Three-dimensional vibrational imaging by coherent anti-Stokes Raman scattering[J]. Phys. Rev. Lett.,1999,82: 4142-4145.

CHAPTER 2

Molecular Spectroscopy

2.1 Jablonski Diagram

For a molecule, besides the nucleus, electrons move, and there are vibrations and rotations between the atoms that make up the molecule, each of which corresponds to a certain amount of energy. Therefore, the energy of the molecule consists of three parts[1]:

$$E = E_e + E_v + E_r, \qquad (2\text{-}1)$$

where, E_e, E_v and E_r correspond to electronic, vibrational and rotational energy, respectively. And usually, such energy can be represented as S, v and J, respectively (Fig. 2-1). The adjacent energy level spacing relationships of the three are as follows

$$\Delta E_e \gg \Delta E_v \gg \Delta E_r.$$

In molecular spectroscopy, a Jablonski diagram is used to illustrate the electronic states of a molecule and the transitions between them. As shown in Fig. 2-2, S_0, S_1, S_2, $\cdots$, S_n represent the ground state of electrons, the first-, second-and n-excited states of electrons in the molecule, respectively, and T represents the triplet state. Nonradiative transitions including internal conversion(IC), intersystem crossing(ISC) and vibrational relaxationare etc., are represented by curved arrows, while radiative transitions, including fluorescence and phosphorescence emission, by straight arrows. The so-called

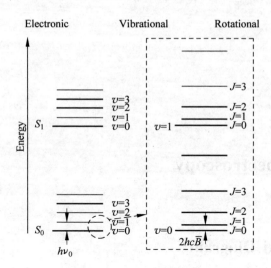

Fig. 2-1 The energy level diagram of a molecule

vibration relaxation, which isn't shown on Fig. 2-2, refers to the excited state relaxes to its own lowest vibration level. And this process involves energy dissipation from molecules to the surrounding environment, so it is impossible for isolated molecules. The thick and thin lines represent vibration ground state and higher vibration states of each electron, respectively[2].

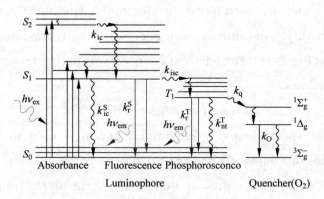

Fig. 2-2 Energy level diagram of the Jablonski
(For colored figure please scan the QR code on page 1)

When molecules absorb photons, the electrons may transit from the ground states(S_0) to the excited states($S_1, S_2, \cdots, S_n$). The instable electron in excited state returns to ground state(S_0) from the just excited state($S_1, S_2, \cdots, S_n$) and emits fluorescence (being the source of fluorescence). Of course, it does not have to emit fluorescence only. It can generate heat or other forms of energy. If the electron transits from the excited state (S_1) to the excited states of the electron T_1 through the ISC, and transfers to T_2 through the IC, and then from the excited state T_2 to S_0, it emits phosphorescence. This is the fundamental difference between phosphorescence and fluorescence.

2.2 Electronic State Transition

2.2.1 Ultraviolet-visible-near IR absorption spectroscopy

The electron transition energy level difference is about 1—20 eV, tens of times larger than the molecular vibration energy level difference. The wavelength of the absorbed light is about 1.25—0.06 μm, mainly in the vacuum ultraviolet to near IR (NIR) region, so it is called ultraviolet-visible (UV-Vis) absorption spectroscopy or electronic spectroscopy.

In the analysis of inorganic solid films, it follows the following light absorption law

$$I = I_0 (1 - R)^2 e^{-\alpha x}, \qquad (2\text{-}2)$$

where, α is the optical absorption coefficient related to the nature of absorbent substance and the wavelength of incident light. Note that studying optical absorption in solids, including the intrinsic, the exciton, the impurity-defect absorption and so on, can directly obtain the band structure analysis, such as electron gap, and impurity-defect level. For our research field, it is more necessary to study the relative contents with the organic compounds. Absorption of organic solutions follows Lambert-Beer's law[3]

$$A = -\lg T = -\lg \frac{I}{I_0}. \qquad (2\text{-}3)$$

When a parallel monochromatic light passes through a dilute solution containing an absorbent, the absorbance of the solution is proportional to the product of the concentration of the absorbent and the thickness of the liquid layer:

$$A = \alpha cl = \varepsilon cl, \tag{2-4}$$

where, c is the absorbent concentration, l is the thickness of transparent layer, and ε is the molar absorption coefficient, equal to the absorbance of 1 mol/L absorbent solution in 1 cm absorption path.

Now, the electronic transitions about organic compounds will be introduced. There are three types of valence electrons: the single bond called σ electron, the double bond π electron, and the unbounded called n electron. The permissible transitions between them are shown in Fig. 2-3, but there are only four types of transitions in organic compounds: $n \to \pi^*$、$\pi \to \pi^*$、$n \to \sigma^*$、$\sigma \to \sigma^*$, and the energy they required increases in turn.

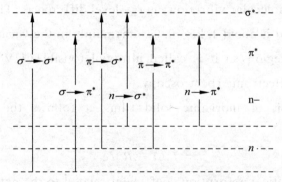

Fig. 2-3　Permissible transitions between electron level orbits

2.2.2　Two-photon absorption spectroscopy

Simultaneous two-photon absorption(TPA) is a nonlinear optical process. This phenomenon was first predicted by Gppert-Mayer in 1931 who calculated the transition probability for a two-quantum absorption process[4]. TPA can create excited states with photons of half the nominal excitation energy, which can provide improved penetration in absorbing or scattering media. TPA is a third-order nonlinear optical process and is the I^2-dependence on the process, which allows for excitation of chromophores with a high degree of spatial selectivity in three dimensions through the use of a tightly focused laser beam[5]. TPA is in great demand for a variety of applications, such as two-photon-excited fluorescence microscopy[6-8], 3D optical memory[9], nanofabrication[10], upconversion lasing[11], solar cell[12-13], nondestructive imaging of biological tissues[14-15], and photodynamic therapy[16].

Theoretical analysis of TPA have been implemented to interpret experimental measurements, very useful to understand structure-property relations. Coupled with finite field techniques, the ab-initio methods are widely used to calculate off-resonant NLO responses[17], the more general method is the application of time-dependent perturbation theory. The sum-over-states(SOS)model is an essentially practical methods, involving calculating both ground and excited-states wavefunctions and the transition dipole moments between them[18-19]. Theese moments in TPA include two parts as three-state and two-state models[18], respectively. M. Richter, and S. Mukamel, et al. demonstrated that the TPA process is a two-step one in which an intermediate state must be experienced during two-photon excitation[19-20]. The complete TPA probability is the sum of the processes that traverse all intermediate states[21], i. e.

$$\delta_{tp} = 8 \sum_{\substack{j \neq g \\ j \neq f}} \frac{|\langle f | \mu | j \rangle \langle j | \mu | g \rangle|^2}{\left(\omega_j - \frac{\omega_f}{2}\right)^2 + \Gamma_f^2} (1 + 2\cos^2 \theta_j) +$$

$$8 \frac{|\Delta\mu_{fg}|^2 |\langle f | \mu | g \rangle|^2}{\left(\frac{\omega_f}{2}\right)^2 + \Gamma_f^2} (1 + 2\cos^2 \phi), \quad (2\text{-}5)$$

where $|g\rangle$, $|f\rangle$ and $|j\rangle$ are the ground state wave function, final state wave function, and the intermediate state of the two-photon transition, respectively, the two Dirac brackets in the first term of the formula are the transition dipole moments from the ground to the intermediate and the intermediate to the final states. The corresponding angle θ_j is between the two transition dipole moment vectors, the $\Delta\mu_{fg}$ in the second term is the difference between the final-and the ground-state permanent dipole moments; the ϕ is the angle between of the vectors of

$\Delta\mu_{fg}$ and $|\langle f|\mu|g\rangle|$; ω_j and ω_f are the energy of intermediate and final states, respectively; the Γ_f is the lifetime of the excited state.

Equation(2-5) describes the TPA cross section of a two-step process. In Fig. 2-4 (a), there are two one-step processes in one-photon absorption. In Fig. 2-4(b), a and b represent two different intermediate states in TPA. This diagram illustrates that the same final TPA state may experience different intermediate states, which may be quite large for a real system. When studying the two-photon process, we not only care about the shape of the absorption spectrum (TPA cross section), but also the electronic motion behavior (such as charge transfer characteristics) in TPA. Therefore, it is advantageous to use Eq. (2-5) to calculate TPA. This is because it reflects the two-step process and can focus

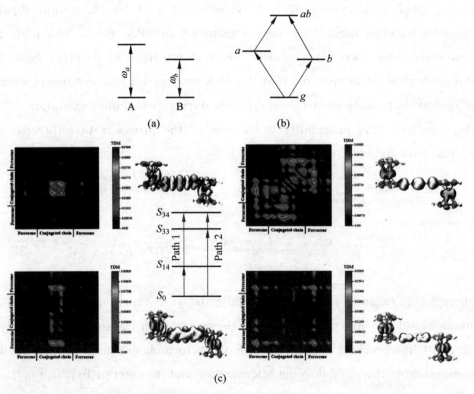

Fig. 2-4　The schematic of (a) One-step process in one-photon absorption, (b) Two-step processes in TPA, and (c) Charge transfer in each step on each channel in TPA

(For colored figure please scan the QR code on page 1)

on the nature of the intermediate state. It is also a factor that plays a crucial role in two-photon excitation.

For the three-state model, in Eq. (2-5), the first part is the transition from the ground to the intermediate states; and the second part is the transition from the intermediate to the final states. For the two-state model, it includes the transition from the ground to the final states; and the difference of permanent dipoles between final excited state and ground state. For the symmetric system, the two-state model can be ignored, since the contribution in Eq. (2-5) is vanished, compared with the three-state model[22].

2. 2. 3 Fluorescence spectroscopy

In 1560 s, Sahagún and Monardes described early fluorescence observations in the infusion named as kidney wood extracted from the wood of Eysenhardtia polystachya and Pterocarpus indicus[23-26]. Clarke[27] and Haüy depicted fluorescence in fluorite in 1819 and 1822, respectively. Brewster described chlorophyll in 1833[28], and Herschel also described quinine in 1845[29]. In 1852, Stokes published a paper on the "refractivity" of light, describing the ability of fluorite and uranium glass to convert invisible light at the ultraviolet end of visible spectrum to blue light. He used "fluorescence" to describe this phenomenon and claimed that "I almost tend to use fluorite to create a word and call it exterior fluorescence, because the similar term" opalescence "derives from the name of minerals"[30]. By the way, the name was the origin from a mineral fluorite called calcium difluoride.

Fluorescence is the light emitted by substances that absorb light or other kinds of electromagnetic radiation. In fact, it belons to a kind of the luminescence. In general, the light of emission has a marked characteristic that it has longer wavelength than the absorbed radiation, so the energy is lower. Interestingly, fluorescence occurs in ultraviolet region of spectrum where the absorbed radiation is invisible to human eye, while usually, the emission light lies in the region of visible, which makes fluorescent substance putting a unique color so that it can be seen only when exposed to ultraviolet light. When removing the radiation source, the fluorescent material almost immediately stops emitting light, unlike the phosphorescent material, which continues to emit light for a period of time.

As shown in Fig. 2-2, S_0, S_1, S_2, $\cdots$, S_n represent the ground state of electrons, the first-, second- and n-excited states of electrons in the molecule, respectively. When molecules absorb photons, the electrons may transit from S_0 to S_1, S_2, $\cdots$, S_n. The electron instability in the excited state returns to S_0 from S_1, S_2, $\cdots$, S_n and emits fluorescence (which is the source of fluorescence). Of course, it does not have to emit fluorescence. It can generate heat or other forms of energy. If the electron transits from S_1 to T_1 through the ISC, and transfers to T_2 through the IC, and then from the excited state T_2 to S_0, it emits phosphorescence. This is the fundamental difference between phosphorescence and fluorescence. By the way, the difference between the excited states of S_1 and T_1 is mainly about the direction of electron spin (singlet and triplet state).

Moreover, the necessary condition for a compound to produce fluorescence is that it absorbes less energy than needed to break the weakest chemical bond when the photons undergo a transition with invariable multiplicity. In addition, to be able to emit light, a compound must have fluorescent groups in its structure. Such groups all contain unsaturated bonds. When these groups are part of the conjugated system of molecules, the compound may produce fluorescence.

Nothing can be accomplished without norms or standards. The excitation and relaxation of electrons in molecules absorbed light need to satisfy different major laws. The excitation process satisfies the Franck-Condon rule while the de-excitation satisfies the Kasha rule. Franck-Condon rule means that the transition process of electrons is very fast, and the relative position of nuclei can not change in this process, which can be simply understood as vertical transition. The Kasha rule stipulates that in the process of electron relaxation recombination, the electrons first relax to the electron excited state's lowest level, and then return to the ground state.

Fluorescence quantum yield is a basic parameter of fluorescent substances, which indicates the fluorescence ability of substances, and its value is between 0 and 1. Fluorescence quantum yield is the result of competition between fluorescence radiation and other radiative and non-radiative transitions. The quantum yield is defined as[31-32]

$$\Phi = \frac{\text{Number of photons emitted}}{\text{Number of photons absorbed}} = \frac{k_f}{k_f + \sum k_i}, \qquad (2-6)$$

where, k_f refers to rate constant of the fluorescence emission process, and $\sum k_i$ represents the sum of the rate constants of other related processes[33]. In fact, to excited state, besides the emission of fluorescence back to ground state, the molecule will return to the ground state through some other processes and these other rates of the decay of excited states are contributed by mechanisms other than photon emission, so they are usually called "non-radiative rates". They may include dynamic collision quenching, resonance energy transfer, ISC and IC. Therefore, if the rate of any path takes palce changing, the fluorescence quantum yield will be affected usually. What's more, the quantum yield is measured by comparing with standard substances, such as the quinine sulfate when it soluted in sulfuric acid.

So-called fluorescence lifetime is the time required when the fluorescence intensity of the molecule decreases to $1/e$ of the maximum fluorescence intensity I_0 while the excitation light is removed (Fig. 2-5), which is commonly expressed as τ. The decay of fluorescence intensity accords with exponential law[34]:

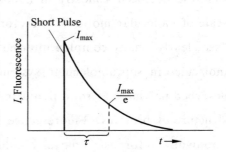

Fig. 2-5 The diagram of fluorescence lifetime

$$I_t = I_0 e^{-kt},\qquad(2\text{-}7)$$

where, I_t is the fluorescence intensity at time t, I_0 is the maximum fluorescence intensity, and k is constant of decay. Fluorescence lifetime is the reciprocal of decay constant k[34]. But it is worth mentioning that transition time is different from the lifetime, the former is the reciprocal of transition frequency.

As we mentioned in the last section, there are some non-radiative processes involved in the de-excitation process. As a result, the process that the excited state molecule returning to ground state is accelerated and the fluorescence lifetime is reduced. The lifetime is related to the rate constants of these processes:

$$\tau = \frac{1}{k_f + \sum k_i}.\qquad(2\text{-}8)$$

The fluorescence lifetime can also be understood as the statistical mean residence time of fluorescent substance in the excited state. Due to the complexity of practical systems, fluorescence attenuation is often described by multi-exponential or non-exponential attenuation equations and fitted by least squares method.

The fluorescence lifetime is related to the polarity and viscosity of the microenvironment in which the substance is located. The changes of the studied system can be directly understood by fluorescence lifetime analysis. Fluorescence occurs mostly in nanosecond order, which is precisely the time scale of molecular motion. Therefore, fluorescence technology can be used to "see clearly" many complex intermolecular processes, such as the clustering of molecules in supramolecular systems, the conformational rearrangement of adsorbed polymers at solid-liquid interfaces, and the changes in the advanced structure of proteins. Fluorescence lifetime analysis has been widely used in photovoltaic, forensic analysis, biological molecules, nanostructures, quantum dots, photosensitization, lanthanides, photodynamic therapy and other fields.

The fluorescence lifetime measurement techniques include time-correlated single-photon counting(TCSPC), phase modulation and flash method. TCSPC is the most popular amony them, which has the advantages of high sensitivity, accurate results and small system error.

Usually, the so-called fluorescence spectrum actually means the relation diagram about the fluorescence energy and the wavelength. In general, there are two kinds of fluorescence spectrum: emission spectrum(PL) and excitation spectrum(PLE). PL can be obtained by fixing the wavelength of excitation light, and recording the fluorescence intensities at different emission wavelengths with the change of emission wavelength. PLE can be obtained by fixing the wavelength of the emitted light, changing the wavelength of the excitation light, and recording the fluorescence intensity changes with the excitation wavelength. Both emission and excitation fluorescence spectra record the variation of emission fluorescence intensity with wavelength. So in fluorescence spectrum, the longitudinal coordinate is intensity and the horizontal ordinate is wavelength. We can obtain many useful information from fluorescence spectrum. Firstly, peak position and half-peak width can be obtained from the graph. The direct expression of peak position is the color of fluorescence, and the half-peak width is the purity of fluorescence.

As we all known, fluorescence spectra often appear at the same time as absorption spectra. So it can be compared with the absorption spectra of molecules. Figure 2-6 shows the ultraviolet-visible (UV-Vis) absorption spectrum absorption spectra, fluorescence emission spectra (PL) and fluorescence excitation spectra(PLE)of the same substance[35]. It is not difficult to find that the excitation spectrum is very similar to the absorption spectrum. But they are fundamentally different. Absorbance is the ordinate of absorption spectrum, and it reflects the

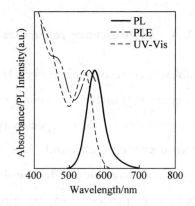

Fig. 2-6 PL, PLE and UV-Vis spectra of InP nanocrystals[16]

absorption of light. Intensity is the ordinate of fluorescence spectrum, which is related to not only the absorptivity of substances, but also the quantum yield. Actually, in many research systems, they are often combined to analyze scientific problems. Generally speaking, fluorescence spectrum refers to fluorescence emission spectrum. The fluorescence emission spectra of most substances may occur red shift, called Stokes shift. Why does Stokes shift occurs? As shown in Fig. 2-2, electrons in the excited state (e. g. S_1) do not radiate directly to the ground state (S_0), but first undergo vibration relaxation and IC. All these processes consume part of the energy. At the same time, a series of vibrational relaxations are required for electrons transiting to the ground state. These causes result in energy loss, which is reflected in the red shift of the spectrum in the spectrogram. Moreover, the shape of the fluorescence emission spectrum is independent of the excitation wavelength, but keeps a certain mirror symmetry with the absorption spectrum. The former is due to the Kasha rule[36]. Whether the electron is excited to that excited state, it will relax to S_1 first, then radiation transits to ground state to emit fluorescence, so the emission spectrum is independent of the wavelength of excitation[37]. The latter is symmetrical because the vibrational energy levels of the first excited singlet state and the ground state are similar, and the vibrational transition probabilities are also similar.

2.2.4 Fluorescence resonance energy transfer

Fluorescence resonance energy transfer (FRET) is the mechanism describing the energy transfer between donors and the acceptors of photosensitive chromophores[38]. More specifically, first, the donor molecule is excited to the electron-excited state, and then the donor transfers energy to the acceptor molecule through non-radiative dipole-dipole coupling when FRET occurs[39]. The above process can be proved by Jablonski diagram (Fig. 2-7)[40]. Moreover, FRET happens when the following conditions are met: first, the

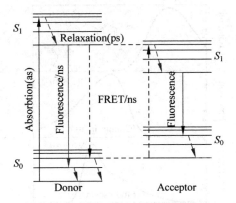

Fig. 2-7 Jablonski diagram for FRET at different timescales[39]
(For colored figure please scan the QR code on page 1)

emission spectra of donors and the absorption spectra of acceptors must overlap; second, the donor and the acceptor must be close enough; third, the direction of transition dipole moment of donor and acceptor can't be perpendicular.

The efficiency of FRET is sensitive to distance because of its physical description[41]:

$$k_t = \left(\frac{1}{\tau_D}\right)\left(\frac{R_0}{r_{DA}}\right)^6, \qquad (2-9)$$

where, k_t is the rate constant about energy transfer from donors to acceptors, r_{DA} is distance between point dipoles, $\frac{1}{\tau_D}$ is the rate of de-excitation including all other de-excitation ways except the way of transferring energy to acceptor, and R_0 represents the distance of molecules pair[41]. There are many common methods for measuring the FRET efficiency, such as acceptor photobleaching FRET, FLIM-FRET, and sensitized emission FRET.

The recent experimental result on the FRET can be seen from Fig. 2-8, where the FRET system of donor and acceptor, and FRET at different ratio between donor and acceptor at 10^{-4} M[23]. The fluorescence of donor, acceptor and FRET can be clearly observed in the figure. Detailed results and discussion can be seen from Ref. [42]. Moreover, the sp can significantly enhance the FRET, physical mechanism and experimental results can refer to Ref. [40].

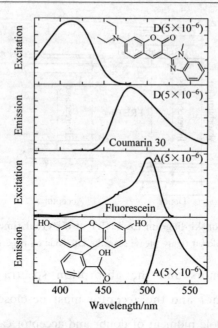
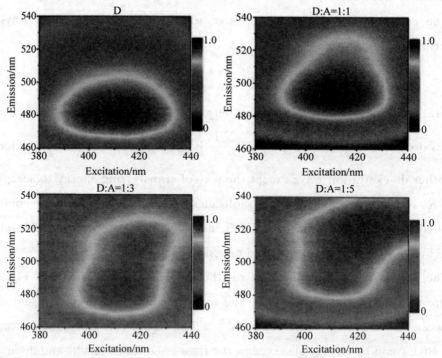

Fig. 2-8 The FRET system of donor and acceptor (left spectra), and FRET at different ratios between donor and acceptor at 10^{-4} M
(For colored figure please scan the QR code on page 1)

CHAPTER 2 Molecular Spectroscopy

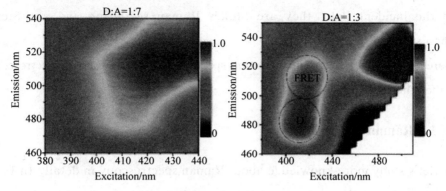

Fig. 2-8(Continued)

2.3 Vibration spectroscopy

The energy level of vibrational transition spectra associated with the vibrational states are shown in Fig. 2-9. Infrared (IR) spectroscopy can selectively absorb certain wavelengths of IR rays and cause the transition of vibrational and rotational levels in molecules. The IR absorption spectra of substances can be obtained by detecting the absorption of IR rays, also known as molecular vibration spectroscopy or vibrational rotation spectroscopy. When the energy of incident light and scattering light are the same, it is Rayleigh scattering. When the energy of scattering light is smaller or larger

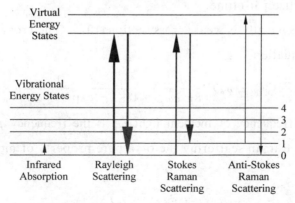

Fig. 2-9 The energy level transitions corresponding several spectra
(For colored figure please scan the QR code on page 1)

than the incident light, they are Stokes Raman scattering and anti-Stokes Raman scattering, respectively.

Here we introduce two main vibration spectra. The one is Raman spectroscopy, and the other is IR spectroscopy.

2.3.1 Raman spectroscopy

Now, let's study more knowledge about Raman spectroscopy in detail. In 1923, Smekal predicted the inelastic scattering of light, but the Raman effect was named after Raman, an Indian scientist who, with Krishnan, observed this effect by sunlight in organic liquids in 1928. In 1930, Raman's discovery earned him Nobel Prize in Physics. Specifically, Raman spectroscopy is the spectroscopic technique which is usually used to detect the system's low-frequency modes, such as vibrational, rotational and others. The magnitude of Raman effect is related to the polarizability of electrons in molecules. The Raman of molecular and ROA could be expressed as

$$I \propto \left| \sum_{k \neq ji} \frac{\langle \varphi_j | \mu_e | \varphi_k \rangle \langle \varphi_k | \mu_e | \varphi_i \rangle}{\omega_{ji} - \omega_0 + i\Gamma} \right|^2 E^4, \quad (2\text{-}10)$$

where, ω_{ji} represents the transition energy, and $i\Gamma$ represents an imaginary number term proportional to the width of the state j and inversely proportional to itself lifetime.

Raman signals usually is weak (cross section) 10^{-3}. The cross section satisfies the following equation[43]:

$$\frac{d\sigma}{d\Omega} = \frac{\pi^2}{\varepsilon_0^2}(\omega_{in} - \omega_p)^4 \frac{h}{8\pi^2 c \omega_p} S_p \frac{1}{45[1 - \exp(-h\omega_p/k_B T)]}, \quad (2\text{-}11)$$

where ω_{in} is the frequency of incident light, ω_p is the frequency p th vibrational mode, and S_p is Raman scattering factor, a pure property of molecular and can be written as[44]:

$$S_p = 45\left(\frac{\partial \alpha}{\partial Q_p}\right)^2 + 7\left(\frac{\partial \gamma}{\partial Q_p}\right)^2, \quad (2\text{-}12)$$

where α is the isotropic polarizability and γ is the anisotropic polarizability.

The Raman spectrum of PATP in methonal is shown in Fig. 2-10.

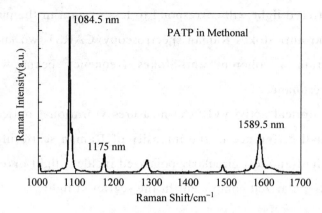

Fig. 2-10 The Raman spectroscopy of PATP in methonal
(For colored figure please scan the QR code on page 1)

Raman shifts are typically reported in wavenumbers, which have units of inverse length, as this value is directly related to energy. In order to convert between the spectral wavelength and the wavenumbers of shift in the Raman spectrum, the following formula can be used:

$$\Delta w(\text{cm}^{-1}) = \left(\frac{1}{\lambda_0(\text{nm})} - \frac{1}{\lambda_1(\text{nm})}\right) \times \frac{10^7 \text{nm}}{\text{cm}}. \qquad (2\text{-}13)$$

There are several kinds of Raman spectroscopy:

(1) Spontaneous Raman spectroscopy. A term used to describe Raman spectroscopy without enhancement of sensitivity.

(2) Resonance Raman spectroscopy. The excitation wavelength is matched to an electronic transition of the molecule or crystal, so that vibrational modes associated with the excited electronic state are greatly enhanced. This is useful for studying large molecules such as polypeptides, which might show hundreds of bands in "conventional" Raman spectra. It is also useful for associating normal modes with their observed frequency shifts.

(3) Stimulated Raman spectroscopy(SRS)-a pump-probe technique, where a spatially coincident, two-color pulse (with polarization either parallel or perpendicular) transfers the population from ground to a revibrationally

excited state. If the difference in energy corresponds to an allowed Raman transition, scattered light will correspond to loss or gain in the pump beam.

(4) Coherent anti-Stokes Raman spectroscopy(CARS)-two laser beams are used to generate a coherent anti-Stokes frequency beam, which can be enhanced by resonance.

(5) Raman optical activity(ROA)-measures vibrational optical activity by means of a small difference in the intensity of Raman scattering from chiral molecules in right-and left-circularly polarized incident light or, equivalently, a small circularly polarized component in the scattered light.

2.3.2 Infrared spectroscopy

Infrared(IR) spectroscopy can selectively absorb certain wavelengths of IR rays and cause the transition of vibrational and rotational levels in molecules. The IR absorption spectra of substances can be obtained by detecting the absorption of IR rays, also known as molecular vibration spectroscopy or vibrational rotation spectroscopy. There are near-, mid-and far-IR, named for their relation to the visible spectrum. The higher-energy near-IR, approximately $14000 \sim 4000$ cm^{-1} ($0.8 \sim 2.5$ μm wavelength)can excite overtone or harmonic vibrations. The mid-IR, approximately $4000 \sim 400$ cm^{-1} ($2.5 \sim 25$ μm) may be used to study the fundamental vibrations and associated rotational-vibrational structure. The far-IR, approximately $400 \sim 10$ cm^{-1} ($25 \sim 1000$ μm), lying adjacent to the microwave region, has low energy and may be used for rotational spectroscopy.

When a continuous wavelength IR light passes through the substance, the group's vibration or rotation frequency in the molecule of matter is the same as that of the IR light. The molecule absorbs energy to the higher energy vibration(rotation) level from the original ground state vibration(rotation) level. After the molecule absorbs IR radiation, vibration and rotational level transitions occur, and that wavelength light is absorbed by matter. IR spectroscopy can be regarded as an analytical method for defining the structure of molecular of substances and detecting compounds based on the information of relative intramolecular atomic vibration and molecular rotation. The IR spectra can be obtained by recording the absorption of IR light by molecules. IR spectroscopy usually uses wave number(σ)or wavelength(λ)as abscissa in order to indicate the location of the absorption peak, and uses absorbance(A) or transmittance($T\%$)as ordinate to indicate absorption intensity. Figure 2-11 is a typical IR spectroscopy.

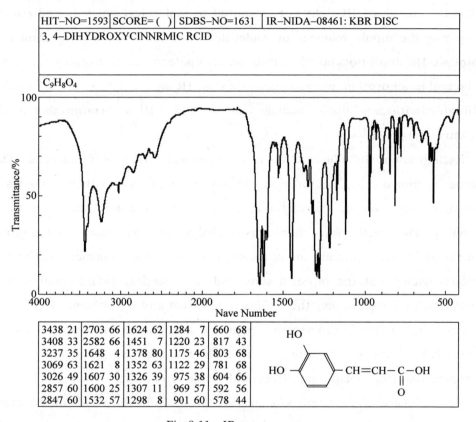

Fig. 2-11　IR spectroscopy

When external electromagnetic wave irradiates the molecule, for example, the energy difference between the irradiated electromagnetic wave and the two levels of the molecule is equal, the electromagnetic wave of this frequency is absorbed by the molecule, which causes the corresponding energy level transition of the molecule, and the macroscopic performance is that the transmitted light intensity decreases. The difference between the energy of electromagnetic wave and the two levels of molecule is one of the conditions that must be satisfied for the material to produce IR absorption spectrum, which decides the location of absorption peak.

The second condition for the generation of IR absorption spectra is that a dual effect between molecules and IR light exists. Therefore, the dipole moment of the molecule must change when it vibrates. This actually ensures that the energy of IR light is transmitted to molecules, which is realized by changing the dipole moment of molecular vibration. Not all vibration can produce IR absorption, but the vibration of dipole moment changing can cause observable absorption phenomenon, that is, IR active vibration; molecular vibration with zero dipole moment can not create IR absorption, that is, IR inactive vibration.

Each group of the molecule has its own particular IR absorption peak. The same functional group's absorption vibration always occurs at the narrow wave number range even in the different compounds, but not in the fixed wave number. The specific wave number is related to the environment of the group in the molecule. There are many factors that cause the frequency shift of the group, among that the physical state and the chemical environment of the molecule are dominations: the temperature effect and the solvent effect. For the internal factors leading to the frequency shift of groups, there are known electrical effects of substituents in molecule, such as induction effect, conjugation effect, mediation effect, and dipole field effect.

The intensity of IR band is a measure of the probability about vibration transition, that is related to the change of dipole moment when molecule vibrates[26]. The intensity of band will increase with the increasing of the dipole moment's change. The change of dipole moment is associated with the inherent dipole moment of the group itself[26]. Therefore, the bigger polarities of the group, the move obvious change of the dipole moment and the greater absorption band. The higher the symmetry of the molecule, the smaller change of the dipole moment and the weaker absorption band[45].

2.3.3 Modes of molecular vibration

The vibration of molecular has two types: the stretching one and the bending one. The former refers to reciprocating motion of atoms along bond axis, and the bond length changes during the vibration process. The latter refers to the vibration of atoms perpendicular to chemical bonds. Generally, different symbols are used to express the different vibrational forms. Stretching vibration is separated to the symmetrical and the antisymmetrical ones, represented by Vs and Vas respectively (Fig. 2-12). Bending vibration is separated to in-plane bending vibration including scissoring bending and in-plane rolling vibration and out-of-plane bending vibration including out-of-plane rolling vibration and twisting vibration (Fig. 2-13). In theory, each fundamental vibration can absorb IR light at the same frequency, and there is an absorption peak at the corresponding position of the IR spectrum. In fact, some vibrating molecules are inactive in IR without dipole moment change. Others have the same frequency and degenerate. Some vibrating frequencies are beyond the scope of the instrument, which makes absorption peaks' number in actual IR spectra much poorer than the value of theories.

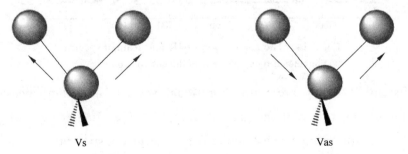

Fig. 2-12 The types of stretching vibration
(For colored figure please scan the QR code on page 1)

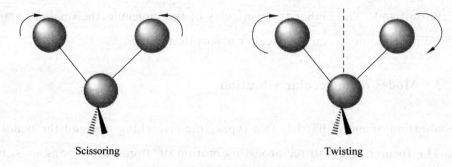

Fig. 2-13 The types of bending vibration
(For colored figure please scan the QR code on page 1)

2.3.4 The difference between Raman and spectra

1. Vividly speaking, the IR spectrum is concave and the Raman spectrum is convex. The two complement each other, that is to say, IR is strong, Raman is weak, and vice versa. But in some cases, the information detected by the two methods is the same (Fig. 2-14).

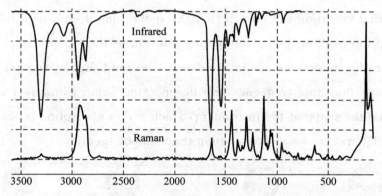

Fig. 2-14 The comparison of IR and Raman spectra
(For colored figure please scan the QR code on page 1)

2. Essentially, they are both vibrational spectra, and they measure the excitation or absorption of the ground state, with the same energy range. But IR is the absorption spectrum, Raman is the scattering spectrum.

3. The selectivity of spectra is different. IR requires that the dipole moment of the molecule change before it can be detected, and Raman is that the polarizability of the molecule changes before it can be detected.

4. IR is easy to measure, and the signal is very good, while the Raman signal is very weak.

5. The wavelength range used is different. IR uses IR light, especially mid-IR, which limits the use. Raman can choose a lot of wavelengths, from visible light to (NIR) spectroscopy.

6. IR can not detect less than 400 wave number. IR is more suitable for organic matter, Raman is more suitable for inorganic matter. IR radiation is greatly disturbed by water.

2.4 Rotational State

Rotational spectroscopy involves measuring the energy of the transition between molecular rotational states of gas molecular weight. The spectra of polar molecules can be measured by far-IR spectroscopy or microwave spectroscopy[46]. The nonpolar molecules'rotational spectrum can't be observed by these methods, but Raman spectra can be used to observe them. Rotational spectroscopy is usually called pure rotational spectroscopy. It is used to distinguish rotational vibration spectroscopy in which rotational energy changes with the change of vibrational energy, or in which electronic, vibrational and rotational energy changes occur at the same time.

In the mechanics quantum, free rotation of molecules is quantized, and therefore, the angular momentum and rotational energy can only take the fixed value, which is only related to the molecule's rotary inertia (I). Three rotary inertia I_A, I_B and I_C for any molecule, and the origin is at the center of the system's mass. This book defines axes $I_A \leqslant I_B \leqslant I_C$ as the general convention, and axis A corresponds to the minimum moment of inertia. For rotational spectra, molecules can be divided into spherical top, linear top and symmetric top according to their symmetry. All three inertia moments of the spherical top (spherical rotor) are equal, that is, $I_A = I_B = I_C$. Such as P_4, CH_4 and CCl_4

and other hexahalides. For the linear molecule, the rotary inertias follow the relationship: $I_A \ll I_B = I_C$. In most cases, I_A can set zero value. Illustrations of the linear molecules are O_2, N_2, OH and CO_2 and so on. If the two inertia's moments are equal: $I_A = I_B$ or $I_B = I_C$, which are the characteristics of the symmetric top. By definition, the symmetrical top must have three times or higher orders rotation axis. For convenience, spectroscopists divided the molecules into two types of symmetrical tops including the prolate symmetric tops with $I_A < I_B = I_C$ and the oblate with $I_A = I_B < I_C$. Their corresponding spectra are very different and can be easily recognized. Common symmetric tops are benzene, ammonia, C_6H_6, CH_3Cl, chloromethane and propyne, and the first three belong to the oblate and the latter three belong to the prolate.

For microwave spectra, it is not able to be detected in the centrosymmetric linear molecules because the rule that transitions between the rotational states are able to be detected in molecules with the permanent electric dipole moments[28]. Tetrahedral molecules, for example, methane have zero-isotropic polarizability and dipole moments, and will not have pure rotation spectra except for the centrifugal distortion effect. When the molecules rotate around a three-fold symmetric axis, the small dipole moment will be generated, so that weak rotation spectra can be observed in microwave spectra[47]. For the symmetric top, the selection rules of pure rotation transition allowed by electric dipole are $\Delta K = 0, \Delta J = \pm 1$. Since these transitions are caused by the emission (or absorption) of a spin-1 single photon, the angular momentum's conservation shows that angular number can change only one unit at most[48]. In addition, the value of quantum number K is restricted with the values between $+J$ and $-J$[49].

For Raman spectroscopy, a molecule undergoes a transition from absorbing incident photons to emitting scattered photons. The general rule of selection for such transition is that molecule's polarizability must be anisotropic, that is, it varies in all directions[47]. The polarized ellipsoids of the spherical apex are actually spherical, so these molecules do not rotational Raman spectra. For all

CHAPTER 2 Molecular Spectroscopy

the other molecules, Stokes lines and anti-Stokes lines should be both detected, because some rotational states are usually thermally filled, they have similar strength. For linear molecule, the rule of selection is $\Delta J = 0, \pm 2$. That is because the polarizabilities return to the same values two times during rotation[49]. The value of $\Delta J = 0$ doesn't associate with the molecular transition, but corresponds to Rayleigh scattering that the incident photon only changes its direction[30]. The rule of selection for symmetric top is

$$\Delta K = 0;$$
$$\text{If } K = 0, \text{then } \Delta J = \pm 2; \quad (2\text{-}14)$$
$$\text{If } K \neq 0, \text{then } \Delta J = 0, \pm 1, \pm 2.$$

2.5 Electronic and Vibrational Spectroscopy by Circularly Polarized Light

The circularly polarized light, which describes the circular trajectory at the end of the rotating electric vector, is the special situation of elliptically polarized light[50]. When the propagation direction is the same, the circularly polarized light with regular change of electric vector can be synthesized by superposition of two teams planar-polarized light whose vibration direction is vertical to each other and whose phase difference is the constant as $\varphi = \left(2m \pm \dfrac{1}{2}\right)\pi$. Circularly polarized waves have two possible states: one is the right circular polarization, where the electric field vector rotates in the right-hand direction relative to the spreading direction, and the other is the left circular polarization, where its vector rotates in the opposite direction, that is, left-hand direction.

2.5.1 Electronic circular dichroism

Electronic circular dichroism (ECD) is refers to differential absorption between the left-handed and the right-handed circularly polarized light.

Circular dichroism spectroscopy is a fast, simple and accurate method widely used to determine the proteins' secondary structure and the conformation in dilute solution. Note that the electronic states absorption of the circularly polarized light and ECD spectroscopy can expressed as[51]

$$I \propto |\langle \varphi_j | \mu_e | \varphi_i \rangle E|^2 + |\langle \varphi_j | \mu_e | \varphi_i \rangle \langle \varphi_j | \mu_m | \varphi_i \rangle B|^{2'}, \quad (2\text{-}15)$$

where, E and B are the electric and magnetic field, respectively; μ_e and μ_m are the electric and magnetic transition dipole moments, respectively; $|\varphi_i\rangle$ and $|\varphi_j\rangle$ are the initial and final wave functions, respectively. In this equation, the

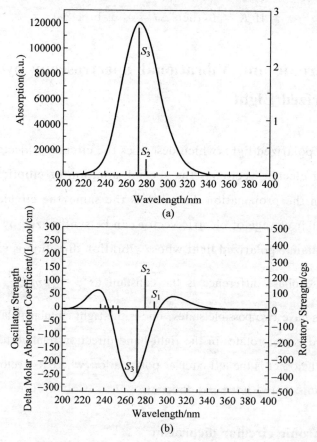

Fig. 2-15 (a) Optical absorption spectra and (b) ECD spectra of the biphenylene groups of n-$C_{12}H_{25}$O-BP-CO-Ala-Ala dipeptides

(For colored figure please scan the QR code on page 1)

former term represents the absorption spectra and the latter term stands for the ECD spectra[52], which reveals the difference between the absorption excited by linear and the circularly polarized light[52].

2.5.2 Raman optical activity

Raman optical activity (ROA)[53-55] is the vibration spectroscopy technique depending on the difference between the intensities of the left and right circularlies polarized light of Raman scattering caused by molecular chirality[56-58]. Therefore, the circular intensity difference was proposed by Buckingham and Barron[53], which is dimensionless and can provide the experimental observation for this phenomenon:

$$\Delta = \frac{|I_L - I_R|}{|I_L + I_R|}, \qquad (2\text{-}16)$$

where I_L and I_R represent the intensites of the Raman scattering about the left-and right-handed circularly polarized light, respectively. Moreover, ROA could be expressed as[59]

$$I \propto \left| \sum_{j \neq i} \frac{\langle \varphi_j | \mu_e | \varphi_k \rangle \langle \varphi_j | \mu_m | \varphi_i \rangle}{\omega_{ji} - \omega_0 + i\Gamma} + \sum_{j \neq i} \frac{\langle \varphi_j | \theta_e | \varphi_i \rangle \langle \varphi_j | \mu_e | \varphi_i \rangle}{\omega_{ji} - \omega_0 + i\Gamma} \right|^2 E^4, \qquad (2\text{-}17)$$

where, μ_e and μ_m are the electric and magnetic transition dipole momens; respectively; θ_e is the quadrupole moment of electric transition; $|\varphi_i\rangle$ and $|\varphi_j\rangle$ are the initial and final wave functions, respectively; ω_{ji} represents the transition energy; $i\Gamma$ represents an imaginary number term proportional to the width of the state j and inversely proportional to itself lifetime[41]. Sun et al. using the ROA(Fig. 2-16) to judge if the selected experimental subject is the excellent candidate for studying remote excitation ROA with chiral plasmon propagation of Ag NW[60].

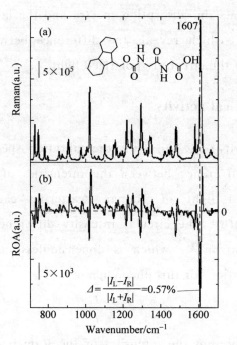

Fig. 2-16 (a) Raman spectrum; (b) ROA spectrum[60]

(For colored figure please scan the QR code on page 1)

References

[1] CRUZ E M, LOPEZ X, AYERBE M, et al. G_2 study of the triplet and singlet $[H_3, P_2]^+$ potential energy surfaces. Mechanisms for the reaction of P^+ ($^1D, ^3P$) with PH_3[J]. The Journal of Physical Chemistry A, 1997, 101: 2166-2172.

[2] ATKINS P, DE PAULA P, ATKINS J. Physical chemistry[M]. New York: Oxford University Press, 2002.

[3] BALL D W. [Bouguer]-Lambert-Beer's law [J]. Spectroscopy-Springfield then Eugene then Duluth, 1999, 14: 16-17.

[4] GOPPERT-MAYER M. Über elementarakte mit zwei quantensprüngen[J]. Ann. Phys. Lpz., 1931, 9: 273-294.

[5] ALBOTA M, BELJONNE D, BRÉDAS J, et al. Design of organic molecules with large two-photon absorption cross section[J]. Science, 1998, 281: 1653-1656.

[6] DENK W, STRICKLER J H, WEBB W W. Two-photon laser scanning fluorescence microscopy[J]. Science, 1990, 248: 73-76.

[7] KÖHLER R H, CAO J, ZIPFEL W R, et al. Exchange of protein molecules through connections between higher plant plastids[J]. Science, 1997, 276: 2039-2042.

[8] LI R, ZHANG Y J, XU X F, et al. Optical characterizations of two-dimensional materials using nonlinear optical microscopies of CARS, TPEF, and SHG [J]. Nanophotonics, 2018, 7: 873-881.

[9] PARTHENOPOULOS D A, RENTZEPIS P M. Three-dimensional optical storage[J]. Science, 1989, 245: 843-845.

[10] CUMPSTON B H, ANANTHAVEL S P, BARLOW S, et al. Two-photon polymerization initiators for three-dimensional optical data storage and microfabrication[J]. Nature, 1999, 398: 51-53.

[11] BAUER C, SCHNABEL B, KLEY E B, et al. Two-photon pumped lasing from a two-dimensional photonic bandgap structure with polymeric gain material[J]. Adv Mater, 2002, 14: 673-676.

[12] LI Y Z, XU B B, SONG P, et al. D—a—Π—a system: light harvesting, charge transfer, and molecular designing[J]. J Phys Chem C, 2017, 121: 12546-12561.

[13] LI Y Z, PULLERITS T, ZHAO M Y, et al. Theoretical characterization of the PC60BM: PDDTT model for an organic solar cell[J]. J. Phys. Chem. C, 2011, 115: 21865-21873.

[14] MILLER M J, WEI S H, PARKER I, et al. Two-photon imaging of lymphocyte motility and antigen response in intact lymph node[J]. Science, 2002, 296: 1869-1873.

[15] LI R, WANG X X, ZHOU Y, et al. Advances in nonlinear optical microscopy for biophotonics[J]. Nanophotonics, 2018, 12: 033007.

[16] KIM S, OHULCHANSKYY T Y, PUDAVAR H E, et al. Organically modified silica nanoparticles co-encapsulating photosensitizing drug and aggregation-enhanced two-photon absorbing fluorescent dye aggregates for two-photon photodynamic therapy[J]. J. Am. Chem. Soc., 2007, 129: 2669-2675.

[17] KANIS D R, RATNER M A, MARKS T J. Design and construction of molecular assemblies with large second-order optical nonlinearities [J]. Quantum chemical aspects. Chem. Rev., 1994, 94: 195-242.

[18] BREDAS J L, ADANT C, TACKX P, et al. Third-order nonlinear optical response in organic materials: theoretical and experimental aspects[J]. Chem. Rev., 1994, 94: 243-278.

[19] RICHTER M, MUKAMEL S. Collective two-particle resonances induced by photon entanglement[J]. Phys. Rev. A, 2011, 83: 063805.

[20] DORFMAN K E, SCHLAWIN F, MUKAMEL S. Nonlinear optical signals and spectroscopy with quantum light[J]. Rev. Mod. Phys, 2016, 88: 045008.

[21] SUN M T, CHEN J N, XU H X. Visualizations of transition dipoles, charge transfer, and electron-hole coherence on electronic state transitions between excited states for two-photon absorption[J]. J. Chem. Phys, 2008, 128: 064106.

[22] MIKHAYLOV A, UUDSEMAA M, TRUMMAL A, et al. Spontaneous symmetry breaking facilitates metal-to-ligand charge transfer: a quantitative two-photon absorption study of ferrocenephenyleneethynylene oligomers [J]. J. Phys. Chem.

Lett.,2018,9: 1893-1899.
[23] ACUÑA A U,AMAT-GUERRI F,MORCILLO P,et al. Structure and formation of the fluorescent compound of lignum nephriticum[J]. Organic Letters,2009,11(14): 3020-3023.
[24] SAFFORD W E, LIGNUM N. Annual report of the board of regents of the smithsonian institution[M]. Sunda: Suda Press,1916,271-298.
[25] VALEUR B, BERBERAN-SANTOS M R N. A brief history of fluorescence and phosphorescence before the emergence of quantum theory[J]. Journal of Chemical Education,2011,88(6): 731-738.
[26] MUYSKENS M ED V. The fluorescence of lignum nephriticum: a flash back to the past and a simple demonstration of natural substance fluorescence[J]. Journal of Chemical Education,2006,83(5): 765.
[27] CLARKE E D. Account of a newly discovered variety of green fluor spar, of very uncommon beauty, and with remarkable properties of colour and phosphorescence [J]. The Annals of Philosophy,1819,14: 34-36.
[28] BREWSTER D. On the colours of natural bodies[J]. Transactions of the Royal Society of Edinburgh,1834,12(2): 538-545.
[29] HERSCHEL J. On a case of superficial colour presented by a homogeneous liquid internally colourless[J]. Philosophical Transactions of the Royal Society of London, 1845,135: 143-145.
[30] STOKES G G. On the change of refrangibility of light[J]. Philosophical Transactions of the Royal Society of London,1852,142: 463-562.
[31] LAKOWICZ J R. Principles of fluorescence spectroscopy[M]. 3rd Edition. New York: Springer Science,2006.
[32] VALEUR B, BERBERAN-SANTOS M. Molecular fluorescence: principles and applications[M]. Weinheim: Wiley-VCH,2012,64.
[33] ATTA D. Time resolved single molecule fluorescence spectroscopy on surface tethered and freely diffusing proteins [M]. Verlag: Forschungszentrum Jülich GmbH Zentralbibliothek,2012.
[34] JIANG J,YU J,CAO S. Au/PtO nanoparticle-modified g-C_3N_4 for plasmon-enhanced photocatalytic hydrogen evolution under visible light[J]. Journal of Colloid and Interface Science,2016,461: 56-63.
[35] BATTAGLIA D,PENG X. Formation of high quality InP and InAs nanocrystals in a noncoordinating solvent [J]. Nano Lett.,2002,2: 1027-1030.
[36] LOUIS J, KUMAR A S. Unusual autofluorescence characteristic of cultured red-rain cells[C]. SPIE Conference 7097,2008.
[37] NARAYANAN J. Spectroscopy in foundamental and biophotonics in dentistry[M]. Singapore: World Scientific,2006.
[38] CHENG P C. Handbook of biological confocal microscopy[M]. New York: Springer, 2006: 162-206.

[39] HELMS V. Principles of computational cell biology[M]. New Jersey: John Wiley & Sons, 2008.

[40] ZONG H, WANG X, MU X, et al. Plasmon-enhanced fluorescence resonance energy transfer[J]. Chem. Rec., 2019, 19: 818-842.

[41] CLEGG R. Förster resonance energy transfer—FRET: what is it, why do it, and how it's done[J]. Elsevier, 2009, 33: 1-57.

[42] ZONG H, MU X, WANG J, et al. The nature of photoinduced intermolecular charger transfer in fluorescence resonance energy transfer[J]. Spectrochimica Acta Part A: Molecular and Biomolecular Spectroscopy, 2019, 209: 228-233.

[43] ZHAO X, LIU S, LI Y, et al. DFT study of chemical mechanism of pre-SERS spectra in Pyrazine-metal complex and metal-Pyrazine-metal junction[J]. Spectrochimica Acta Part A Molecular & Biomolecular Spectroscopy, 2010, 75(2): 794-798.

[44] SUN M, XIA L, CHEN M. Self-assembled dynamics of silver nanoparticles and self-assembled dynamics of 1, 4-benzenedithiol adsorbed on silver nanoparticles: surface-enhanced Raman scattering study[J]. Spectrochimica Acta Part A Molecular & Biomolecular Spectroscopy, 2009, 74(2): 509-514.

[45] MA H, LI X, YE R, et al. Crystal structure and color point tuning of β-Sr1.98-yMgySiO$_4$-1.5xNx: 0.02Eu^{2+}: a single-phase white light-emitting phosphor[J]. Journal of Alloys and Compounds, 2017, 703: 486-499.

[46] GORDY W A. Microwave molecular spectra in technique of organic chemistry[M]. IX. New York: Interscience, 1970.

[47] HOLLAS M J. Modern spectroscopy[M]. 3rd ed. Hoboken: Wiley, 1996: 95-111.

[48] ATKINS P W, DE PAULA J. Physical chemistry [M]. 8th ed. New York: Oxford University Press, 2006: 431-469.

[49] BANWELL C N, MCCASH E M. Fundamentals of molecular spectroscopy[M]. 4th ed. New York: McGraw-Hill, 1994.

[50] KUCH W, RUDOLF S, PETER F, et al. Magneto-optical effects[M]. New York: Springer, 2015.

[51] WARNKE I, FURCHE F. Circular dichroism: electronic[J]. WIREs Comput. Mol. Sci., 2012, 2: 150-166.

[52] MU X, WANG J, DUAN G, et al. The nature of chirality induced by molecular aggregation and self-assembly [J]. Spectrochimica Acta Part A: Molecular and Biomolecular Spectroscopy, 2019, 212: 188-198.

[53] BARRON L D, BUCKINGHAM A D. Rayleigh and Raman scattering from optically active molecules[J]. Mol Phys, 1971, 20: 1111-1119.

[54] BARRON L D, BOGAARD M P, BUCKINGHAM A D. Raman scattering of circularly polarized light by optically active molecules[J]. J Am Chem Soc, 1973, 95: 603-605.

[55] HUG W, KINT S, BAILEY G F, et al. Raman circular intensity differential spectroscopy. Spectra of (2)-a-pinene and (1)-a-phenylethylamine[J]. J Am Chem Soc, 1975, 97:

5589-5590.
[56] SEBESTIK J, BOUR P. Raman optical activity of methyloxirane gas and liquid[J]. J. Phys. Chem. Lett. ,2011,2: 498-502.
[57] HOPMANN K H, SEBESTI´K J, NOVOTNA J, et al. Determining the absolute configuration of two marine compounds using vibrational chiroptical spectroscopy[J]. J Org Chem,2012,77: 858-869.
[58] HAESLER J, SCHINDELHOLZ I, RIGUET E, et al. Absolute configuration of chirally deuterated neopentane[J]. Nature,2007,446: 526-529.
[59] LAURENCE A. Encyclopedia of spectroscopy and spectrometry[M]. 3rd edition. Elsevier: Elsevier Science and Technology,2017: 891.
[60] SUN M, ZHANG Z, WANG P, et al. Remotely excited Raman optical activity using chiral plasmon propagation in Ag nanowires[J]. Light: Science & Applications,2013, 2: e112.

CHAPTER 3

Photonics and Plasmonics

3.1 Introduction

Electronic circuits provide us with the ability to control the transport and storage of electrons. However, their performance is now becoming rather limited when digital information needs to be sent from one point to another. Photonics offers an effective solution to this problem by implementing optical communication systems based on optical fibers and photonic circuits. Unfortunately, the micrometer-scale bulky components of photonics have limited the integration of these components into electronic chips, which are now measured in nanometers. Surface plasncon-based circuits, which merge electronics and photonics at the nanoscale[1], may offer a solution to this size-compatibility problem. So, plasmonics can be considered as a kind of nanophotonics. In this chapter, we introduce the excitation kinds of plasmons and plasmonics, which are mainly focused on the exciton and polariton.

3.2 Exciton

3.2.1 Brief introduction of excitons

In 1931, Frenkel (Fig. 3-1) first proposed the concept about excitons, he indicated that it was possible for neutral excitation of a crystal through light where electrons remained bound to the holes generated at the lattice sites, the lattice positions were determined to be quasiparticles, i.e. exciton[2]. Excitons were electrically neutral and were mostly found in insulators, semiconductors and certain liquids[2-3]. When photons were absorbed by a semiconductor, excitons were generated, causing electrons to transition from the valence band to the conduction band. The decay of exciton, in other words, the recombination about hole

Fig. 3-1 Yakov Il'ich Frenkel

and electron, was limited by the resonance stability owing to the overlap of hole and electron wave functions, causing a prolonged lifetime about the exciton.

Since the discovery of the excitons, their lifespan has been very short, only 10 microseconds, and the energy extremely low, which made the study of exciton movement very difficult. However, due to the development of laser, low-temperature and electronic technologies, extremely favorable conditions had been provided for the study of surface excitons. Because the laser could emit a monochromatic beam of intense energy with a large intensity, the intensity of the light indicated that the number of photons contained in the beam increases, which caused a large number of excitons to be generated in the crystal. In 1966, Haynes first observed electron-hole droplets, that is, a large number of exciton condensates, which led to a new phase of exciton research. Later, the researchers studied fluid quantum mechanics. In addition, the formation of crystal defects and motion and exciton were linked, and certain results had been achieved. Therefore, the study of excitons was currently a hot area in solid state physics. Moreove addition, excitons were divided into many types, such as the Frenkel[4-14], the Wannier-Mott[15], the charge-transfer[16-23], the surface[24-26], and the atomic and molecular excitons[27-28], as well as the self-trapping of excitons[29-30].

3.2.2 Exciton classification

Frenkel excitons

The Frenkel excitons named after him, has a binding energy of 0.1 eV to 1 eV. And it had considerable interaction cross sections having molecular vibrations. Particularly, about coherently coupled Frenkel excitons, the rate of exciton scattering was significantly enhanced. Furthermore, the application of Frenkel excitons was very extensive. For example, in previous study, the J band in the PIC aggregates in the glass and Langmuir-Blodgett (LB) films could be described by the disordered Frenkel exciton band, it was proved that resonance light scattering was an important tool for studying exciton dynamics in polymers and aggregates. And in organic semiconductor microcavities, the studies of photon-mediated hybridization of Frenkel excitons was demonstrated. Microcavities were artificial structures of wavelength dimension in which coupling could be excited by their mutual interaction with restricted photon modes. Energy storage, capture and transfer between excitations of coherently coupled molecules could be studied. Studies had shown that it was possible to generate mixed exciton materials consisted of coherently coupled excitons in a microcavity. These novel hybrid exciton-photon structures had probable research interests as model systems for studying energy capture and storage and energy transfer between excitations of coherently coupled molecules was of great help. Especially the bright and dark states might allow for the study of energy capture and storage functions. If energy transfer was indeed feasible, the Frenkel-Wannier hybrid exciton device will be more widely used.

There had some progress in the study about the interference between the charge-transfer (CT) state and the Frenkel exciton for the perylene pigments crystallochromy. For compounds having smaller holes and electron transfer, such as diindenoperylene (DIP) and 3,4,9,10-perylene tetracarboxylic dianhydride (PTCDA), their molecular structural formulas were shown in

Fig. 3-2. For materials with relatively large electron and hole transport, the exciton model allowed the energy about the CT transition to be determined down the stacking direction with an accuracy better than 0.05 eV, which was much lower than the systematic deviations produced by the general calculation. In addition, for the breakdown about the Frenkel exciton model, the exciton band structure about pentacene molecular solids was applied here. Studies had shown that the exciton dispersion behavior of pentacene measured from electron energy-loss spectroscopy(EELS) was inconsistent with Frenkel-like electron-hole pairs, and further research is needed. And, there were of course many other applications, such as in-band relaxation about the fluorescence decay time about 1D Frenkel excitons, and low-energy spectra about 1D local Frenkel excitons of Hidde structures.

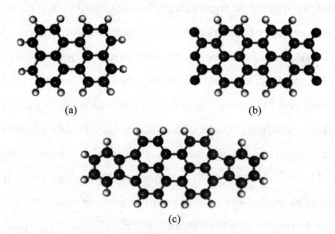

Fig. 3-2 The structures of (a) Perylene, (b) PTCDA and (c) DIP[8]
(For colored figure please scan the QR code on page 1)

Wannier-Mott excitons

Excitons were the basic optical excitation of dielectric solids. Two excitons were created in nature, namely the Frenkel exciton with smaller radius and the Wannier exciton with larger radius. Among them, Wannier excitons were formed in materials used in semiconductor technology. Because of the different radii of the two excitons, they had completely different properties. Wannier exciton interaction density is low, resulting in a multi-particle effect which is related to laser action and optical nonlinearity; while the Frenkel excitons, due to their small radius, barely saw each other but produce huge features in the spectrum[15].

Wannier-Mott excitons had a large dielectric constant, therefore, the electric field shielding effect weakened the Coulomb interaction between holes and electrons, i. e. , Wannier excitons, whose radius was greater than the lattice spacing. The small effective electron mass typical for semiconductors favors excitons with larger radii. Compared with hydrogen atoms, owing to its small mass and Coulomb interaction, its binding energy was small, about 0.01 eV.

In recent years, inorganic and organic semiconductor nanostructures had been the subject of much research. There was a need to synthesize composite inorganic or organic semiconductor heterostructures for using in the synthesis of novel nanostructures of optics, electrons, and transport to further understand their size-dependent physical properties. Hybrid excitons could be generated from the resonant mixing about Frenkel-Wannier-Mott excitons in inorganic or organic quantum wells, which was a new elementary state produced by optical excitation. This hybrid exciton exhibited unique physical properties while providing a foundation for new electro-optic technology. This unique mixed exciton had the properties of two excitons. In addition, the resonator of the Frenkel exciton was quite strong, and the Wannier-Mott exciton had a large radius and was delocalized.

In previous studies, the authors had research the nonlinear and linear optical properties about inorganic-organic nanostructure caused by resonance interactions between Wannier-Mott excitons in semiconductor quantum wells (QWs) and Frenkel excitons in organic QWs. For the Coulomb dipole-dipole interacts and through the cavity photons in the microcavity, this coupling resulted in the mixing of Frenkel-Wannier excitons. Studies had shown that enhancement of resonance optical nonlinearity, relaxation processes, and the fluorescence efficiency could be achieved by adjusting this hybrid state and the plotted dispersion curve.

It was known that in the physical system, there were two quantum wells, the inorganic semiconductor and the molecular solid. Therefore, the authors considered a novel kind of electronic excitation owing to Frenkel and Wannier exciton hybridization. The physical system contained a 2D layer separated by a distance z_0, which was a tightly bound Frenkel excitons (FE) and a loosely bound Wannier excitons (WE), and they had the same energy. These two were referred to as organic quantum wells and inorganic quantum wells, respectively, and it was worth noting that this model was equally applicable to any heterostructure having a layer that maintains large and small radius resonant excitons. The real excitation in the physical system was a hybrid exciton (HE) owing to the dipole-dipole interaction between WE and FE. When EF and WF were almost completely mixed, 2D HE were most likely to display strong optical nonlinearitiys in some fields. For example, excitons with smaller radii had larger oscillator strength, and excitons with larger radii had a smaller saturation density.

In 1998, the authors had proposed a new type of HF in organic-inorganic heterostructures, and had researched some nonlinear optical properties in theory. For ordinary semiconductor quantum wells, it could be predicted that the linear and nonlinear portions of the susceptibility coefficient will produce strong enhancement. Studies had shown that the above results were consistent with FE with larger oscillator strength and WE with smaller saturation densities, both of which were hybrid resonances. If the above structure could be synthesized successfully, it will show significant nonlinearity. In addition, hE were also important systems for studying the properties of electro-optical and magneto-optical properties. To further demonstrate the above experimental results, the authors also studied resonant inorganic and organic quantum wells in microcavities. Although only a hypothesis, the authors believed that in this structure the exciton relaxation time could be reduced to a state with a larger radiation width and a shorter fluorescence decay time. The authors also hypothesized that the combination of the electrical pumping about excitons with fast relaxation in inorganic quantum wells and the fluorescence about excitons in organic quantum wells would provide new ways to study the exciton process in microcavities, which will also be applied to more devices.

Charge-transfer excitons

Between the WE and the FE, there was a special excitons called charge-transfer excitons (CT excitons)[16]. When an electron and a hole separatelg occupy adjacent molecules, the exciton was generated. CT excitons were mainly present in ionic crystals. Unlike the WEs and the FE, they were capable of exhibiting electrostatic dipole moments. CT excitons could be generated at the organic semiconductor surface and interface, i.e. the donor and acceptor interfaces in an organic heterojunction solar cell. According to the classic dielectric constant about organic semiconductors and the size about conjugated molecules, it was estimated that the binding energy about CT excitons at the acceptor and donor interfaces will be an order of magnitude larger than $k_B T$ in condition of room temperature, where T was temperatureand k_B was the Boltzmann constant.

The use of CT excitons was widespread. For example, in organic photovoltaic (OPV) devices, the generation of photocurrents relied on the dissociation about excitons into free electrons and holes at the acceptor/donor heterointerface. Since the organic semiconductors dielectric constant was low, the strong interaction between electrons and hole pairs effectively prevented the generation of free charges, and electrons and holes will overcome the capture of Coulomb interactions to some extent. Studies had shown that thermal CT excitons played a key role in this process. The authors used femtosecond nonlinear spectroscopy and non-adiabatic hybrid quantum mechanics to generate thermal CT excitons using the phthalocyanine-fullerene model OPV system. Studies had shown also that when the reaction was carried out for 10~

13 s, that is, the phthalocyanine just started to excite, it will generate thermal CT excitons, then reduce the energy and shorten the distance between electron-holes. At the same time, as mentioned above, for typical molecular size and dielectric constant, the interfacial CT exciton binding energy of several hundred millivolts of electrons was estimated to be an order of magnitude higher than the thermal energy at the condition of room temperature.

Another important application was the use of the FE with the lowest energy and CT excitons in the quasi-1D structure for N,N'-dimethylperylene-3,4,9,10-dicarboximide(MePTCDI) and PTCDA crystals. The authors proposed a Hamiltonian that included some vibrating FE and a nearest neighbor CT exciton that could describe the polarization direction, peak intensity, and energy position of a 1D crystal for a molecule that contained only one molecule. The intrinsic oscillator strength of crystal and CT excitons was very small. Electro-absorption measurement was the most ideal direct observation tool. It proved that CT excitons had large internal transition dipole moments in quasi-1D crystals having strong orbital overlap, which seriously affected the polarization direction about the mixed excitons. However, the CT transition dipole reached a polarization ratio for a three-dimensional (3D) crystal containing two molecules and having a weak coupling between 1D stacks. This mechanism directly demonstrated the mixing about Frenkel and CT excitons in MempCDI, which was not directly related to electroabsorption measurements[17].

Other applications of CT excitons included polymer/fullerene mixtures, bulk heterojunctions of polyfluorene copolymers and fullerene derivatives, van der Waals interfaces, polymer semiconductor heterojunctions, organic solar cells, narrow band gap polymer-based volume heterojunctions, and organic solids in first principle[18-22]. Among them, the molecular donor/acceptor and the van der Waals interface of the graphene-based 2D semiconductor were the key to the photo-electrical mutual conversion, including photodetectors, light-emitting diodes and solar cells. A distinguishing characteristic of the two van der Waals interfaces was the poorly shielded Coulomb potential, resulting in bound electron-hole pairs, i. e. CT excitons. For organic solar cells, the dissociation efficiency about the CT state of the weakly bonded interface was very significant for organic heterojunction solar cells. And the authors examined a variety of donor polymers and acceptors through photoluminescence (PL) quenching to observe that the dissociation is not different from CT excitons and FE. Studies had shown that the field-dependent photocurrent about pure polymers was related to the quenching effect. However, the correlation between CT exciton quenching and photocurrent was not significant. It was worth noting that for pure polymers, PL and electroluminescence were the same, but red shift of blend electroluminescence could still be observed. The above indicated that the energy in the blending was low and was not visible in PL. Thus, the luminescent state of the PL-detected blend was produced by photocurrent[23].

Surface excitons[24-26]

In previous studies, the theory about surface excitons in molecular crystals showed that the presence of surface excitons and their states on or below the exciton state were based on the nearest neighboring environmental displacement term and the exciton transfer term determines. In addition, the surface exciton's localization energy was not sensitive to the thickness of crystal.

Surface excitons were widely used. For example, ionic crystals and rare gases. Here was a brief introduction to the application of surface excitons in ZnO crystals, which were n-type semiconductors with upwardly band bending on the surface, and generated by negative acceptor-like surface states. When the light was excited, the surface electric field separated the photocarrier from the surface, and the photocarrier was accompanied by movement to the bulk of the electrons and holes. Note that the surface excitons studied here had very specific time behaviors. For bound excitons, their decay time was very short.

The study of the radiation and non-radiative decay mechanisms about surface excitons needed to require more in-depth research. Therefore, the authors suggested that the surface-bound excitons decay might be related to low temperatures.

Surface excitons could still be observed in rare gases. The first evidence showed that surface excitons were observed only under ultrahigh vacuum (HHV) conditions, and their strength was quickly reduced even in very small ranges, such as 10^{-10} Torr. More than twice as much as half an hour. The residual gas adsorbed by the single layer causes the surface excitons to disappear. When covering different rare gas films, the surface excitons will disappear, while the surface excitons of the overlay will emerge.

For studying the dependence of surface excitons on film thickness, since the transmittance of each film was constant, the transmittance of the multilayer film decreased with increasing thickness. The authors speculated that surface excitons caused the sample to be limited to 1 layer to 2 layers. In addition, studies had shown that the position of excitons, splitting, and the strength of the oscillator needed to be considered. Compared to bulk excitation, surface excitation was caused by changes in the spatial environment close to the surface.

Another important application was volume and surface excitons in solid neon. Rare gas solids (RGSs) were prototype materials for insulators because RGS had a relatively simple electronic structure at the ground state. The valence band was produced by the relatively weak van der Waals force in the crystal and the outermost closed p-shell electron[24].

Atomic and molecular excitons

In previous studies, molecular exciton models were used to deal with excited-state resonance interactions in weakly coupled electronic systems, which was seen as an important tool for studying the photochemistry and spectroscopy of complex molecules. For composite molecules, loosely bound light absorbing units were bonded together by van der Waals forces or hydrogen bonds.

Molecular excitons had some special characters, one of which was Förster resonance energy transfer. If the energy between the molecular excitons and the spectral absorption of the second molecule matches, the excitons could jump from one molecule to another. The process was related to the intermolecular distance between species in solution, so molecular excitons could be widely used in molecular scales and sensing.

Förster (also known as fluorescence) resonance energy transfer (FRET) involved non-radiative transfer about excitation energy from an excited donor fluorophore to a ground-state acceptor fluorophore, and the process was to absorb high-energy photons first, the low energy photons were then irradiated. This energy transfer process was driven by dipole-dipole interactions. Wherein, the non-radiative energy transfer formula could be expressed as

$$k_{DA} = \frac{B \times Q_D I}{\tau_D r^6} = \left(\frac{1}{\tau_D}\right) \times \left(\frac{R_0}{r}\right)^6, \qquad (3-1)$$

where, τ_D was the excited state radiation lifetime of the donor. N_A was the Avogadro constant. The fluorescence separation distance R_0 could be expressed as

$$R_0 = \left(\frac{9000(\ln 10)\kappa_P^2 Q_D}{N_A 128\pi^5 n_D^4} I\right)^{1/6}. \qquad (3-2)$$

The efficiency of FRET was E, which could be described as

$$E = \frac{k_{DA}}{k_{DA} + \tau_D^{-1}} = \frac{R_0^6}{R_0^6 + r^6}. \qquad (3-3)$$

In the process of FRET, energy transfer was manifested by a decrease in the fluorescence intensity of the donor and a shortened lifetime of the excitons, which was combined with an increase about the acceptor fluorescence intensity and an extension about the exciton lifetime. In addition, FRET was used in an extensive range of other applications, such as chemical analysis, biological analysis, and photophysical studies. Biological studies included *in vivo* and *in vitro*, for example, detection of DNA sequencing and hybridization, protein conformation, detection of cell membrane dynamics and intracellular receptor-ligand binding. This was mainly due to the fact that the FRET technique had a relatively high sensitivity to subtle changes in orientation and separation distance between the acceptor and donor dipoles, so FRET was called a spectral scale[27].

As a unique fluorescence technology, FRET had been studying the life process of living cells for more than a decade, and its spatial resolution at the nanometer scale was very high. In other respects, FRET had some additional applications, such as molecular sensor, material chemistry, polymer chemistry, host-pathology interactions, biomolecular interactions, drug and ligand screening, folding dynamics and conformations, and lipid membrane[28].

For atomic excitons, most of them occurred in undoped semiconductors, and the Coulomb interaction between holes and electrons causes exciton peaks in the spectrum about semiconductors and metals to appear. For undoped semiconductors, a photonic excitation electron and a hole form a bound state, which caused an atomic exciton spectral peak to appear.

Self-trapping of excitons

The lattice vibration in the crystal, that is, the exciton interacts with the phonon. If this coupling was very weak in a classic semiconductor, such as Si or GaAs, then the excitons were scattered through the phonons. Conversely, when a hole or electron was strongly coupled to the crystal lattice, the carrier, in the lattice distortion field, could self-trap into a small polariton. This bound electron-hole pair was often called self-trapping excitons. Such excitons could significantly affect energy transfer, luminescence, and lattice defects in the crystal. Exciton self-trapping was a very common phenomenon, especially in metal halides and rare gas crystals. Studies had shown that exciton self-trapping could produce a dense virtual phonon cloud, which led to a weakening of the ability of excitons to pass through the crystal, and the process was inhibited[29].

CHAPTER 3 Photonics and Plasmonics

For the problem of exciton trapping in rare gas solids, the author cavefally analyzes it from the perspective of dynamics. The ultraviolet emission spectrum emitted from a rare gas solid or liquid could be used to explain the radiation decay of the excited state dimer. This dimer could be produced in a variety of ways, such as by electron trapping of trapped holes, and by dynamically capturing excitons into an excimer state[30].

3.3 Polariton

3.3.1 Brief introduction of polariton

Generally, for the exciton state, it was generally got from the light absorption related to the excitation that excitons could be watched under the band gap. When a photon interacts with an exciton, a polariton, also called an exciton-polariton, was produced[31]. In physics, a polariton was a quasiparticle produced by the strong coupling of an electric or magnetic dipole with electromagnetic wave. Polaritons expressed a common quantum phenomenon, the level repulsion phenomenon, or the principle of avoiding intersection. Polarization was used to describe the crossing about light scattering with any interacting resonance. A polariton was also an electron plus an attached phonon cloud. Polaritons were produced by the mixing about the excitation of the polarity between the material and a photon. In addition to exciton polaritons[31-36], there were many other types of polaritons, including phonon polaritons[37-39], inter-sub-polaritons[40], Bragg polaritons[41,42], SPPs[43,44], etc. They were formed in different ways.

3.3.2 Polariton types

Exciton polaritons

In 1925, Einstein (Fig. 3-3) predicted that Bose-Einstein condensation (BEC) would occur in any perfect gas, and this gas corresponds to non-interacting boson particles. The key to understand superfluid physics and BEC was the special excitation spectrum and particle-particle interaction. In 1947, Bogoliubov(Fig. 3-4) came up with a set of quantum field theory formulas for weakly interacting Bose condensation systems, which predicted the existence of a lower momentum phonon-like excitation spectrum. The authors used a two-photon Bragg scattering technique to verify the atomic BEC theoretically.

Fig. 3-3　Albert Einstein　　　　　　Fig. 3-4　Nikolay Nikolayevich Bogolyubov

In semiconductor microcavities having single-or multiple-quantum wells, comparing the decay rate of photons and excitons, when the coupling rate of cavity photon-exciton was large, the eigenstate was transformed into a new normal mode, also called exciton polaritons. In recent years, such polaritons have been used as important solid-state systems for studying dynamic cohesion in solids. Their effective mass was very small, 8 orders about magnitude smaller than hydrogen atoms and 4 orders about magnitude smaller than excitons. Therefore, the temperature about the polariton BEC transition could reach room temperature conditions. Therefore, the exciton-polariton in the semiconductor microcavity was the basic excitation produced by the strong coupling interaction between the microcavity photons and the quantum well excitons. In solid-state systems, it was considered a new BEC candidate. The exciton-polariton was produced by the coupling of excitons and visible light.

In the previous research, the exciton-polariton was theoretically studied, and the interaction effect between the excitation spectrum and the exciton-polariton condensate was proposed, which was consistent with the Bogoliubov theory. The author had used the experimental setup shown in Fig. 3-5 to study exciton-polariton traps. In this device, three stacks about four GaAs QWs are embedded in the three antinodes of the center of the planar microcavity of the AlAs/AlGaAs distributed Bragg reflector. As could be seen from Fig. 3-5, there was a microcavity between the two distributed Bragg reflectors, this microcavity consisting of $\lambda/2$ AlAs, where λ was the cavity wavelength[34].

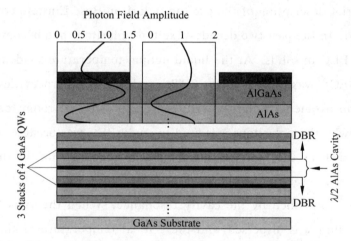

Fig. 3-5　Schematic of a polariton trap formed by a relatively thin metal film on top of a microcavity structure with a circular hole[34]

(For colored figure please scan the QR code on page 1)

Previous studies had shown that, under certain conditions, polaritons condensed in real space rather than formed Bose-Einstein condensates. The authors analyzed the different mechanisms about the polariton-polariton interactions, containing the van der Waals, the direct exchange, and the average field electrostatic interactions as well as the indirect exchange interactions through double exciton states and dark excitons. The exciton-polariton in the planar semiconductor accompanying the quantum well consists of cavity photons and heavy hole excitons. The authors concluded that the magnitude of the exciton-polariton interaction was related to spin, and the spin-parallel excitons attract each other. Conversely, the spin-parallel excitons repel each other. Compared to negative photon detuning, when attraction was more exclusive, the exciton-polariton could form a Bose-Einstein condensation in the planar microcavity. In addition, the authors also found that the attraction was stronger than the repulsive force for certain regions about a particular sample, which suggested that for the same sample, there might be this special phenomenon, that is, in real space, classical coagulation condensation and Bose-Einstein condensates, in the reciprocal space of exciton-polaritons, coexist[31].

Here, a brief description of the exciton-polariton Bose-Einstein condensation is introduced. In the past two decades, exciton-polaritons had become the latest prospect of BEC in solids. At the liquid helium temperature, evidence of some polariton BECs was obtained in CdTe and GaAs microcavities. Due to technical improvements, polaritons BECs were possible at room temperature. The reason why the polaritons was used to study BEC was because it has many advantages, such as the light effective mass of the polariton in the $k_{//} \ll k_c$ region, where $k_{//}$ was the in-plane wave number and k_c was the longitudinal wave number controlled by the cavity resonance. When the value of $k_{//}$ was high, the exciton was dispersed, and the critical temperature of the polariton condensation gradually increased from a few Kelvin to room temperature then to a higher temperature. This significantly reduced the requirements of the polariton laboratory device, making the polariton a practical candidate for quantum device applications. Another important advantage was that the effective mass of the polariton was small, so its density of state (DOS) and energy were small, and the quantum-degenerate seed about the polariton was easier to implement. Therefore, the final state of the boson was stimulating, so that the energy relaxation about the polariton was accelerated. In addition, some characters about the polariton gas could be measured by the leaked photon field, which had advanced spectral and quantum optical techniques such as population distribution in the coordinates, coherence function, vortices, configuration space and transmission properties. The authors also studied the nonlinear interaction about polaritons and this was the core of polaritons condensation. After years of research, the polariton BEC was limited by thermal depletion. For more efficient cooling schemes or microcavities with higher cavity quality factor Q values, the quantum properties of the BEC ground state will be explained. For the applications of polariton BEC, the largest was as a low threshold source for coherent light[31]. In addition, for the study of exciton polaritons there were many aspects, for example, mediated superconductivity[36], spin switches[33], and amplifiers.

Phonon polaritons

Phonon polaritons (PhPs) were quasiparticles produced by the strong coupling interaction of optical phonons and infrared (IR) photons, among them, optical phonons belong to polar crystals. PhPs differ from SPP, which typically needs to span a fairly wide range of energies. PhP could provide spectrally selective responses in relation to optical phonon modes over a range of spectra, from IR to terahertz spectroscopy, and others. Moreover, PhP had a powerful space limitation, which could discover potential applications to store high-density IR data, enhance IR light-substance interaction, coherent heat emission, and develop of metamaterials, as well as potential applications such as frequency-coordinated terahertz waves[39]. In previous studies, the authors controlled and excited these quasiparticles, also regarded as PhPs, in two-dimensional (2D) materials, providing the chance to transmit and limit nanoscale light.

Here was a brief introduction to hexagoxal boron nitride (hBN), the structure was shown in Fig. 3-6, a relatively suitable model system for learning PhPs in ultra-thin materials. This was due to the fact that hBN flakes of different thicknesses were used to make materials without dangling bonds and higher quality, and thus could be used as a typical 2D van der Waals material. In addition, for nanotechnology applications, hBN was the most promising 2D

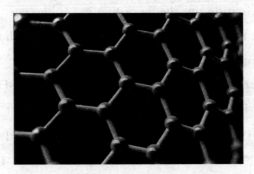

Fig. 3-6 The structure of hexagonal boron nitride
(For colored figure please scan the QR code on page 1)

material due to its layered structure, and its thermal stability and chemical properties were very good, most important was its electrical insulation was also very good. At the same time, hBN was a substrate of graphene with high mobility and a perfect spacer and dielectric layer for 2D heterostructures. However, hBN was a normal hyperbolic material. For bulk hBN, there were many waveguide models about PhPs. It was necessary to fully understand the out-of-plane and in-plane dielectric constants. In the limitation of 2D materials, the thickness of the hBN flakes might be only a few tens of nanometers or only a few atomic layers. Studies had shown that the only mode observed with near-field microscopy was the fundamental. And the authors observed that the PhPs in-plane mode frequency was higher, the response was stronger, and the measured wavelength of the polariton was proportional to the thickness of the hBN flake, while the PhPs out-of-plane mode response was weaker[39].

The above research results have brought nano-scale IR light engineering to a new platform, which has played a key role in promoting new photonic nanodevices[39].

Surface phonon polaritons

A surface polariton was an electromagnetic wave that was capable of coupling to the charge oscillation of a material, and capable of collecting and directing optical energy lower than the diffraction limit. Subsequently, the surface phonon polaritons(SPhPs)included electrons oscillations and was available in precious metals of near-infrared(NIR)and visible wavelengths. However, for SPhPs, it was related to phonon resonance in polar materials, and was active in the mid-IR region. Research showed that the use of precious metal SPhPs in the mid-IR region was limited, while for SPhPs, the polariton wavelength was usually longer at the flat interface and could provide reasonable field confinement or field enhancement. A few years ago, the authors demonstrated that SPhPs could be propagated in 1D materials. This material consists of boron nitride nanotubes with mid-IR wavelengths. Studies have shown that SPhPs exhibit high field enhancement and field limitations, and the effective index was very high. The results had shown that the modal characteristics and propagation length characteristics about SPhPs were controlled by adjusting the size of the substrate and nanotubes, thereby realizing the application in the mid-IR region[40].

The IR counterparts of SPPs and SPhPs were produced by the coupling between optical phonons about polar materials and electromagnetic waves. So, the characteristics of SPPs and SPhPs were very similar, including: ①The high field density at the surface; ②The propagation along the boundary; ③Guided wavelength reduction compared to the free space; ④The coincidence of the light in the charge oscillation space. In general, metal plasmon were difficult to obtain in the mid-IR spectrum. In previous studies, specially designed structured surfaces and lines were shown to extend the application of SP to the mid-IR region, but, the cost was too high and too complicated. Therefore, this had always been an imagination[40].

The hBN had strong phonon resonance and supports mid-IR SPhPs. Therefore, SPhPs could develop nanophotonics to the mid-IR range. Previously, the authors studied the multiwall boron nitride nanotubes (BNNTs), which were 1D forms of boron nitride, whose geometry was variable and mainly regulated by chemical synthesis. BNNTs had mechanical strength and chemical inertness. Studies have shown that 1D SPhPs in BNNTs had a fairly high effective index, greater than SPhPs in planar materials and SPPs in precious metals. And the propagation properties of the 1D SPhPs were determined by the geometry of the BNNTs and the structure of the substrate. The local geometry and dielectric properties of the substrate could control the propagation and generation of SPhPs in BNNTs. For example, when the surface of the metal substrate was rough, the loss of SPhPs could be appropriately reduced[40].

Intersubband polaritons

IR or terahertz photons were coupled with excitons between subbands to produce intersubband (ISB) polaritons. They exhibited bosonic properties at very low densities, allowing the observation of excited polariton scattering[40].

Bragg polaritons

In previous studies, the Bragg polariton was often used as an important tool for customizing the coupling of light materials. The authors detected the Bragg polariton branch by a white light reflectance measurement experiment. And the Bragg polariton was generated by the coupling of the Bragg photon mode. In addition, the Bragg modes appeared on the high-energy side about the photonic band gap (PBG)[41-42].

3.4 Plasmon and surface plasmons

3.4.1 Plasmons

In physics, a plasmon is a quantum of plasma oscillation. Just as light (an optical oscillation) consists of photons, the plasma oscillation consists of plasmons. The plasmon can be considered as a quasiparticle since it arises from the quantization of plasma oscillations, just like phonons from the mechanical vibrations. Thus, plasmons are collective(a discrete number)oscillations of the free electron gas density. For example, at optical frequencies, plasmons can couple with a photon to create another quasiparticle called a plasmon polariton.

The plasmon was initially proposed in 1952 by Pines and Bohm[43] and was shown to arise from a Hamiltonian for the long-range electron-electron correlations[44]. Since plasmons are the quantization of classical plasma oscillations, most of their properties can be derived directly from Maxwell's equations[45]. Plasmons can be described in the classical picture as an oscillation of electron density with respect to the fixed positive ions in a metal. To visualize a plasma oscillation, imagine a cube of metal placed in an external electric field pointing to the right. Electrons will move to the left side (uncovering positive ions on the right side)until they cancel the field inside the metal. If the electric field is removed, the electrons move to the right, repelled by each other and attracted to the positive ions left bare on the right side. They oscillate back and forth at the plasma frequency until the energy is lost in some kind of resistance or damping. Plasmons are a quantization of this kind of oscillation.

Plasmons play a major role in the optical properties of metals and semiconductors. Light of frequencies below the plasma frequency is reflected by a material because the electrons in the material screen the electric field of the light. Light of frequencies above the plasma frequency is transmitted by a material because the electrons in the material cannot respond fast enough to screen it. In most metals, the plasma frequency is in the ultraviolet, making them shiny(reflective)in the visible range. Some metals, such as copper and gold, have electronic interband transitions in the visible range, whereby specific light energies (colors) are absorbed, yielding their distinct color. In semiconductors, the valence electron plasmon frequency is usually in the deep ultraviolet, while their electronic interband transitions are in the visible range, whereby specific light energies (colors) are absorbed, yielding their distinct color which is why they are reflective. It has shown that the plasmon frequency may occur in the mid-IR and NIR region when semiconductors are in the form of nanoparticles with heavy doping. The plasmon frequency can often be estimated in the free electron model as

$$\omega_P = \sqrt{\frac{Ne^2}{\varepsilon_0 m}}, \tag{3-4}$$

where, N is the conduction electron density, e is the elementary charge, m is the electron mass, and ε_0 is the permittivity of free space.

3.4.2 Surface plasmons

Surface plasmons(SPs) are those confined to surfaces and interacted strongly with light resulting in a polariton. They occur at the interface of a material exhibiting positive real part of their relative permittivity, i.e. dielectric constant, (e.g. vacuum, air, glass and other dielectrics), or a material whose real part of permittivity is negative at the given frequency of light, typically a metal or heavily doped semiconductors. In addition to opposite sign of the real part of the permittivity, the magnitude of the real part of the permittivity in the negative permittivity region should typically be larger than the magnitude of the permittivity in the positive permittivity region, otherwise the light is not bound to the surface(i.e. the SPs do not exist), as shown in the famous book by Raether. At visible wavelengths of light, e.g. 632.8 nm, provided by a HeNe laser, interfaces supporting SPs are often formed by metals like silver or gold(negative real part permittivity) in contact with dielectrics such as air or silicon dioxide. The particular choice of materials can have a drastic effect on the degree of light confinement and propagation distance due to losses. SPs can also exist on interfaces other than flat surfaces, such as particles, rectangular strips, v-grooves, cylinders, or other structures. Many structures have been investigated due to the capability of SPs to confine light below the diffraction limit of light.

SPs can play a role in surface-enhanced Raman spectroscopy and in explaining anomalies in diffraction from metal gratings (Wood's anomaly), among other things. SP resonance is used by biochemists to study the mechanisms and kinetics of ligands binding to receptors (i.e. a substrate binding to an enzyme). Multi-parametric SP resonance can be used not only to measure molecular interactions, but also nanolayer properties or structural changes in the adsorbed molecules, polymer layers or graphene, for instance.

3.4.3 Surface plasmon polaritons

The surface plasmon polaritons (SPP) was a propagating transverse magnetic (TM) polarized optical surface wave. It was worth noting that it reached a maximum at the interface and then decayed exponentially. SPPs could be used on many metal structures, such as thin metal films and stripes, gaps, slits, holes, corrugations/grooves in metal films, and metal nanoparticles of different sizes and shapes[46].

SPPs was produced by the coupling of light and, where the wavelength was determined by the substance and its geometry. SPPs had a number of important properties, such as the asymptote of energy in the dispersion curve, localization, field enhancement and resonance, subwavelength limitation, and sensitivity of bulk and surface area. Because of these special properties, SPP was widely used in nanophotonics, spectroscopy, circuits, biosensing, imaging, etc[46].

SPPs dissipated energy by interacting with metals. In general, for a metal defined by a perfect dielectric, its loss was caused by the scattering of free electrons in the metal, and the wavelength was very short, absorbed by the interband transition. This was the basis of the loss mechanism of SPP. If the wavelength could be reasonably selected, the absorption of the transition between bands could be appropriately avoided. However, the energy consumption caused by the scattering of free electrons was reduced to some extent by technical improvement, but could not be completely eliminated. In addition to this, there was an additional energy loss, for example, scattering the SPPs to the bulk wave, resulting in a rough metal interface energy loss[46].

However, in real life, the phenomenon of excessive energy loss in SPPs was existed. It was precisely because of the excessive loss of energy that the application of SPPs in life was limited. First, because the loss was too large, the propagation length of SPPs was not large. In the previous researchers, the authors once mentioned the method of loss amplification and compensation, that is, adding optical gain to the dielectric of the delimited metal. Research had shown that SPPs was affected by the greater the limit, the greater the attenuation. There was no limit to the attenuation balance in the dielectric waveguide. Of course, for a well-made structure, its attenuation could be neglected. SPPs amplification was a good application. For example, SPPs amplifiers were used as stand-alone components or integrated with biosensors and plasmon components to compensate for energy loss and improve performance. SPPs oscillators, also known as SPPs lasers, or "spasers", were produced because SPPs could be limited to deep sub-wavelength dimensions, so this application was achieved in nanoscale structures. This showed that the SPPs oscillator or amplifier was very different in performance and form from the traditional amplifier[46].

In the last few years, SPPs amplifiers have continued to develop. In order to adapt to many high-performance works at present, SPPs amplifiers needed to be further improved, mainly from the following seven aspects: ①Pumping, it was best to use electric pump instead of optical pumping; ② Resonators; ③Power dissipation, how to reduce power consumption was a major problem that researchers need to solve; ④Efficiency, try to increase the efficiency of the SPPs amplifier; ⑤Signal-to-noise ratio, it was necessary to reduce the noise in the SPPs amplifier; ⑥Operating temperature, the temperature was best close to room temperature, which helped the work to proceed smoothly; ⑦Stability, reduced oscillation, and increased stability of the SPPs amplifier[46].

In addition, the surface plasmon polariton absorption modulator (SPPAM) had also been widely used, which was a very small device (only a few micrometers), very space-saving and convenient to use. The device consists of a superposition of insulator, metal, metal layers, and metal oxide. The principle of this device was to achieve absorption modulation by modulating electrically the density of free carrier in the metal oxide. It was worth noting that, unlike the previous traditional SPPAM, the device was more suitable for NIR operation[47].

3.5 Plasmon-Exciton Coupling: Plexciton

Plexcitons are polaritonic modes[48] that result from the strong coupling between excitons and plasmons. Plexcitons aid direct energy flows in exciton energy transfer. SP modes can coherently hybridize with molecular excitons, and its large oscillator strength can lead to a so-called strong coupling regime. In such a regime, a coherent coupling between LSPs and excitons overwhelms all losses and results in two new mixed states of light. In the weak coupling regime, the resonant molecules can be treated as dielectric materials with a complex refractive index (RI) in dispersion. Through the Kramers-Kronig relations, the wavelength dependence on the real part of the RI is related to the molecular absorption resonance as described by the imaginary part of the complex refractive index. As a result, switching the molecular resonance can change the real part of the RI and thus cause LSPR peak shifts due to the sensitivity of the LSPR to the RI of the surroundings.

References

[1]　OZBAY E. Plasmonics: merging photonics and electronics at nanoscale dimensions[J]. Science, 2006, 311: 189-193.
[2]　FRENKEL J. On the transformation of light into heat in solids[J]. Phys. Rev., 1931, 37: 17-44.

[3] YABLONOVITCH E. Inhibited spontaneous emission in solid-state physics and electronics[J]. Phys. Rev. Lett.,1987,58: 2059-2062.

[4] LIU G B, XIAO D, YAO Y, et al. Electronic structures and theoretical modelling of two-dimensional group-VIB transition metal dichalcogenides[J]. Chem. Soc. Rev., 2015,44: 2643-2663.

[5] FIDDER H, TERPSTRA J, WIERSMA D A. Dynamics of Frenkel excitons in disordered molecular aggregates[J]. J. Chem. Phys.,1991,94: 6895.

[6] SCHWEIZER K S. Electronic absorption of Frenkel excitons in topologically disordered systems[J]. J. Chem. Phys.,1986,85: 4638.

[7] LIDZEY D G, BRADLEY D D C, ARMITAGE A, et al. Photon-mediated hybridization of Frenkel excitons in organic semiconductor microcavities[J]. Science, 2000,288: 1620-1623.

[8] BLUMSTENGEL S, SADOFEV S, XU C, et al. Converting Wannier into Frenkel excitons in an inorganic/organic hybrid semiconductor nanostructure[J]. Phys. Rev. Lett.,2006,97: 237401.

[9] GISSLÉN L,SCHOLZ R. Crystallochromy of perylene pigments: interference between Frenkel excitons and charge-transfer states[J]. Phys. Rev. B,2009,80: 115309.

[10] SCHUSTER R,KNUPFER M. Exciton band structure of pentacene molecular solids: breakdown of the Frenkel exciton model[J]. Phys. Rev. Lett.,2007,98: 037402.

[11] MALYSHEV V,MORENO P. Hidden structure of the low-energy spectrum of a one-dimensional localized Frenkel exciton[J]. Phys. Rev. B,1995,51: 14587.

[12] TVINGSTEDT K,VANDEWAL K,ZHANG F L,et al. On the dissociation efficiency of charge transfer excitons and Frenkel excitons in organic solar cells: a luminescence quenching study[J]. J. Phys. Chem. C,2010,114: 21824-21832.

[13] PEARLSTEIN R M. Impurity quenching of molecular excitons. (i). kinetic comparison of forster-dexter and slowly quenched Frenkel excitons in linear chains[J]. J. Chem. Phys.,1972,56: 2431.

[14] BEDNARZ M, MALYSHEV V A, KNOESTER J. Intraband relaxation and temperature dependence of the fluorescence decay time of one-dimensional Frenkel excitons: the Pauli master equation approach[J]. J. Chem. Phys.,2002,117: 6200.

[15] AGRANOVICH V M,LA ROCCA G C,BASSANI F,et al. Hybrid Frenkel-Wannier-Mott excitons at interfaces and in microcavities [J]. Optical Materials, 1998, 9: 430-436.

[16] ZHU X Y,YANG Q, MUNTWILER A M. Charge-transfer excitons at organic semiconductor surfaces and interfaces [J]. Acc. Chem. Res., 2009, 42 (11): 1779-1787.

[17] JAILAUBEKOV A E, WILLARD A P, TRITSCH J R, et al. Hot charge-transfer excitons set the time limit for charge separation at donor/acceptor interfaces in organic photovoltaics[J]. Nature Materials,2013,12: 66-73.

[18] HOFFMANN M, SCHMIDT K, FRITZ T, et al. The lowest energy Frenkel and

charge-transfer excitons in quasi-one-dimensional structures: application to MePTCDI and PTCDA crystals[J]. Chemical Physics,2000,258: 73-96.

[19] HALLERMANN M, KRIEGEL I, COMO E D, et al. Charge transfer excitons in polymer/fullerene blends: the role of morphology and polymer chain conformation [J]. Adv. Funct. Mater.,2009,19: 3662-3668.

[20] ZHU X Y, MONAHAN N R, GONG Z Z, et al. Charge transfer excitons at van der Waals interfaces[J]. J. Am. Chem. Soc.,2015,137: 8313-8320.

[21] GELINAS S, PARE-LABROSSE O, BROSSEAU C N, et al. The binding energy of charge-transfer excitons localized at polymeric semiconductor heterojunctions[J]. J. Phys. Chem. C,2011,115: 7114-7119.

[22] JARZAB D, CORDELLA F, GAO J, et al. Low-temperature behaviour of charge transfer excitons in narrow-bandgap polymer-based bulk heterojunctions[J]. Adv. Energy Mater.,2011,1: 604-609.

[23] SHARIFZADEH S, DARANCET P, KRONIK L, et al. Low-energy charge-transfer excitons in organic solids from first principles: the case of pentacene[J]. J. Phys. Chem. Lett.,2013,4: 2197-2201.

[24] BRANDT O, PFÜLLER C, CHÈZE C, et al. Sub-meV linewidth of excitonic luminescence in single GaN nanowires: direct evidence for surface excitons[J]. Phys. Rev. B,2010,81: 045302.

[25] SAITE V, KOCH E E. Bulk and surface excitons in solid neon[J]. Phys. Rev. B, 1979,20: 784.

[26] KAMBHAMPATI P. On the kinetics and thermodynamics of excitons at the surface of semiconductor nanocrystals: are there surface excitons[J]. Chemical physics,2015, 446: 92-107.

[27] CLAPP A R, MEDINTZ I L, MATTOUSSI H. Forster resonance energy transfer investigations using quantum-dot fluorophores[J]. Chem Phys Chem,2006,7: 47-57.

[28] SAHOO H. Förster resonance energy transfer-a spectroscopic nanoruler: principle and applications[J]. Journal of Photochemistry and Photobiology C: Photochemistry Reviews,2011,12: 20-30.

[29] WILLIAM R T, SONG K S. The self-trapped exciton[J]. J. Phys. Chem. Solids., 1990,51: 679-716.

[30] MARTIN M. Exciton self-trapping in rare-gas crystals[J]. J. Chem. Phys.,1971, 54: 3289.

[31] DENG H, HAUG H, YAMAMOTO Y. Exciton-polariton Bose-Einstein condensation [J]. Rev. Mod. Phys.,2010,82: 1489.

[32] LAGOUDAKIS K G, WOUTERS M, RICHARD M, et al. Quantized vortices in an exciton-polariton condensate[J]. Nature Physics,2008,4: 706-710.

[33] AMO A, LIEW T C H, ADRADOS C, et al. Exciton-polariton spin switches [J]. Nature Photonics,2010,4: 361-366.

[34] UTSUNOMIYA S, TIAN L, ROUMPOS G, et al. Observation of Bogoliubov

excitations in exciton-polariton condensates[J]. Nature Physics,2008,4: 700-705.

[35] VLADIMIROVA M, CRONENBERGER S, SCALBERT D. Polariton-polariton interaction constants in microcavities[J]. Phys. Rev. B,2010,82: 075301.

[36] LAUSSY F P,KAVOKIN A V, SHELYKH I A. Exciton-polariton mediated superconductivity[J]. Phys. Rev. Lett.,2010,104: 106402.

[37] CALDWELL J D,LINDSAY L,GIANNINI V,et al. Low-loss,infrared and terahertz nanophotonics using surface phonon polaritons[J]. Nanophotonics,2015,4: 44-68.

[38] XU X G, GHAMSARI B G, JIANG J H, et al. One-dimensional surface phonon polaritons in boron nitride nanotubes[J]. Nature Communications,2014,5: 4782.

[39] SHI Z W,BECHTEL H A, BERWEGER S, et al. Amplitude-and phase-resolved nanospectral imaging of phonon polaritons in hexagonal boron nitride [J]. ACS Photonics,2015,2: 790-796.

[40] ZANOTTO S,BIASIOL G,DEGLINNOCENTI R,et al. Intersubband polaritons in a one-dimensional surface plasmon photonic crystal [J]. Appl. Phys. Lett., 2010, 97: 231123.

[41] ASKITOPOULOS A, MOUCHLIADIS L, IORSH I, et al. Bragg polaritons: strong coupling and amplification in an unfolded microcavity[J]. Phys. Rev. Lett.,2011, 106: 076401.

[42] SEDOV E S,IORSH I V, ARAKELIAN S M, et al. Hyperbolic metamaterials with Bragg polaritons[J]. Phys. Rev. Lett.,2015,114: 237402.

[43] DAVID P,DAVID B. A collective description of electron interactions: Ⅱ. collective vs individual particle aspects of the interactions[J]. Phys. Rev.,1952,85: 338.

[44] DAVID B,DAVID P. Coulomb interactions in a degenerate electron gas[J]. Phys. Rev. A,1953,92(3): 609-625.

[45] JACKSON J D. Plasma oscillations. Classical electrodynamics [M]. 2nd ed. New York: John Wiley & Sons.,1970.

[46] BERINI P,LEON I D. Surface plasmon-polariton amplifiers and lasers[J]. Nature Photonics,2012,6: 16-24.

[47] MELIKYAN A, LINDENMANN N, WALHEIM S, et al. Surface plasmon polariton absorption modulator[J]. Optics Express,2011,19: 8855-8869.

[48] CAO E,LIN W, SUN M T, et al. Exciton-plasmon coupling interactions: from principle to applications[J]. Nanophotonics,2017,7: 145-167.

CHAPTER 4

2D Borophene excitons

4.1 Introduction

Borophnene has attracted widespread attention in the field of two-dimensional (2D) materials since it was first synthesized in 2015[1]. It is a 2D material made entirely of boron atoms, which have important potential applications in electronics[2-4], energy storage[5-8], transport[9], catalysis[10-12], plasmonics[13-14], superconductivity[15-18], sensor[19-21],etc. In addition,the 2D metallic properties[1,22],Dirac Fermions[23],ideal flexibility and strength[24], antiferromagnetism[25] and excellent electronic[26-27], gas-sensing[28] and transport[9] properties of borophene have also attracted widespread attention from researchers[29].

In 1997, Boustani proposed the Aufbau principle to construct the 2D boron sheets, that is, a stable 2D quasi-planar boron sheet can be obtained by combining two basic structural units, one is the pentagonal pyramid unit B_6, and the other is the hexagonal pyramid unit B_7[30]. In 2007, Tang et al. proposed a boron sheet composed of triangular and hexagonal boron structures, and explained that since the boron atom has only three valence electrons, some strong in-plane sp^2 bonding states are not occupied, which explains the instability of boron sheet, if electrons are available from other sources, the sheet readily accepts electrons to increase its stability[31]. This also provides an idea for the subsequent synthesis of borophene on metal substrates.

Boron atom is of two electrons in the $2s$ orbital and one electron in the $2p$ orbital, so it can form sp^2 hybridization similar to carbon atom, which is conducive to the production of two-dimensional borophene[32]. The structure of borophene is comprised of a layer of atoms arranged in a solid or hollow hexagonal lattice[29,33-34]. Boron is lighter than carbon and has a variety of bonding methods[31], which make it more attractive than graphene for some applications[31,34-37]. Like graphene, borophene also exists in 0-dimensional and 1-dimensional allotropes (all-boron fullerenes[38] and boron nanocones[39]/tubes[40-42]). Borophene is an excellent conductor of electricity[15-16,43-44]. The efficient electrical conductivity renders borophene ideal for use in electronics. The strength and lightweight properties of borophene, which surpass even graphene's exceptional strength, make it an attractive material for use in advanced composites and flexible devices. Additionally, borophene's flexibility means that it can be bent and stretched without breaking, making it a promising material for use in flexible electronics[24]. And one of the most promising areas of borophene is in electronics. The excellent conductivity, combined with its strength and flexibility, make it an ideal candidate for high-performance electronic devices. Researchers are also exploring the use of borophene in transistors. Some special borophene phases and hydrogenated borophene exhibit semiconducting properties[45-46], which make borophene transistors a reality[2-4,29,47-48], and borophene could be utilized to produce faster and more efficient transistors than those currently available. Borophene is also being studied for use in batteries, due to the large surface zone and exceptional conductance making it an ideal material for use in electrodes, which are critical components of batteries and other energy storage devices[5]. In addition to electronics and energy storage, borophene is being explored for use in catalysis[10]. And the excellent photoresponse properties of borphene make it a suitable candidate material for photodetectors[3]. Topological Dirac semimetals display a unique electronic structure and has broad application prospects in the fields of electronic devices and semiconductors. Graphene is a typical lightweight Dirac material, and many novel physical phenomena and electronic properties are caused by Dirac cones. As the nearest neighbor of the carbon atom, borophene is also of similar Dirac cone structure. In 2017, Feng et al. found that the β_{12}-phase borophene can be decomposed into two triangular sublattices in a manner similar to a honeycomb lattice, thereby hosting Dirac cones. Due to the interaction between the borophene layer and the substrate, these Dirac cones is divided into two pairs, which motivates the investigation of superconductivity, topological order, and high-speed electron transport and switching in single-layer borophene[23]. And they used high-resolution ARPES to observe the first two-dimensional anisotropic Dirac cone in χ_3-phase borophene. The Dirac cone is centered on the X and X' point, and the interaction between borophene and Ag(111) is very weak, so the Dirac cone is preserved[49]. The vacancy density and coordination number of borophene are two important parameters used to distinguish the characteristics of different borophene, and borophene with different vacancy densities and coordination numbers also have different physical properties. Vacancy density ν = (No. of hexagon holes)/(No. of atoms in the original triangular sheet). According to different coordination numbers, borophene can be divided into different types, among which the δ-type borophene is characterized by a single coordination number. The α-type borophene is of a coordination number of 5 and 6. The coordination numbers of β-type borophene are 4, 5 and 6. The coordination numbers of χ-type borophene are 4 and 5[33]. And the ν of σ_3, σ_4, σ_5 and σ_6 phase are 1/3, 1/4, 1/7 and 0, respectively; the ν of α, α_1, α_2 and α_3 phase are 1/9, 1/8, 1/8 and 1/9, respectively; the ν of χ_1, χ_2, χ_3 and χ_4 phase are 3/17, 1/6, 1/5 and 1/6, respectively; the ν of β_1, β_2, β_3 and β_{12} phase are 1/8, 1/7, 1/6 and 1/6, respectively. And in 2014, Zhou et al. used the Aufbau principle to propose the 8-pmmn phase borophene whose energy is much lower than that of the planar structure and the structure was first confirmed to have a distorted Dirac cone[50].

Despite its many promising properties, borophene is still a relatively new material[51]. Currently, most monolayer borophene are produced by molecular beam epitaxy[52-53], depositing boron atoms on substrates in a vacuum chamber such as, Au[54], Cu[4,55], Al[56], Ir[57] and Ag[1,22,58-60]. Most of them are metallic, but in particular, Hydrogenated borophene obtained by thermally decomposing NaOH under H_2 exhibits semiconducting properties[45]. The charge redistribution between the metal film and borophene facilitates its synthesis[51]. However, the researchers focused on exploring the properties of multilayered borophene because the twisting[61-62] and heterojunction properties of multilayered two-dimensional materials often bring novel physical properties, but the synthesis of multilayered borophene is often prevented by boron clusters rather than the formation of flat multilayered borophene. However, single-layer borophene is easily oxidized and has poor stability, which adversely affects the fabrication of devices. Therefore, the synthesis of bilayer borophene is very important. In 2022, Wu et al. and Liu et al. successfully synthesized bilayer borophene on Cu(111) and Ag(111) substrates, respectively[63-64]. The stability and antioxidant of BL borophene are greatly improved, which is beneficial to the fabrication of devices. Bilayer borophene on Ag(111) is of larger work function and stronger charge transfer than monolayer layer and shows excellent optical and thermoelectric properties[44]. The bilayer borophene on Cu(111) also undergoes greater charge transfer and is more resistant to oxidation. The double-dirac cones and higher Fermi velocity than graphene[65], antiferromagnetism[25], nodal line fermions[66], and ultrahigh critical strain[67] of bilayer borophene open up new applications for borophene-based materials.

In this review, we introduce several monolayer borophene synthesized by molecular beam epitaxy on metal substrates, which generally display 2D metallicity and excellent electronic properties. Then we reviewed two successfully synthesized bilayer borophene and the theoretical studies on α-phase bilayer borophene, and finally we introduced four borophene-based heterostructure. This electronically abrupt metal/semiconductor interface provides a new solution for the integration of borophene into nanodevice applications.

4.2 Monolayer borophene

4.2.1 Monolayer borophene on Ag(111)

Although a series of theoretical studies have proved the existence of borophene[33,35,41,68-69], but before 2015, borophene has not been synthesized experimentally. In 2013, Liu et al. predicted that the Ag(111) substrate was favorable for the synthesis of borophene[70]. In 2015, Mannix et al. first synthesized 2D borophene under the UHV on Ag(111) film[1], which maintained between 450℃ and 700℃. STM images show an anisotropic planar structure. At a boron coverage of 1.0 ML, the substrate is fully enveloped by boron sheets and clusters (Fig. 4-1(a)). Two distinct phases can be observed in the STM image: the homogeneous phase and the striped phase as shown with red and white arrows in Fig. 4-1(a). The difference can be seen from the high-resolution STM images of the two different phases. The homogeneous phase has the characteristics of periodic atomic chains, short-range rhombic Moiré patterns and long-range one-dimensional Moiré patterns. The striped phase contains a rectangular lattice corresponding with the stripe areas. Through XPS, we can find that this monolayer borophene is easily oxidized, but the silicon/silicon oxide can delay this oxidation to several weeks. The monolayer borophene on Ag(111) substrate simulated by DFT is shown in Fig. 4-1(b). The symmetries and lattice constants are in good agreement with experimental values, and the simulation and experimental STM are also very consistent (Fig. 4-1(c)).

The energy band and DOS of borophene is shown in Fig. 4-2(a), which shows its metallicity property at G-X and Y-S high symmetry points and shows a bandgap at G-Y and S-X high symmetry points. Borophene's metallic characteristics can be examined through I-V curves (Fig. 4-2(b)) and dI/dV curves (Fig. 4-2(c)) obtained via scanning tunneling spectroscopy (STS).

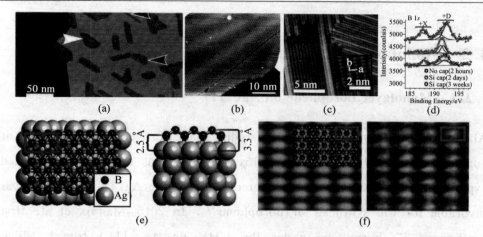

Fig. 4-1 (a) STM topography of borophene. High-resolution STM images of (b) homogeneous phase and (c) striped phase. (d) XPS B 1s spectra. (e) DFT model of borophene. (f) simulation and experimental STM images[1]. This figure has been adapted from Ref. [1] with permission from American Association for the Advancement of Science, copyright 2015

(For colored figure please scan the QR code on page 1)

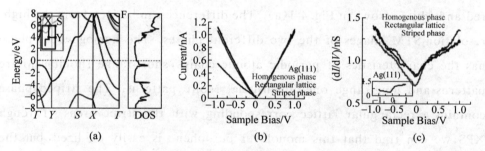

Fig. 4-2 (a) Energy band and DOS of the lowest-energy borophene structure in Fig. 4-1(e). (b) I-V curves (c) dI/dV curves[1]. This figure has been adapted from Ref. [1] with permission from American Association for the Advancement of Science, copyright 2015

(For colored figure please scan the QR code on page 1)

Feng et al. synthesized monolayer borophene under UHV using molecular beam epitaxy (MBE) in 2016[22]. At 570 K, type1 phase borophene is formed (Fig. 4-3(a)). When the sample is annealed to 650 K, type 1 phase is transformed into type 2 phase (Fig. 4-3(b)). Type 2 phase can coexist with type 1 in the temperature region of 650 ~ 800 K, and when the temperature of Ag(111) is 680 K. Type 2 borophene could also be synthesized straight. The

β_{12} borophene exhibits a pattern of pore chains that are separated by hexagonal boron rows and the lattice constants are 3.0 Å and 5.0 Å (Fig. 4-3(c)), which is consistent with the experimental value. Due to lattice mismatch, a Along the boron row direction, a 1.5 nm periodicity Moiré structure arises. The type 2 stage is χ_3 borophene, where boron row is narrower than type 1 (Fig. 4-3(d)). Once a boron coverage of 1 ML is reached, an abundance of 3D clusters will appear on the surface, and no multilayer borophene will be obtained. Because the interaction of the Ag-boron interface is necessary for the synthesis of borophene. When boron coverages above 1 ML, the interfacial interplay is saturated, leading to the inability to form multilayer borophene. The total forming energy of type 1 is slightly less than that of type 2, indicating the

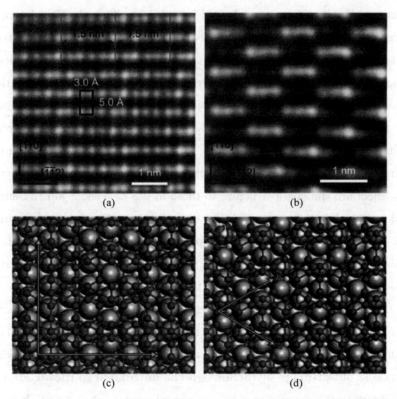

Fig. 4-3 (a) and (b) 1 nm-resolution STM image of S1 and S2 phases, respectively. (c) and (d) DFT model of (a) and (b)[22]. This figure has been adapted from Ref. [22] with permission from Springer Nature, copyright 2016

(For colored figure please scan the QR code on page 1)

Ag(111) substrate as being more conducive to the stability of type 2. The lattice parameters of β_{12} and χ_3 borophene are consistent with Ag (111), which can effectively reduce the strain and facilitate the synthesis of Ag (111). The vacancy density of β_{12} and χ_3 borophene are very close (1/6 and 1/5), which makes the transformation of the two structures easy.

In 2018, Liu et al. used UHV STM, spectroscopy and DFT to study the structural properties of borophene. They found that under certain conditions, borophene phases can intermix and form new phases due to line defects, which is the characteristic high anisotropy and similar polymorphs of borophene. The research also revealed the presence of a charge density wave that is moderated by line defects[58]. The STM image show boron particles and borophene islands (Fig. 4-4(a)), and the 1 nm resolution STM image shows two different phases with vacancy density of 1/6 and 1/5 (Fig. 4-4(b)). The phase of borophene is influenced by the growth temperature, resulting in a slow change from β_{12}-phase (1/6) to χ_3-phase (1/5) with increasing temperature. DFT calculations show that the β_{12} and χ_3 phases have comparable chemical potentials and stabilities. And the STS spectra show that both phases of borophene are metallic two-dimensional materials (Fig. 4-4(c)). The STM topography show that that bigger borophene consists of both β_{12} and χ_3 structure with corresponding line defects (showed by blue and red, respectively). The line defects in the β_{12} sheet are similar in structure to those in the χ_3. The 2 nm resolution STM images show that line defects resemble with each other (Fig. 4-4(e)), as the defect of the β_{12} sheet resembling that of χ_3-phase, and the defect of the v1/5 sheet resembling that β_{12}-phase, with the line distance matching those of defect corresponding χ_3 and β_{12} phases, respectively.

The DFT calculation shows that the substrate had a templating effect on the growth of these structures, resulting in an epitaxial growth with parallel rows. The structures were found to have an atomically smooth phase boundary and no observable out-of-plane distortion, which contributed to the steadiness. The calculated structures successfully fit the experimental results, and there is good consistency between the simulated STM results and the experimental STM measurements as shown in Fig. 4-5.

CHAPTER 4 2D Borophene excitons

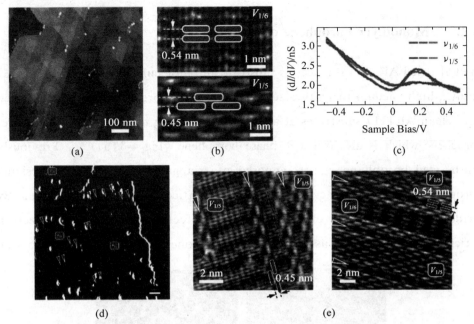

Fig. 4-4 (a) STM topography (b) 1 nm-resolution STM images of two phases. (c) STS spectra (d) The derivative image of (a) and (e) 2 nm-resolution STM image[58]. This figure has been adapted from Ref. [58] with permission from Springer Nature, copyright 2018

(For colored figure please scan the QR code on page 1)

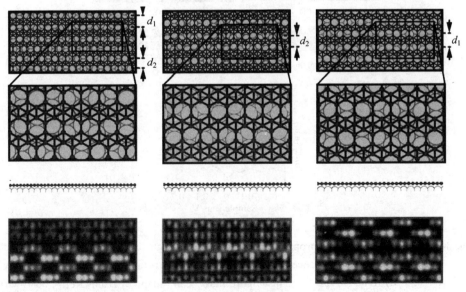

Fig. 4-5 DFT model and Simulated STM images[58]. This figure has been adapted from Ref. [58] with permission from Springer Nature, copyright 2018

(For colored figure please scan the QR code on page 1)

4.2.2 Monolayer borophene on Al(111)

Li et al. used Al(111) surfaces as substrate and molecular beam epitaxy (MBE) in ultra-high vacuum to synthesize borophene, and boron atoms are evaporated onto clean Al(111) substrates at about 500 K forming the honeycomb monolayer borophene which is also called δ_3-phase borophene (Fig. 4-6(a))[56]. Previously synthesized borophene were composed of triangular lattice, so synthesizing borophene with graphene-like honeycomb lattice became a challenge. By virtue of its three free electrons, Al can effectively balance the electron deficiency of borophene, as evidenced by the honeycomb lattice structure seen

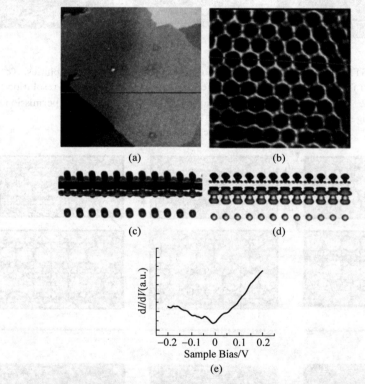

Fig. 4-6 (a) STM image of borophene on Al (111), (b) high resolution STM, (c) and (d) charge density difference of borophene on Al (111) and Ag (111), respectively. (e) dI/dV curves[56]. This figure has been adapted from Ref. [56] with permission from Elsevier, copyright 2018

(For colored figure please scan the QR code on page 1)

in high-resolution STM images (Fig. 4-6(b)). The lattice constant is slighter than that for honeycomb borophene in vacuum (2.9 Å compared to 3 Å), and is similar to that of Al(111) (2.86 Å). For freestanding borophene, the honeycomb lattice is unstable[31]. The The stability of borophene is greatly influenced by the Al(111) substrate. A comparison of the formation energies on Ag(111) (0.81 eV) and Al(111) (0.31 eV) suggests that borophene is more stable on the Al substrate. Charge transfer from the first layer of the substrate to the borophene layer can be inferred from a significant electronic charge at the interface between them (Fig. 4-6(c)). In contrast, for the Ag substrate and borophene, there is almost no electron accumulation at the interface (Fig. 4-6 (d)). The high electron density of Al and the small lattice mismatch with borophene provide a possibility for the synthesis of honeycomb borophene. And Fig. 4-6(e) indicates that this type borophene is metallicity and has the depression around the Fermi energy level (Fig. 4-6(e)).

4.2.3 Monolayer borophene on Ir (111)

Vinogradov et al. synthesized χ_6 borophene with a vacancy density of 1/6 on Ir (111) substrates by direct exposure to B-atom fluxes at a wide range of high temperatures[57]. The borophene phase is shown in Fig. 4-7(a), which is a 6×2 superlattice structure and is a triplet of equivalent 120° rotational domains. Figure 4-7(b) shows the STM image of the sample with three 120° rotational domains with more than hundred nanometers in size so that the borophene is growing with only one structural phase on metal film. To match two Ir-Ir distances and three B-B distances, the B-B bond along [0 1] is stretched by roughly 7% from 1.69 Å. Along [1 0], the B-B bond dimension reduces by approximately 3.6%, matching six Ir-Ir distances with 10 B-B distances, partially relieving the strain. The borophene is a χ_6 structure with a vacancy density of 1/6. DFT calculations indicate that interfacial interplay from substrate to borophene is very vital for the stability borophene, as shown in the Fig. 4-7(c), although the unsupported I_4 structure is more stable than the χ_6 structure, interactions with the support favor the stability of the χ_6 structure. The calculated average B-B bond is about 1.8 Å, that is in good agreement with the hypothesis of borophene lattice expansion. In the interfacial region between borophene and metal, a significant redistribution of the charge density occurs, which indicates the formation of chemical bonds (Fig. 4-7(d)).

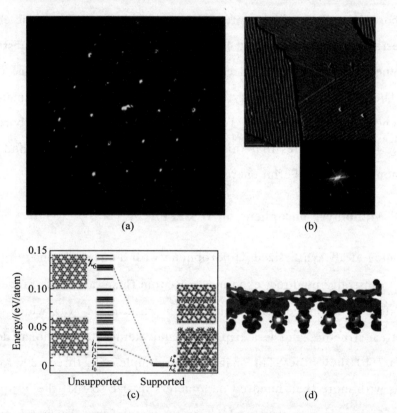

Fig. 4-7 (a) and (b) LEED pattern and STM of borophene on Ir (111), respectively. (c) DFT calculation of stability (d) charge density difference[57]. This figure has been adapted from Ref. [57] with permission from American Chemical Society, copyright 2019

(For colored figure please scan the QR code on page 1)

4.2.4 Monolayer borophene on Au(111)

Borophene on Au (111) substrates was successfully synthesized by Kiraly in 2018[54]. Unlike the synthesis process on Ag(111) previously studied, B atoms diffuses into Au at high temperatures and separates to the surface forming islands of borophene when the substrate cools. The synthesis of borophene also changes the surface reconstruction of the substrate, forming a triangular network that can be templated for growth at low coverage. When the substrate temperature is 550℃, the boron clusters are disappeared and a transition from the conventional herringbone reconstruction to a triangular network is observed on the Au surface, where small islands of nanoscale borophene appear, as shown in Fig. 4-8(a). Figure 4-8(b) shows the borophene model on Au with a vacancy density of 1/12. Au surfaces with distributed boron particles show clear peaks at the B $1s$ core energy level (Fig. 4-8(c)). However, the reduced peak intensity of the triangular network sample shows a much lower surface coverage of boron in these cases, suggesting that boron diffuses into the Au sublayer region at higher substrate temperatures. There is no significant peak shift in the B $1s$ core energy level, indicating little valence change. The growth of borophene is caused by the separation of boron from the native surface as the sample cools (as shown in Fig. 4-8(d)), which is very different from the growth of borophene on other metal surfaces.

DFT calculations show that the topmost subsurface site is the steadiest location for isolated boron atoms, which is about 0.3 eV higher along the Au (111) step edge than the planar subsurface site (Fig. 4-9(a)). Subsurface boron reduces the spatial constraints of surrounding Au, which in turn allows for stronger Au-B interactions. And there is a tendency to form minor B clusters in the Au(111) film (Fig. 4-9(b)). The increased depth of the dimer and trimer potential wells suggests a tendency to form boron-boron bonds in this subsurface. In addition, steadiest construction includes the expulsion of Au

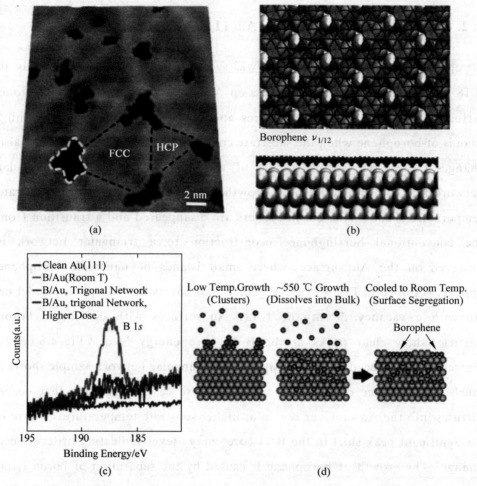

Fig. 4-8 (a) STM image of borophene on Au (111), (b) DFT borophene model, (c) B 1s core-level spectra, (d) Schematic diagram of the dynamics of borophene growth[54]. This figure has been adapted from Ref. [54] with permission from American Chemical Society, copyright 2019

(For colored figure please scan the QR code on page 1)

atoms from the surface to make room for the boron trimer cluster, which relieves interfacial stress. As the boron trimer docks onto the surface and excess Au atoms are removed, it nucleates and spreads to a two-dimensional island. A magnified images of the borophene islands (Fig. 4-9(c)) reveals the periodicity, with herringbone stripes visible under the borophene (black arrows), indicating its atomic thinness. Borophene can be broken down into rhombic parts, and clustering multiple rhombic structural units can explain larger borophene islands image (Fig. 4-9(d)). Energy bands and density of states reveal that the borophene structure on Au (111) shows metallic.

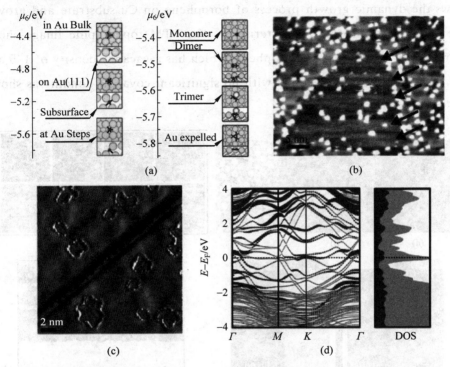

Fig. 4-9 (a) Free energies of B on the Au (111) surface and bulk. (b) 5 nm-STM image of larger borophene island (c) the magnified image of borophene (d) energy band and DOS of borophene[54]. This figure has been adapted from Ref. [54] with permission from American Chemical Society, copyright 2019

(For colored figure please scan the QR code on page 1)

4.2.5 Monolayer borophene on Cu(111)

In 2019, Wu successfully synthesized micron-sized borophene on Cu(111) surfaces[55]. This borophene has a new crystal structure, as shown in the atomic scale STM image. In this borophene, the hollow hexagonal lattice presents a periodic broken line distribution, which is not observed in the previous borophene structure. It may be because of this periodic broken line structure that the borophene is easy to extend and form a large-scale crystal. And the calculated STM structure is very similar to the experimental results, and the structure model is shown in Fig. 4-10(c). Because Cu is more inert than Ag and somewhat reactive, it is expected to produce large-area borophene. Figure 4-10(a) shows the dynamic growth process of borophene on Cu substrate and growth rate exhibits anisotropic characteristics. The STM topographic image shows formation of large areas of borophene, which has a vacancy density of 1/5 and was found to be electron-doped with no significant covalent bonding as shown in Fig. 4-10(d).

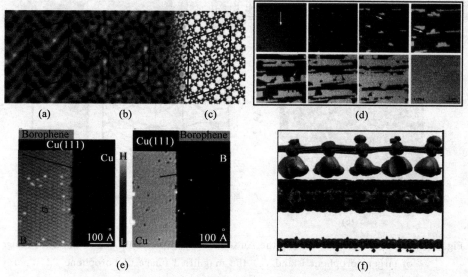

Fig. 4-10 (a) atomic scale STM image. (b) simulated data of (a). (c) borophene structure (d) the dynamic growth process of borophene on Cu substrate. (e) The large-scale STM topographic. (f) charge density difference of borophene and Cu[55]. This figure has been adapted from Ref. [55] with permission from Springer Nature, copyright 2019

(For colored figure please scan the QR code on page 1)

4.3 Bilayer borophene

4.3.1 Bilayer borophene on Ag(111)

Due to the relatively poor stability of SL borophene, which does not meet the needs of borophene-based devices, and BL borophene is predicted to have excellent properties such as double-Dirac cones[65] and ultra-high critical strains[67]. Therefore, the synthesis of BL boronene has become a research hotspot in the field of two-dimensional materials.

In 2022, BL borophene was successfully synthesized on Ag (111) substrate for the first time by Liu et al.[63] Research has found that the key to synthesize bilayer bilayers is using flat micron-sized atomically Ag (111) terraces. The successfully synthesized BL borophene is shown in Fig. 4-11 (a), which is surrounded by two SL borophenes with vacancy densities of 1/5 and 1/6. Through CO-STM to improve spatial resolution, we can find obvious honeycomb structure in STM images as shown in Fig. 4-11(b). DFT calculations yielded a lattice constant of 5.7 Å consistent with STM measurements and reproduced the moiré superlattice with a period of 2.2 nm, and the model of BL borophene is shown in Fig. 4-11(c), which is α phase and the opposite atoms in the solid hexagonal lattice in the bottom and top layers are close to each other to form covalent bonds. And due to the formation of interlayer bonds, bilayer borophene shows stronger stability, which meets the needs of fabricating borophene-based devices. The simulated CO-STM and CO-AFM images are very similar to the experimental results.

The FER spectra taken on BL, SL borophene and Ag(111) substrate show that BL-α borophene has sharper image-potential states (IPS) and longer lifetimes compared to 1/6 phase SL borophene and substrate. And the scattering-induced finite lifetime is calculated by the full-width-at-half-maximum (Γ_n) of the nth IPS peak. The larger n has the longer lifetime leading to the sharper the IPS as shown in Fig. 4-12(b). The work function of BL borophene is about 0.41eV higher than that of Ag (111) substrate, so charge transfer from Ag substrate to BL borophene will occur. In 2022, Yang et al. theoretically investigated the electronic, optical and thermoelectric properties of BL-α borophene[44]. The charge density difference shows that significant charge transfer from metal film to BL borophene layer as shown in Fig. 4-12(c). And due to the interaction of substrate and BL borophene, the

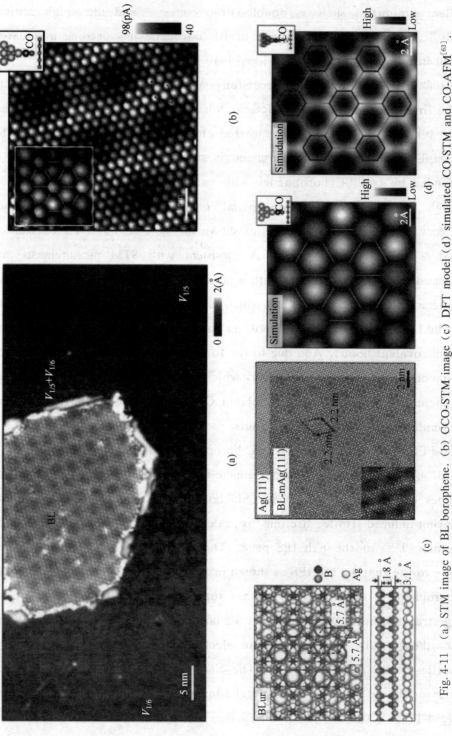

Fig. 4-11 (a) STM image of BL borophene. (b) CCO-STM image (c) DFT model (d) simulated CO-STM and CO-AFM[63]. This figure has been adapted from Ref. [63] with permission from Springer Nature, copyright 2022
(For colored figure please scan the QR code on page 1)

surface Ag atoms fluctuate. Figure 4-12(d) shows the bonding analysis of the interlayer chemical bonds, which reveals that a typical covalent bond is formed between the layers.

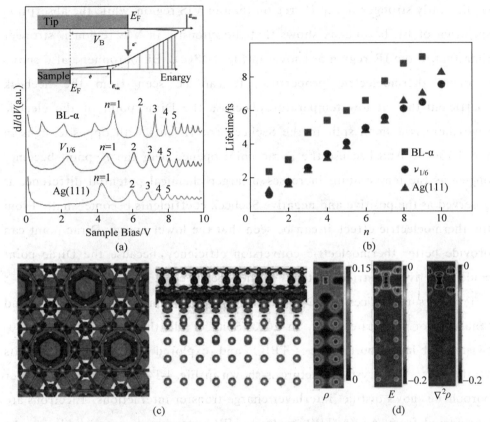

Fig. 4-12 (a) FER spectra of BL-α borophene, SL ν1/6 borophene and Ag(111). (b) lifetimes of the image-potential states in (a)[63]. (c) charge density difference (d) bonding analysis[44]. This figure has been adapted from Refs. [44] and [63] with permission from Elsevier and Springer Nature, copyright 2022
(For colored figure please scan the QR code on page 1)

The energy band and density of states of BL-α borophene reveal the metallicity of this two-dimensional material and demonstrates the strong coupling between the upper and lower layers as shown in Fig. 4-13(a). In Fig. 4-13(b), the enlarged energy band shows two Dirac points near the K point and at the K points. But they cannot be observed in monolayer α borophene, which also proves that the formation of interlayer chemical bonds leads to the generation of Dirac points. And BL borophene and SL borophene exhibit a significant difference in their absorption spectra, which is caused by the difference in energy bands. The absorption spectrum of SL borophene is significantly stronger in the IR region than in VIS region, while the absorption spectrum of BL borophene shows that the intensity in VIS region is stronger than that in the IR region as shown in Fig. 4-13(c). BL borophene also shows excellent thermoelectric properties. It can be seen from the Seebeck coefficient that at low temperature, because the Dirac point of the electric cage, there is a small split in the Seebeck coefficients with opposite signs in Fig. 4-13(d). And because the Dirac point opened near the K point becomes bigger, as the temperature increases, a larger chemical potential difference is observed as the positive and negative Seebeck coefficients become small. From the thermoelectric effect, it can be seen that the lower energy Dirac point can provide better thermoelectric conversion efficiency, because the Dirac point could effectively restrict electrons in the electric cage.

Twisted superlattices of 2D materials often bring intriguing properties and tunability of properties[61-62]. In 2023, Song et al. theoretically studied 30° twisted α-bilayer borophene (TBB) and explored the possibility of its synthesis[71]. The crystal structure is shown in Fig. 4-14(a). Twisted α-bilayer borophene shows distinct interlayer charge-transfer interactions. Electrons are transferred from Ag to TBB, making TBB synthesized more stably on Ag substrate. Twisting can significantly improve the thermoelectric performance of bilayer borophene. The inversion of the Seebeck coefficient is due to the splitting of the Dirac point near point Γ. In Fig. 4-14(b), the twisted superlattice increases the Seebeck coefficient from 300 μV/K to 500 μV/K and increases the optimum temperature to around 110 K. Due to the multiple quantum well of the superlattice, the high conductivity and large Seebeck coefficient are tuned to the same range by twisting as shown in Fig. 4-14(d). Large S and σ in the same region lead to the increase in ZT as shown in Fig. 4-14(e), which shows significant structure-property correlation.

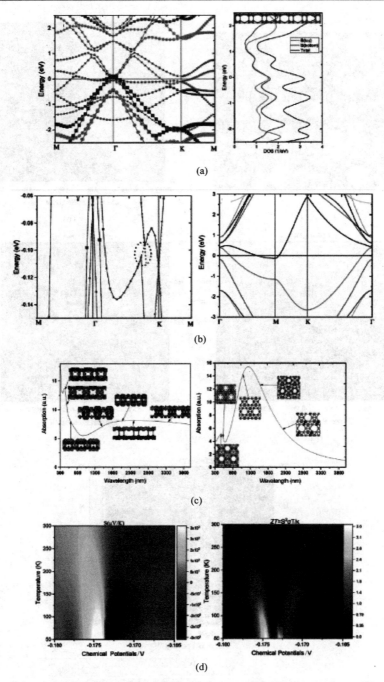

Fig. 4-13 (a) energy band and DOS of BL borophene (b) Dirac cone near K points and energy band of SL borophene. (c) Optical absorption spectra (d) Seebeck coefficient and thermoelectric effect[44]. This figure has been adapted from Ref. [44] with permission from Elsevier, copyright 2022

(For colored figure please scan the QR code on page 1)

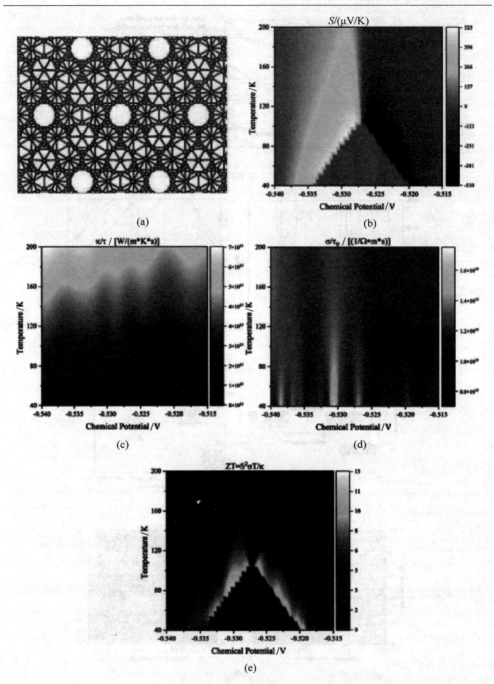

Fig. 4-14 （a）30° twisted α-bilayer borophene，Chemical dependent （b）Seebeck coefficient,（c）thermal conductivity,（d）electrical conductivity, and （e）thermoelectric effect[71]. This figure has been adapted from Ref. [71] with permission from John Wiley and Sons，copyright 2023

（For colored figure please scan the QR code on page 1）

In 2023, Cao et al. studied the Fano-resonant propagating plexcitons and Rabi-splitting local plexcitons in the TERS-BL borophene system as shown in Fig. 4-15 (a)[72]. Figure 4-15 (b) shows the electric field enhancement calculated by STM-TERS with two localized surface plasmon resonance peaks at 488 nm and 633 nm with electric fields up to 10^5. There are two strong absorption peaks and weak absorption peaks around 488 nm and 633 nm of borophene and the dielectric function along z of BL borophene is negative near 488 nm. The wavelength match between STM-TERS and BL borophene indicates the presence of plasmon-exciton interactions. The interaction between the broad LSPR mode of TERS and the narrow resonance mode of BL borophene results in Fano resonance behavior from 550 nm to 700 nm. Because there are both strong absorption near 488 nm in TERS system and BL borophene and the real part of the dielectric function along z of BL borophene is negative. Therefore, both TERS and BL borophene have plasmon properties near 500 nm. The strong plasmon exciton coupling interaction near 488 nm in TERS leads to two stronger plexcitonic peaks and the peak at 448 nm can be enhanced to 10^7 magnitude.

Plexciton waveguides on the surface of BL borophene are investigated as shown in Fig. 4-16. It can be found that the plexciton waveguides can propagate along the surface of BL borocephe at specific wavelengths and are strongly controlled by different wavelengths. The reason is that in the 616 nm to 622 nm region, the plasmon energy is strongly transferred to BL borophene, and the Fano resonance propagating plexcitons can efficiently propagate longer distances with less loss. From 624 nm to 630 nm, the plexciton gradually resemble the plasmons in Fig. 4-16(d). The edge of propagating and localized plexcitonic modes is found at 624 nm. At 626 nm, the plexciton mode becomes progressively closer to localization. At 628 nm and 630 nm, the plexciton modes are clearly localized plexciton resonances (LPER). The plexciton mode at 600 nm is the LPER mode, which is hardly affected by the Fano resonance. At 606~614 nm, plasmons gain energy from excitons, and there is not enough

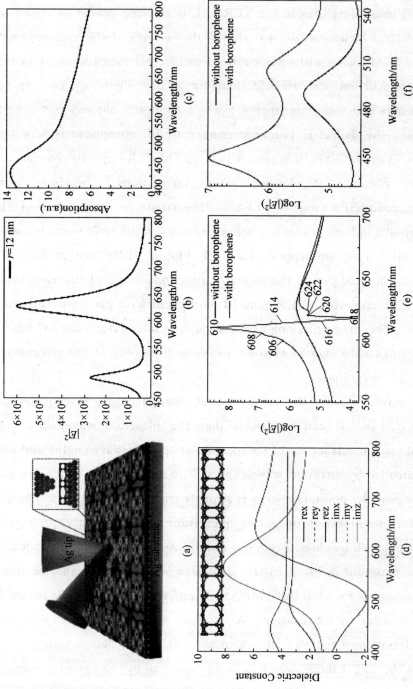

Fig. 4-15 (a) Schematic diagram of the TERS system, (b) Calculated electric field of (a), (c) The absorption spectrum of BL borophene, (d) Dielectric function of BL borophene, (e) Weak and (f) Strong coupling of plasmon and exciton.[72] This figure has been adapted from Ref. [72] with permission from American Institute of Physics, copyright 2023

(For colored figure please scan the QR code on page 1)

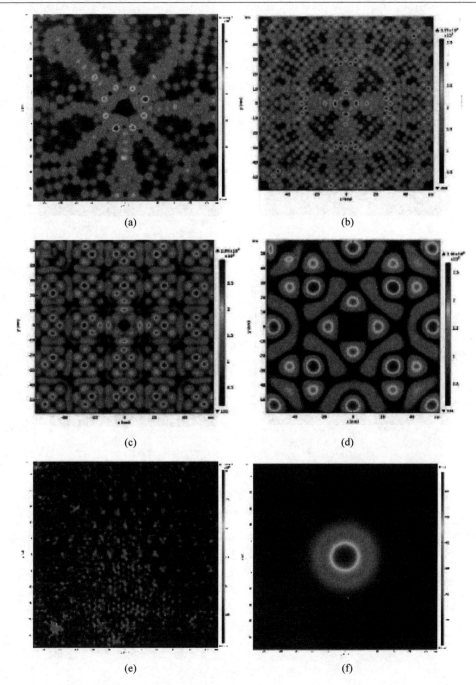

Fig. 4-16 Fano resonance propagating plexcitons at (a) 616 nm, (b) 618 nm, (c) 620 nm, (d) 622 nm. Plexcitonic modes at (e) 624 nm, (f) 626 nm, (g) 628 nm, (h) 630 nm, (i) 600 nm, (j) 606 nm, (k) 610 nm, (l) 614 nm[72]. This figure has been adapted from Ref. [72] with permission from American Institute of Physics, copyright 2023
(For colored figure please scan the QR code on page 1)

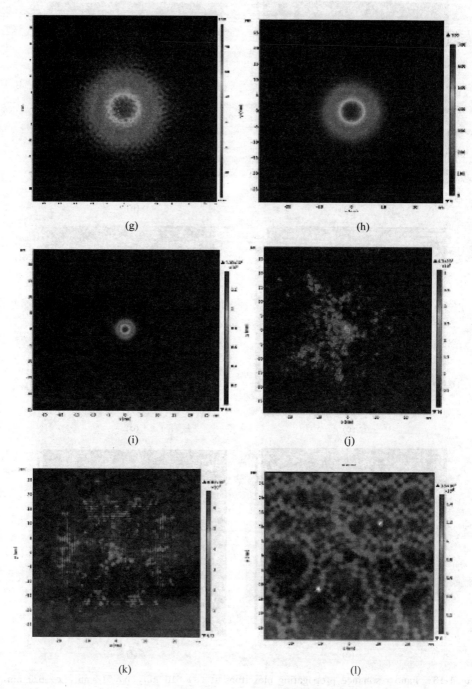

Fig. 4-16(Continued)

CHAPTER 4 2D Borophene excitons

energy to support a plexciton waveguide. Therefore, the quality of the plexciton waveguides is not good enough.

4.3.2 Bilayer borophene synthesis on Cu(111)

In 2022, Chen et al. synthesized β_{12}-like bilayer borophene on Cu(111) film using MBE[64]. The bilayer borophene contains two slightly different monolayers by covalent interlayer bonding, and each monolayer layer has the characteristics of periodic broken line distribution of the hollow hexagonal lattice. And a large transfer and redistribution of charge occurs in bottom borophene and substrate in formation of the bilayer borophene.

The structure is shown in the Fig. 4-17(a), which is a β_{12}-like structure. However, the structure of top and bottom borophene layer is similar but

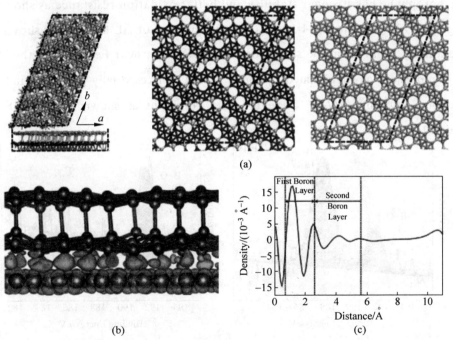

Fig. 4-17 (a) The structure of BL borophene, (b) Charge density difference, (c) Planar-averaged electron density difference[64]. This figure has been adapted from Ref. [64] with permission from Springer Nature, copyright 2022

(For colored figure please scan the QR code on page 1)

different, the vacancy density of bottom layer is 5/36, and that of top layer is 1/6, so the top layer has 6 atoms less than the bottom layer, and the vacancy density of BL borophene is 11/72. Through the charge distribution, we can find that a large number of electrons are transferred to the borophene as shown in Fig. 4-17(b). Large charge transfer provides additional electrons to the borophene and facilitates the formation of interlayer chemical bonds. And the excess electrons in the first layer of boron appear to be transferred to the second layer, causing a relocation of electrons in top and bottom borophene. This redistribution is seen as advantageous for the generation of β_{12}-like BL borophene.

Through XPS, we can find that only a small part of BL borophene in the air is oxidized, while SL borophene is completely oxidized, which also proves that BL borophene has stronger stability and better oxidation resistance as shown in Fig. 4-18(a). And in Fig. 4-18(b), the energy band of BL borophene shows the metallicity. And it can be seen from the PDOS that near Fermi level, the DOS of borophene is mostly donated by the p orbital of B, especially the p_z orbital and the total DOS and dI/dV curve are very consistent as shown in Fig. 4-18(c) and (d).

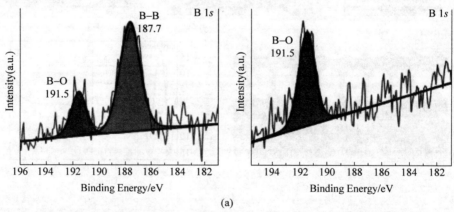

Fig. 4-18 (a) XPS of BL and SL borophene (b) energy band (c) PDOS (d) dI/dV curve[64]. This figure has been adapted from Ref. [64] with permission from Springer Nature, copyright 2022

(For colored figure please scan the QR code on page 1)

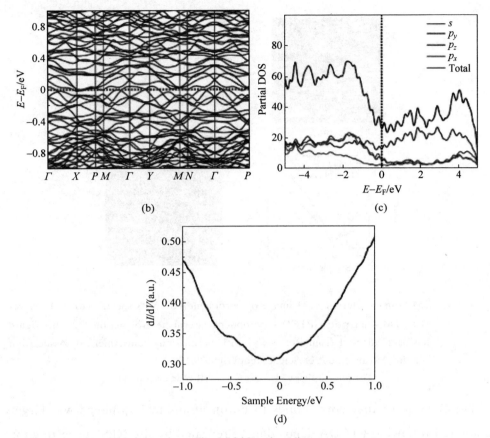

Fig. 4-18(Continued)

4.4 Borophene heterostructure

4.4.1 Borophene-PTCDA lateral heterostructure

In 2017, Liu et al. first synthesized borophene-PTCDA lateral heterostructure[73]. The borophene on the Ag substrate is the brick wall structure as shown in Fig. 4-19(a). And the situ XPS shows the B $1s$ peak at 188 eV, and there is no B-O peak at 192 eV, which proves that borophene is not oxidized or polluted. The I-V curve reveals the metallicity of borophene as shown in Fig. 4-19(c). The PTCDA/borophene lateral heterojunction after PTCDA deposition is shown in Fig. 4-19(d).

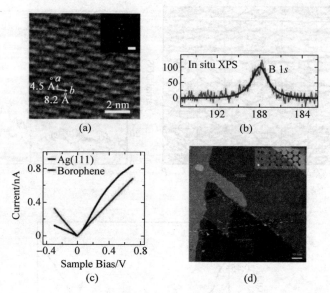

Fig. 4-19 (a) 2 nm-resolution STM image of borophene (b) XPS spectra (c) I/V curves (d) STM image of a PTCDA/borophene lateral heterojunction[73]. This figure has been adapted from Ref. [73] with permission from American Association for the Advancement of Science, copyright 2017

(For colored figure please scan the QR code on page 1)

The B 1s peak that corresponds to boron atoms in borophene was largely unaltered following PTCDA deposition, as revealed by situ XPS, suggesting that there are not PTCDA covered on the borophene in Fig. 4-20(a). The C 1s peak shows an increase tendency after PTCDA deposition, with subpeaks corresponding to perylene core and carbonyl groups in PTCDA. Few PTCDA molecules were found on the borophene surface, and charge transfer between borophene and PTCDA generated a slight shift in the B 1s peak. The study examines the interfacial electronic interaction through the STS spectra taken at different lateral displacements across the Ag/borophene/PTCDA interfaces. The results show an abrupt transition from PTCDA to borophene occurring within 1～2 nm, and weak van der Waals interactions are thought to be responsible for a slight downshift of the LUMO + 1 state as approaching the intersection from PTCDA as shown in Fig. 4-20(c) and (d).

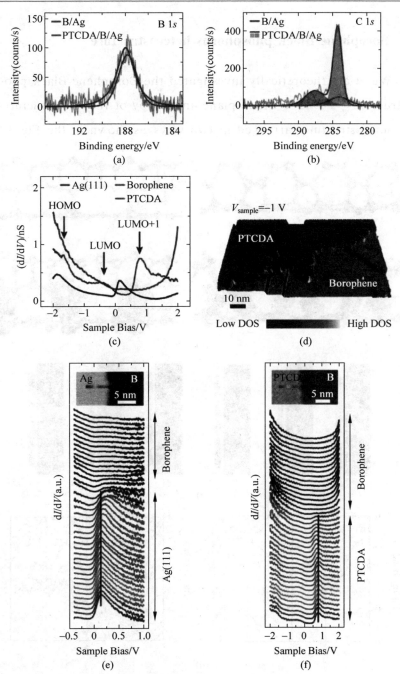

Fig. 4-20 (a) and (b) XPS of the B 1s and C 1s, respectively. (c) and (d) Spatially-resolved STS spectra of borophene/Ag and borophene/PTCDA, respectively[73]. This figure has been adapted from Ref. [73] with permission from American Association for the Advancement of Science, copyright 2017

(For colored figure please scan the QR code on page 1)

4.4.2 Borophene-Black phosphorus heterostructure

In 2023, Wu et al. theoretically investigated the Borophene-Black phosphorus heterostructure[74]. Due to the in-plane anisotropy of the two structures, there are two heterostructures arranged in two ways, as shown in the Fig. 4-21(a).

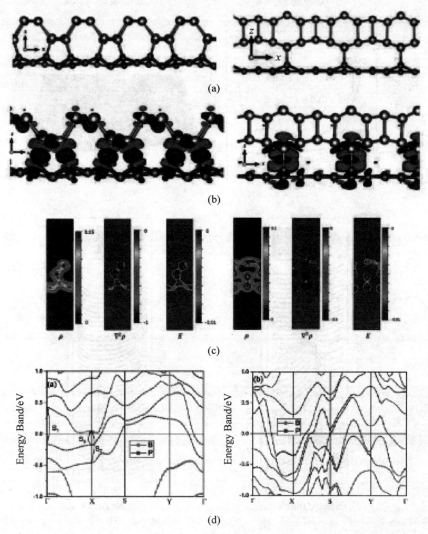

Fig. 4-21 (a) Structure of two Borophene-Black phosphorus heterostructures (b) Charge density difference (c) bonding analysis (d) energy bands[74]. This figure has been adapted from Ref. [74] with permission from Elsevier, copyright 2023
(For colored figure please scan the QR code on page 1)

Chemical bonds are formed between the two structural layers and the charge density difference both exhibit a hole-electron-hole sandwich structure to stabilize the heterostructure. The charges of the latter structure are mostly in bonding track, resulting in the stronger charge attraction force in this direction, so it is more stable. The charge analysis of the interlayer bonds all showed obvious covalent bonding as shown in Fig. 4-21(c). The energy bands of the two structures are shown in the Fig. 4-21(d), both showing metallicity.

The absorption spectrum of BP-borophene-BP sandwich heterostructure shows obvious In-plane optical anisotropy, especially at 1100 nm. The distribution of the strongest electric field with different polarization angles shows a strong polarization feature as shown in Fig. 4-22(b). When the polarization angle of increases from 0° to 90°, the maximum electric field strength progressively declines from 7.220 to 2.315. When the materials are coupled, the electric field strength increases significantly. The Fig. 4-22(c) shows the maximum electric field intensity under different polarization angles when the coupling spacing is 2 nm, showing a stronger polarization dependence. And the maximum electric field strength reaches 10.70.

In 2023, Wu et al. investigated the current, photocurrent and thermocurrent properties of borophene-black phosphorus heterostructure[75]. Due to the anisotropy of the heterostructure, the electrical properties of zigzag and armchair directions are studied as shown in Fig. 4-23(a). The I-V curve of the armchair device shows a better linear relationship than the zigzag device from -0.2 V to 0.2 V, so the armchair device is a good resistance between the voltage region. Borophene and black phosphorus are separated from the heterostructure, and calculated its I-V curves as shown in Fig. 4-23(c) and (d). The I-V curves of borophene show obvious anisotropy, while the difference between the I-V curves of black phosphorus in the two directions are not obvious, indicating that the anisotropy of heterostructure device is mainly from borophene. The photocurrent also shows a strong direction dependence, and the photocurrent of the zigzag device is much stronger than

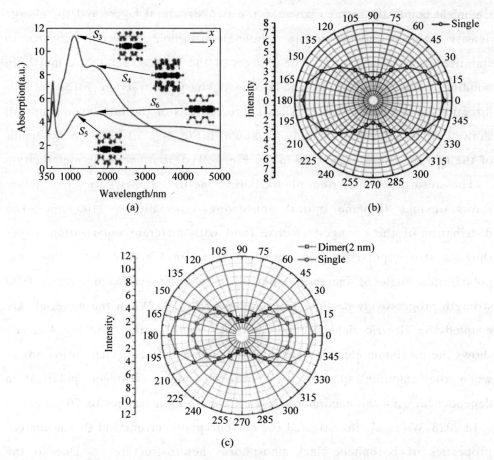

Fig. 4-22 (a) Absorption spectra of sandwich structure. (b) Maximum electric field intensity of different polarization angles at 1100 nm. (c) Maximum electric field intensity of the single and dimer materials with at different polarization angles at 1100 nm[74]. This figure has been adapted from Ref. [74] with permission from Elsevier, copyright 2023

(For colored figure please scan the QR code on page 1)

that of the armchair device, which is because the real part of the dielectric function in the armchair direction is larger than that of zigzag direction. When the wavelength of the incident light is 980 nm, the direction of the photocurrent can be changed by adjusting the polarization angle. Figure 4-23(g) shows that the thermoelectric current under different temperature differences. Under the same temperature difference, the thermoelectric current of the armchair device is larger than that of the zigzag device, and the thermoelectric current of the

CHAPTER 4 2D Borophene excitons

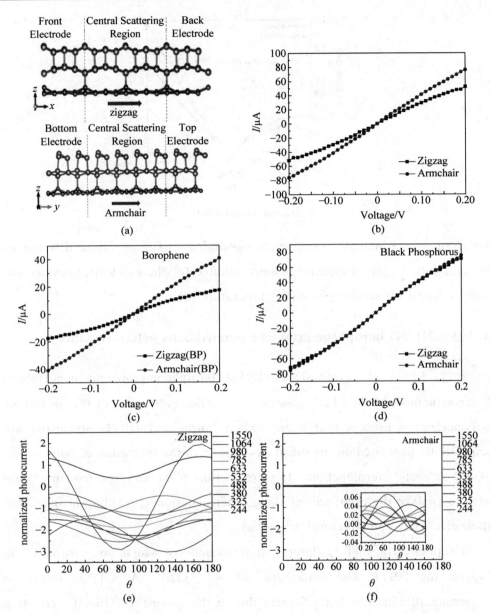

Fig. 4-23 (a) Schematic diagram of borophene-black phosphorus heterostructure device in zigzag and armchair directions (b) I-V curve of heterostructure (c) I-V curve of borophene (d) I-V curve of black phosphorus (e) Photocurrent in zigzag direction (f) Photocurrent in armchair direction (g) The thermoelectric current of heterostructure[75]. This figure has been adapted from Ref. [75] with permission from Elsevier, copyright 2023

(For colored figure please scan the QR code on page 1)

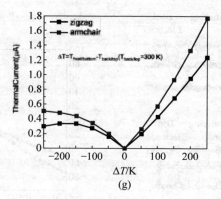

Fig. 4-23(Continued)

armchair device increases faster with the increase of temperature difference, indicating that armchair device is more suitable for thermoelectric devices and more suitable for working in high temperature regions.

4.4.3 2D/1D borophene-graphene nanoribbons heterostructure

In 2021, Li et al. synthesized 2D/1D borophene-graphene nanoribbons heterostructures on Ag (111) substrates[76]. The study explores the on-surface polymerization process that occurs when boron and DBTP precursors are sequentially deposited on the substrate, leading to the formation of borophene/aGNR lateral heterojunction. High-resolution STM and spectroscopy were used to analyze the interfaces of these heterojunctions, providing insights into their electronic and structural properties.

Borophene/Ag (TP)$_n$ lateral heterojunction, which are formed via evaporating DBTP and borophene on Ag (111). Ag (TP)$_n$ refers to organometallic intermediates formed during the on-surface Ullmann coupling reaction of dehalogenated DBTP monomers on Ag(111). Borophene is first synthesized on an Ag(111)/mica substrate by direct electron beam evaporation from pure B rods at a substrate temperature of 450℃. Two distinct borophene phases with v1/6 and v1/5 structures were showed in high-resolution scanning tunneling microscopy (STM) images, with the former possessing a rectangular-shaped unit and parallel striped shapes, while the latter had a diamond-shaped unit with brick-wall shapes such as Fig. 4-24(a) and (b). And DBTP was then deposited on Ag(111) at room temperature. STM images revealed strictly crowded arrangements of ordered DBTP, which self-assembled into two types of structures with brick-wall shapes and different vacancy density as shown in Fig. 4-24(c) and (d). Following annealing at 50℃, the monomers underwent an surface Ullmann coupling reaction, resulting in the formation of more extended nanostructures., which were identified as Ag(TP)$_n$ intermediates. STM images showed ordered Ag(TP)$_n$ chains with bright protrusions allocated to Ag adatoms trapped in the organometallic intermediates. Longer Ag(TP)$_n$ structures were also imaged as rodlike structures or chains with three protrusions decorated inside. Borophene/Ag(TP)$_n$ lateral heterojunction were synthesized by evaporating DBTP on borophene with Ag(111) (Fig. 4-24(f)). STM images showed that DBTP preferentially adsorbed on Ag(111), leading to self-assembly of this lateral heterojunction. The atomic structure of borophene was used as an orientational sign to determine the orientation of the Ag(TP)$_n$ domain with respect to the underlying borophene lattice. After demetallization of Ag(TP)$_n$ by annealing the sample at a higher temperature, borophene/aGNR lateral heterostructures are generated (Fig. 4-24(e)). The aGNRs are 3 carbon atoms wide and have an average distance of (1.0 ± 0.1) nm between adjacent parallel 3-aGNRs. The direction of the 3-aGNRs is at a 30° angle to the boron row direction in borophene, which is equal to the Ag(111) atomic chain orientation. STS measurements showed a band gap opening in the borophene phase upon introduction of the Ag(TP)$_n$ phase, indicating electronic coupling between the two phases (Fig. 4-24(g)). The authors attribute this to the formation of borophene/Ag(TP)$_n$ interfaces, which create states within the band gap of borophene due to hybridization between boron orbitals and Ag(TP)$_n$ orbitals. The 2D/1D borophene-graphene nanoribbons heterostructure have potential applications in optoelectronics and catalysis.

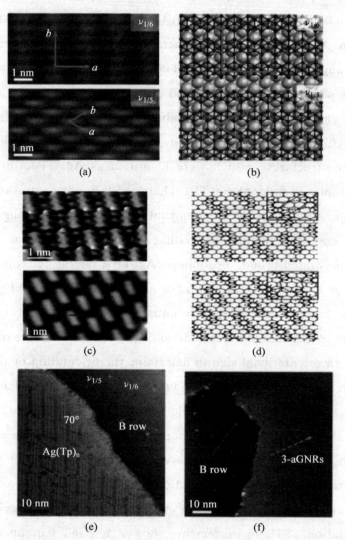

Fig. 4-24 (a) 1 nm-resolution STM images of v1/6 and v1/5 borophene. (b) DFT model of (a). (c) two type of DBTP structures. (d) DFT model of (c). (e) STM image of borophene/Ag(TP)$_n$ lateral heterojunction. (f) STM image of a borophene/3-GNR lateral heterojunction. (g) Spatially resolved STS spectra[76]. This figure has been adapted from Ref. [76] with permission from American Chemical Society, copyright 2021

(For colored figure please scan the QR code on page 1)

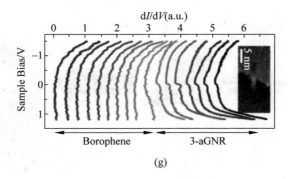

Fig. 4-24(Continued)

4.4.4 Borophene-graphene heterostructure

In 2021, Hou et al. successfully prepared borophene/graphene heterojunction by decomposing $NaBH_4$ and graphene mixture under H_2 environment without metal substratez[19]. The preparation method is shown in Fig. 4-25(a). In Fig. 4-25(b), the SEM shows the synthetic heterojunction. And the XPS of heterojunction proves the existence of borophene and graphene in Fig. 4-25(c). And the B 1s peak consists of two peaks of 187.8 eV and 188.8 eV, which proves that there are two kinds of B-B bonds as shown in Fig. 4-25(d). And the Raman spectrum shows the E_g mode of the B-B cluster the B-H bond, and the G and 2D mode of graphene, which further proves the synthesis of heterojunction as shown in Fig. 4-25(e).

Because Borophene/Graphene heterojunction shows good stability in various solvents. So heterojunction is a candidate material for humidity sensors. Boeophene/Graphene sensor device is shown in Fig. 4-26(a), and calculate its sensitivity through changes in current. In Fig. 4-26(b), when the humidity increases, the sensitivity of the sensor gradually increases. As shown in Fig. 4-26(c) and (d), under the conditions of RH = 85%, the sensitivity of heterojunction is about 700 times higher than graphenne, which is about 27 times higher than borophene's sensitivity, and the highest sensitivity is as high as 4200%, far exceeding other

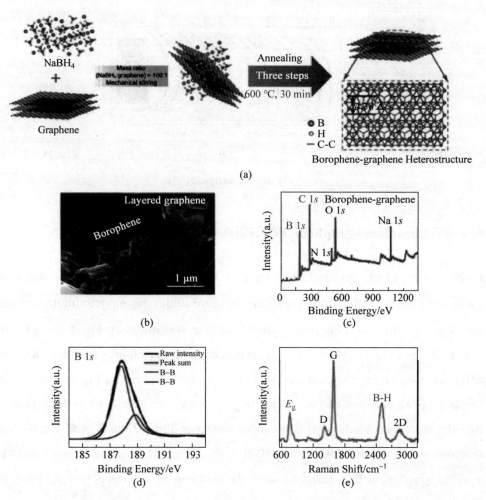

Fig. 4-25 (a) preparation method, (b) SEM images, (c) XPS of borophene-graphene heterojunction, (d) XPS of B 1s peak (e) Raman spectrum[19]. This figure has been adapted from Ref. [19] with permission from Springer Nature, copyright 2020

(For colored figure please scan the QR code on page 1)

two-dimensional materials. There are three reasons for the excellent performance of heterojunction snesor: maximizing the active sites, increased holes concentration, and decreased Schottky barrier height.

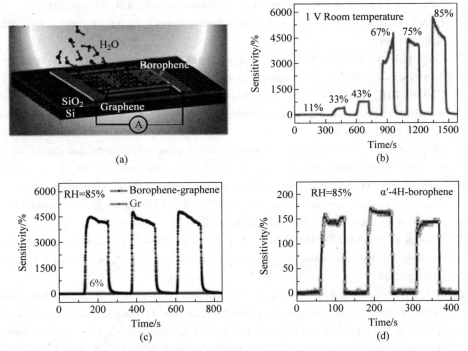

Fig. 4-26 (a) borophene/graphene heterojunction snesor, (b) Sensor performance at different humidities, (c) and (d) the repeatability of borophene/graphene heterojunction, graphene, and borophene sensor at 85% RH[19]. This figure has been adapted from Ref. [19] with permission from Springer Nature, copyright 2020
(For colored figure please scan the QR code on page 1)

References

[1] MANNIX A J, ZHOU X F, KIRALY B, et al. Synthesis of borophenes: Anisotropic, two-dimensional boron polymorphs[J]. Science. ,2015,350: 1513-1516.

[2] SANG P, WANG Q, WEI W, et al. Hydrogenated borophene as a promising two-dimensional semiconductor for nanoscale field-effect transistors: A computational study [J]. ACS Appl. Nano Mater. ,2021,4: 11931-11937.

[3] GUHA S, KABIAJ A, MAHAPATRA S. Discovery of clustered-P1 borophene and its application as the lightest high-performance transistor[J]. ACS Appl. Mater. Inter. , 2023,15: 3182-3191.

[4] TAI G, HU T, ZHOU Y, et al. Synthesis of atomically thin boron films on copper foils [J]. Angew. Chem. Int. Ed. ,2015,54: 15473-15477.

[5] ZHANG X, HU J, CHENG Y, et al. Borophene as an extremely high capacity electrode

material for Li-ion and Na-ion batteries[J]. Nanoscale. ,2016,8: 15340-15347.

[6] ZHAN C,ZHANG P,DAI S,et al. Boron supercapacitors[J]. ACS Energy. Lett. ,2016, 1: 1241-1246.

[7] KABIRAJ A, MAHAPATRA S. High-throughput assessment of two-dimensional electrode materials for energy storage devices[J]. Cell. Rep. Phys. Sci. ,2022,3: 100718.

[8] VISHNUBHOTLA V, KABIRAJ A, BHATTACHARYYA A J, et al. Global minima search for sodium-and magnesium-adsorbed polymorphic borophene[J]. J. Phys. Chem. C. ,2022,126: 8605-8614.

[9] LI D,GAO J,CHENG P,et al. 2D boron sheets: structure,growth,and electronic and thermal transport properties[J]. Adv. Funct. Mater. ,2020,30: 1904349.

[10] LIN H, SHI H, WANG Z, et al. Scalable production of freestanding few-layer β_{12}- borophene single crystalline sheets as efficient electrocatalysts for lithium-sulfur batteries[J]. ACS nano. ,2021,15: 17327-17336.

[11] MIR S H, CHAKRABORTY S, JHA P C, et al. Two-dimensional boron: Lightest catalyst for hydrogen and oxygen evolution reaction [J]. Appl. Phys. Lett. , 2016, 109: 053903.

[12] SHI L,LING C,OUYANG Y,et al. High intrinsic catalytic activity of two-dimensional boron monolayers for the hydrogen evolution reaction[J]. Nanoscale,2017,9: 533-537.

[13] HUANG Y,SHIRODKAR S N,YAKOBSON B I. Two-dimensional boron polymorphs for visible range plasmonics: a first-principles exploration[J]. J. Am. Chem. Soc. , 2017,139: 17181-17185.

[14] LIAN C,HU S Q,ZHANG J,et al. Integrated plasmonics: broadband Dirac plasmons in borophene[J]. Phys. Rev. Lett. ,2020,125: 116802.

[15] PENEV E S,KUTANA A,YAKOBSON B I. Can two-dimensional boron superconduct [J]. Nano Lett. ,2016,16: 2522-2526.

[16] XIAO R, SHAO D, LU W, et al. Enhanced superconductivity by strain and carrier- doping in borophene: A first principles prediction [J]. Appl. Phys. Lett. , 2016, 109: 122604.

[17] GAO M,LI Q Z. Prediction of phonon-mediated superconductivity in borophene[J]. Phys. Rev. B. ,2017,95: 024505.

[18] ZHAO Y,ZENG S,NI J. Superconductivity in two-dimensional boron allotropes[J]. Phys. Rev. B. ,2016,93: 014502.

[19] HOU C,TAI G A,LIU B,et al. Borophene-graphene heterostructure: Preparation and ultrasensitive humidity sensing[J]. Nano Res. ,2021,14: 2337-2344.

[20] HOU C,TAI G,LIU Y,et al. J. Mater. Chem. A. ,2021,9: 13100-13108.

[21] HOU C, TAI G, LIU Y, et al. Borophene gas sensor [J]. Nano Res. , 2022, 15: 2537-2544.

[22] FENG B, ZHANG J, ZHONG Q, et al. Experimental realization of two-dimensional boron sheets[J]. Nat. Chem. ,2016,8: 563-568.

[23] FENG B,SUGINO O,LIU R Y,et al. Dirac fermions in borophene[J]. Phys. Rev.

Lett.,2017,118: 096401.
[24] ZHANG Z,YANG Y,PENEV E S,et al. Elasticity, flexibility, and ideal strength of borophenes[J]. Adv. Funct. Mater.,2017,27: 1605059.
[25] ZHOU X F,OGANOV A R, WANG Z, et al. Two-dimensional magnetic boron[J]. Phys. Rev. B.,2016,93: 085406.
[26] LOPEZ-BEZANILLA A, LITTLEWOOD P B. Electronic properties of 8-Pmmn borophene[J] Phys. Rev. B.,2016,93: 241405.
[27] PENG B,ZHANG H, SHAO H, et al. The electronic, optical, and thermodynamic properties of borophene from first-principles calculations[J].J. Mater.Chem.C.,2016, 4: 3592-3598.
[28] SHUKLA V,WARNA J,JENA N K,et al. Toward the realization of 2D borophene based gas sensor[J].J. Phys. Chem.C.,2017,121: 26869-26876.
[29] MANNIX A J,ZHANG Z, GUISINGER N P, et al. Borophene as a prototype for synthetic 2D materials development[J]. Nat. Nanotechnol.,2018,13: 444-450.
[30] BOUSTANI I. Systematic ab initio investigation of bare boron clusters: mDetermination of the geometryand electronic structures of B n ($n = 2 - 14$)[J]. Phys. Rev. B.,1997,55: 16426.
[31] TANG H,ISMAIL-BEIGI S. Novel precursors for boron nanotubes: the competition of two-center and three-center bonding in boron sheets[J]. Phys. Rev. Lett., 2007, 99: 115501.
[32] KANETI Y V, BENU D P, XU X, et al. Borophene: two-dimensional boron monolayer: synthesis,properties,and potential applications[J]. Chem. Rev.,2021,122: 1000-1051.
[33] WU X,DAI J,ZHAO Y,et al. ACS nano.,2012,6: 7443-7453.
[34] HOU C, TAI G, WU Z. Two-dimensional boron monolayer sheets [J]. ChemPlusChem.,2020,85: 2186-2196.
[35] PIAZZA Z A,HU H S,LI W L, et al. Planar hexagonal B_{36} as a potential basis for extended single-atom layer boron sheets[J]. Nat. Commun.,2014,5: 3113.
[36] ZHOU X F, DONG X, OGANOV A R, et al. Semimetallic two-dimensional boron allotrope with massless Dirac fermions[J]. Phys. Rev. Lett.,2014,112: 085502.
[37] HUANG W,SERGEEVA A P,ZHAI H J,et al. A concentric planar doubly π-aromatic B_{19}-cluster[J]. Nat. Chem.,2010,2: 202-206.
[38] ZHAI H J,ZHAO Y F,LI W L, et al. Observation of an all-boron fullerene[J]. Nat. Chem.,2014,6: 727-731.
[39] WANG X, TIAN J, YANG T, et al. Single crystalline boron nanocones: Electric transport and field emission properties[J]. Adv. Mater.,2007,19: 4480-4485.
[40] LIU F,SHEN C,SU Z, et al. Metal-like single crystalline boron nanotubes: synthesis and in situ study on electric transport and field emission properties[J]. J. Mater. Chem.,2010,20: 2197-2205.
[41] YANG X,DING Y, NI J. Ab initio prediction of stable boron sheets and boron

nanotubes: Structure, stability, and electronic properties[J]. Phys. Rev. B., 2008, 77: 041402.

[42] TIAN J, XU Z, SHEN C, et al. One-dimensional boron nanostructures: Prediction, synthesis, characterizations, and applications[J]. Nanoscale., 2010, 2: 1375-1389.

[43] CHENG C, SUN J T, LIU H, et al. Suppressed superconductivity in substrate-supported β_{12}-borophene by tensile strain and electron doping[J]. 2D Mater., 2017, 4: 025032.

[44] YANG R, SUN M. Superior thermoelectric properties of twist-angle superlattice borophene induced by interlayer electrons transport[J]. Mater. Today Phys., 2022, 23: 100652.

[45] HOU C, TAI G, HAO J, et al. Ultrastable crystalline semiconducting hydrogenated borophene[J]. Angew. Chem. Int. Ed., 2020, 59: 10819-10825.

[46] MORTAZAVI B, MAKAREMI M, SHAHROKHI M, et al. Borophene hydride: a stiff 2D material with high thermal conductivity and attractive optical and electronic properties[J]. Nanoscale, 2018, 10: 3759-3768.

[47] RANJAN P, LEE J, KUMAR P, et al. Borophene: New sensation in flatland[J]. Adv. Mater., 2020, 32: 2000531.

[48] XIE Z, MENG X, LI X, et al. Two-dimensional borophene: properties, fabrication, and promising applications[J]. Research., 2020, 2020: 2624617.

[49] FENG B, ZHANG J, ITO S, et al. Discovery of 2D anisotropic Dirac cones[J]. Adv. Mater., 2018, 30: 1704025.

[50] ZHOU X, DONG X, OGANOV A, et al. Semimetallic two-dimensional boron allotrope with massless Dirac fermions[J]. Phys. Rev. Lett., 2014, 112: 085502.

[51] ZHANG Z, YANG Y, GAO G, et al. Two-dimensional boron monolayers mediated by metal substrates[J]. Angew. Chem., 2015, 127: 13214-13218.

[52] OU M, X WANG, L YU, et al. The emergence and evolution of borophene[J] Adv. Sci., 2021, 8: 2001801.

[53] WANG Z Q, LÜT Y, Wang H Q, et al. Review of borophene and its potential applications[J]. Front Phys., 2019, 14: 33403.

[54] KIRALY B, LIU X, WANG L, et al. Borophene synthesis on Au (111)[J]. ACS Nano., 2019, 13: 3816-3822.

[55] WU R, DROZDOV I K, ELTINGE S, et al. Large-area single-crystal sheets of borophene on Cu (111) surfaces[J]. Nat. Nanotechnol., 2019, 14: 44-49.

[56] LI W, KONG L, CHEN C, et al. Two-dimensional honeycomb borophene oxide: Strong anisotropy and nodal loop transformation[J]. Sci. Bull., 2018, 63: 282-286.

[57] VINOGRADOV N A, LYALIN A, TAKETSUGU T, et al. ACS Nano., 2019, 13: 14511-14518.

[58] LIU X, ZHANG Z, WANG L, et al. Single-phase borophene on Ir (111): formation, structure, and decoupling from the support[J]. Nat. Mater., 2018, 17: 783-788.

[59] ZHANG Z, MANNIX A J, HU Z, et al. Substrate-induced nanoscale undulations of borophene on silver[J]. Nano Lett., 2016, 16: 6622-6627.

[60] XU S, ZHAO Y, LIAO J, et al. The nucleation and growth of borophene on the Ag (111) surface[J]. Nano Res. ,2016,9: 2616-2622.

[61] CAO Y, FATEMI V, DEMIR A, et al. Correlated insulator behaviour at half-filling in magic-angle graphene superlattices[J]. Nature,2018,556: 80-84.

[62] CAO Y, FATEMI V, FANG S, et al. Unconventional superconductivity in magic-angle graphene superlattices[J]. Nature,2018,556: 43-50.

[63] LIU X, LI Q, RUAN Q, et al. Borophene synthesis beyond the single-atomic-layer limit [J]. Nat. Mater. ,2022,21: 35-40.

[64] CHEN C, LV H, ZHANG P, et al. Synthesis of bilayer borophene[J]. Nat. Chem. , 2022,14: 25-31.

[65] MA F, JIAO Y, GAO G, et al. Graphene-like two-dimensional ionic boron with double dirac cones at ambient condition[J]. Nano Lett. ,2016,16: 3022-3028.

[66] XU S G, ZHENG B, XU H, et al. Ideal nodal line semimetal in a two-dimensional boron bilayer[J]. J. Phys. Chem. C. ,2019,123: 4977-4983.

[67] ZHONG H, HUANG K, YU G, et al. Electronic and mechanical properties of few-layer borophene[J]. Phys. Rev. B. ,2018,98: 054104.

[68] ZHAI H J, KIRAN B, LI J, et al. Hydrocarbon analogues of boron clusters—planarity, aromaticity and antiaromaticity[J]. Nat. Mater. ,2003,2: 827-833.

[69] LAU K C, Pandey R. Stability and electronic properties of atomistically-engineered 2D boron sheets[J]. J. Phys. Chem. C. ,2007,111: 2906-2912.

[70] LIU Y, PENEV E S, YAKOBSON B I. Probing the synthesis of two-dimensional boron by first-principles computations[J]. Angew. Chem. Int. Ed. ,2013,52: 3156-3159.

[71] SONG J, CAO Y, DONG J, et al. Superior thermoelectric properties of twist-angle superlattice borophene induced by interlayer electrons transport[J]. Small, 2023, 19: 2301348.

[72] CAO Y, FENG Y, CHENG Y, et al. Fano-resonant propagating plexcitons and Rabi-splitting local plexcitons of bilayer borophene in TERS[J]. Appl. Phys. Lett. ,2023, 122: 23.

[73] LIU X, WEI Z, BALLA I, et al. Self-assembly of electronically abrupt borophene/organic lateral heterostructures[J]. Sci. Adv. ,2017,3: e1602356.

[74] WU Y, SUN M. Polarization-dependent excitons in Borophene-Black phosphorus heterostructures[J]. Spectrochim. Acta A. ,2023,291: 122372.

[75] WU Y, SUN M. Current, photocurrent and thermocurrent of Borophene-Black phosphorus heterostructures[J]. Results in Physics. ,2023,50: 106583.

[76] LI Q, LIU X, Aklile E B, et al. Self-assembled borophene/graphene nanoribbon mixed-dimensional heterostructures[J]. Nano Lett. ,2021,21: 4029-4035.

CHAPTER 5

Surface Plasmons

5.1 Brief Introduction of SPs

Surface plasmons (SPs) were known to be collective oscillations about free electrons at the interface between dielectrics and metals[1] and widely used in a variety of chemical reactions, such as catalyses, dissociation[2-6], photochromic reactions, isomerization, and polymerization. In general, SPs are divided into two types, the local SP (LSP) and the propagating SP (PSP, also known as plasmon waveguide)[7-8]. Their schematic was shown in Fig. 5-1. Among them, the LSP was confined to the surface of the metal nanoparticles (NPs) and belongs to the oscillation of the charge density. LSP resonance (LSPR) was an important mechanism of surface-enhanced Raman spectroscopy, which could greatly enhance the local field[9]. The SP was coupled with a photon as a SP polariton (SPP), which was a collective excitation about conduction electrons, and confined near the interface and could propagate down the metal surface, and the energy gradually dissipated due to radiation or heat loss during propagation[10]. When incident light having a certain wavelength was irradiated onto the metal surface, a large increase in local field and SPR of the electromagnetic (EM) field on the metal surface was caused. SP was widely used in SERS and TERS spectra[11-12].

CHAPTER 5 Surface Plasmons

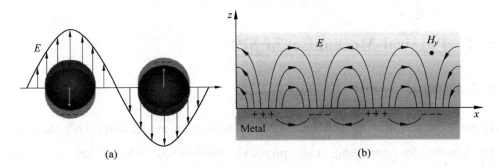

Fig. 5-1 Sketches of the mechanism of LSPs and PSPs. (a) LSPs are the collective oscillation of free electrons confined on the surface of metal NPs and (b) PSPs are the excitation of collective electrons that can only propagate near the vicinity of the interface[13]

(For colored figure please scan the QR code on page 1)

LSPR peaks can be well manipulated by nanostructures, see Fig. 5-2(a). SPP can be propagated with the surface of nanostructures, known as propagating SPP (PSPP), see Fig. 5-2(b).

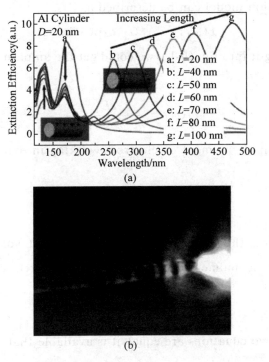

Fig. 5-2 (a) LSPR, and (b) PSPP
(For colored figure please scan the QR code on page 1)

5.2 Physical Mechanism of SPs

5.2.1 Drude model

It was essential to understand a collective oscillation model about free electrons in metals for studying the physical mechanism about SP in metal nanostructure. When operating in an external electric field, the motion about free electrons in the metal could be expressed as

$$m \frac{d^2 x}{dt^2} + m\gamma \frac{dx}{dt} = -eE_0 \exp(-i\omega t), \tag{5-1}$$

where, x is the electronic location, m the electronic quality, g the damping constant, e the electronic charge, E_0 the amplitude of external electric field and ω the circular frequency of external electric field. Here, the electronic collective oscillation model can be described as

$$x(\omega, t) = x_0(\omega) \exp(-i\omega t). \tag{5-2}$$

Then, substituting it into Eq. (5-1), we could get the amplitude of the electronic oscillation as

$$x_0(\omega) = \frac{eE_0}{m(\omega^2 + i\gamma\omega)}. \tag{5-3}$$

So, the induced dipole moment of electron, working in external electric field, can be described as

$$P = N(-ex_0) = -\frac{Ne^2 E_0}{m(\omega^2 + i\gamma\omega)}, \tag{5-4}$$

where N represents the number of electrons in a unit volume. At the same time, induced dipole moment is a function of dielectric constant, it can be described as

$$P = \varepsilon_0 [\varepsilon(\omega) - 1] E_0, \tag{5-5}$$

since the above two equations are equal, it is available that

$$\varepsilon(\omega) = \varepsilon_r + i\varepsilon_i = 1 - \frac{\omega_p^2}{\omega(\omega + i\gamma)}, \tag{5-6}$$

where $\omega_p = \sqrt{\dfrac{Ne^2}{\varepsilon_0 m}}$ is electronic plasmon frequency. Eq. (5-6) represents the collective oscillation model (Drude model) of free electrons in a metal, it can describe the relationship between the metal dielectric constant and the electric field frequency of incident light. The authors suppose $\gamma \ll \omega$, then could get that $\varepsilon(\omega) \approx 1 - \dfrac{\omega_p^2}{\omega^2}$. It follows that dielectric constant was negative number, the refractive index was a ratio and there was a strong interaction between incident EM wave and metal when $\omega < \omega_p$. However, this showed that dielectric constant was positive number, the refractive index was real and metal was just a traditional dielectric material for incident light when $\omega > \omega_p$. Although Drude model had many approximate conditions, it could still be used to give details the physical mechanism about SP and many experimental phenomena.

5.2.2 Relationship between Refractive Index and Dielectric Constant

When visible light incident on a metal, its energy was absorbed by the metal surface layer to excite free electrons, it reached a higher energy state. The photon was emitted when electrons transition from a high energy state to a low energy state. Since the metal was opaque, the absorption phenomenon only occurred in its surface layer having a thickness of about 100 nm, in other words, the sheet metal was "transparent" below 100 nm. Only short-wavelength rays, such as X-rays and g-rays, could pass through a certain thickness of metal. Therefore, the interaction between metal and visible light was mainly reflection, resulting in metallic luster.

When incident light entered the light-transmitting material at an angle, the phenomenon of bending was refraction. The refractive index n was defined as

$$n = \frac{c_{vac}}{c} = \frac{\sin\theta_i}{\sin\theta_r}, \tag{5-7}$$

where θ_i and θ_r represented the incident angle and the refraction angle, respectively. When light entered a denser material from a vacuum, its speed reduced. The ratio of the light between vacuum and material was the refractive index of the material. If light entered material 2 from material 1 through the interface, the relationship between incident angle, refraction angle and refractive index of material could be described as (Fig. 5-3)

$$\frac{\sin i_1}{\sin i_2} = \frac{n_2}{n_1} = n_{21} = \frac{v_1}{v_2}. \tag{5-8}$$

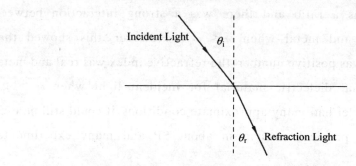

Fig. 5-3 Schematic of the relationship between the angle of incidence, the angle of refraction and the refractive index of a material as it passes from one material to another

(For colored figure please scan the QR code on page 1)

The refractive index of the medium was a positive number that was always greater than 1. Such as for air, $n = 1.0003$; for solid oxide, $n = 1.3 : 2.7$, and for silicate glass, $n = 1.5 : 1.9$. Different compositions or structures, different refractive indices.

There were several aspects that affect the factor of n. One of the important factors the authors first introduced was the ionic radius of the constituent elemental materials. According to Maxwell's theory of EM waves, the propagation speed of light in a medium can be fitted with

$$v = \frac{c}{\sqrt{\varepsilon\mu}}, \tag{5-9}$$

where, μ was the permeability of the medium, c was the speed of light in vacuum and ε was the dielectric constant of the medium, so that

$$n = \sqrt{\varepsilon\mu}. \tag{5-10}$$

In inorganic material media, $\mu = 1$, so $n = \sqrt{\varepsilon}$. It shows that the refractive index about the medium increased with the increase of its dielectric constant, while the dielectric constant was related to the polarization of the medium. Due to the interaction of EM radiation and the electronic system of the atom, the light wave was decelerated. When the ionic radius increased, its dielectric constant and n also increased. Therefore, it was possible to obtain a material having a high refractive index with a large ion, such as for PbS, $n = 3.912$. In contrast, a low refractive index material could be obtained with small ions, such as for $SiCl_4$, $n = 1.412$.

5.2.3 Dispersion relations

In physical science and electrical engineering, dispersion relations describe the effect of dispersion in a medium on the properties of a wave traveling. The dispersion relations describe the relationship between the angular frequency (ω) of a wave and the magnitude of its wave vector (k).

In the dispersion relation for bulk plasmon, the wave equation can be written as

$$\frac{\varepsilon_m}{c^2} \frac{\partial^2 \boldsymbol{E}(r,t)}{\partial t^2} = \nabla^2 \boldsymbol{E}(r,t), \tag{5-11}$$

where the investigated solution of the form is

$$\boldsymbol{E}(r,t) = \mathrm{Re}\{\boldsymbol{E}(r,\omega)\exp(i\boldsymbol{k}\cdot\boldsymbol{r} - i\omega t)\}. \tag{5-12}$$

According to

$$\omega^2 \varepsilon_m = c^2 k^2 \quad \text{and} \quad \varepsilon_m = 1 - \frac{\omega_p^2}{\omega^2}, \tag{5-13}$$

one can get

$$\omega^2 \left(1 - \frac{\omega_p^2}{\omega^2}\right) = \omega^2 - \omega_p^2 = c^2 k^2, \tag{5-14}$$

and then,

$$\omega(k) = \sqrt{\omega_p^2 + c^2 k^2}. \tag{5-15}$$

So, the dispersion relation for bulk plasmon can be seen from Fig. 5-4.

The dispersion relationship of SP, for dielectric-metal boundary can be understand by the model in Fig. 5-5.

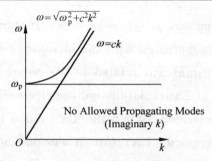

Fig. 5-4 The dispersion relation for bulk plasmon
(For colored figure please scan the QR code on page 1)

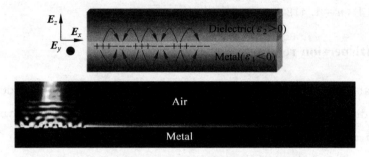

Fig. 5-5 The model of dispersion relationship of SP, for dielectric-metal boundary
(For colored figure please scan the QR code on page 1)

The permittivity of metal can be written as

$$\varepsilon_m(\omega) = 1 - \frac{\omega_p^2}{\omega^2 + \gamma^2} + i\frac{\omega_p^2}{\omega^2 + \gamma^2}\left(\frac{\gamma}{\omega}\right) \approx 1 - \frac{\omega_p^2}{\omega^2}. \tag{5-16}$$

Condition of SPs is

$$\frac{k_{zm}}{\varepsilon_m} = \frac{k_{zd}}{\varepsilon_d}, \tag{5-17}$$

and relation for k_x is

$$k_{xm} = k_{xd}. \tag{5-18}$$

For any EM,

$$k^2 = \varepsilon_i \left(\frac{\omega}{c}\right)^2 = k_x^2 + k_{zi}^2, \tag{5-19}$$

and according to Eq. (5-17) and Eq. (5-19), we can obtain ($k_{SPP} = k_x$) the dispersion relationship

$$k_{SPP} = \frac{\omega}{c}\left(\frac{\varepsilon_d \varepsilon_m}{\varepsilon_d + \varepsilon_m}\right)^{1/2}. \tag{5-20}$$

Detailed information of dispersion relationship can be seen from Fig. 5-6.

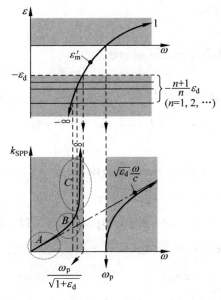

Fig. 5-6 Dispersion relation of SP, for dielectric-metal boundary
(For colored figure please scan the QR code on page 1)

At low frequency, a SPP approaches a Sommerfeld-Zenneck wave, where the dispersion relation (between frequency and wavevector) is the same as in free space. Thus the wave's speed is ω/k, the dispersion relationship becomes

$$k_{SPP} = \frac{\omega}{c} \lim_{\varepsilon_m \to -\infty} \left(\frac{\varepsilon_d \varepsilon_m}{\varepsilon_d + \varepsilon_m} \right)^{1/2} \approx \frac{\omega}{c} \sqrt{\varepsilon_d}. \qquad (5-21)$$

At a higher frequency, the dispersion relation bends over and reaches an asymptotic limit called the "surface plasma frequency" in Eq. (5-20).

When $\varepsilon_m = -\varepsilon_d$, $k_{SPP} \to \infty$,

$$\omega = \omega_{sp} = \frac{\omega_{sp}}{\sqrt{1+\varepsilon_d}}. \qquad (5-22)$$

Along the z direction,

$$k_{zi}^2 = \varepsilon_i \left(\frac{\omega}{c}\right)^2 - k_x^2 \quad \text{and} \quad k_{zi} = \pm \frac{\omega}{c} \left(\frac{\varepsilon_i^2}{\varepsilon_d + \varepsilon_m} \right)^{1/2}. \qquad (5-23)$$

For bound SP mode, k_{zi} must be a pure imaginary number, $\varepsilon_d + \varepsilon_m < 0$, and $\varepsilon_m < -\varepsilon_d$.

5.3 Localized SPs

5.3.1 LSPs in metallic nanosphere

The electric fields inside and outside the particle are E_{in} and E_{out}, respectively. And the potentials are $\phi_{in}(r,\theta)$ and $\phi_{out}(r,\theta)$, respectively, where r is the position vector.

Supposed that the spherical metal particles of radius a were in a uniform electrostatic field of $E = E_0 r \cos\theta$ (Fig. 5-7), the electric fields inside and outside the particle were E_{in} and E_{out}, respectively. The potentials were $\phi_{in}(r,\theta)$ and $\phi_{out}(r,\theta)$, respectively, so that

$$\begin{cases} E_{in} = -\nabla\phi_{in}, & E_{out} = -\nabla\phi_{out}, \\ \nabla^2\phi_{in} = 0 \ (r<a), & \nabla^2\phi_{out} = 0 \ (r>a), \end{cases} \quad (5\text{-}24)$$

Eq. (5-24) satisfied the following boundary conditions

$$\begin{cases} \phi_{in} = \phi_{out} \\ \varepsilon \dfrac{\partial\phi_{in}}{\partial r} = \varepsilon_m \dfrac{\partial\phi_{out}}{\partial r} \end{cases} (r=a), \quad (5\text{-}25)$$

where ε and ε_m were the dielectric constants of spherical particles and surrounding media, respectively. In other words, the boundary conditions were that the potentials at the interface were equal. Specifically, the potentials at infinity were undisturbed, it could be found that the solution satisfying the above partial differential equations and boundary conditions was

$$\phi_{in} = \frac{-3\varepsilon_m}{\varepsilon + 2\varepsilon_m} E_0 r\cos\theta, \quad (5\text{-}26)$$

$$\phi_{out} = -E_0 r\cos\theta + a^3 E_0 \frac{\varepsilon - \varepsilon_m}{\varepsilon + 2\varepsilon_m} \frac{\cos\theta}{r^2}. \quad (5\text{-}27)$$

It was easy to see that the potential energy ϕ_{out} outside the particle could be considered as the sum of the potential of the incident electric field and the potential energy about the other dipole, i.e. the first and second items on the

CHAPTER 5 Surface Plasmons

right side of Eq. (5-27). And the potential energy about the dipole could be described as

$$\phi = \frac{p\cos\theta}{4\pi\varepsilon_m r^2}. \tag{5-28}$$

And its dipole moment was

$$p = 4\pi\varepsilon_m a^3 \frac{\varepsilon - \varepsilon_m}{\varepsilon + 2\varepsilon_m} E_0. \tag{5-29}$$

The polarizability of the dipole could be estimated with

$$\alpha = 4\pi a^3 \frac{\varepsilon - \varepsilon_m}{\varepsilon + 2\varepsilon_m}. \tag{5-30}$$

In other words, when the radius about the particle was much smaller than the incident light wavelength, it could be treated as a dipole approximation. Its polarizability was a function of the dielectric constant and radius about the particle. Further deriving, the scattering and absorption cross sections of the particles were given by

$$C_{abs} = k\operatorname{Im}\{\alpha\} = 4k\pi a^3 \operatorname{Im}\left\{\frac{\varepsilon - \varepsilon_m}{\varepsilon + 2\varepsilon_m}\right\}, \tag{5-31}$$

$$C_{scat} = \frac{k^4}{6\pi}|\alpha|^2 = \frac{8}{3}k^4\pi a^6 \left|\frac{\varepsilon - \varepsilon_m}{\varepsilon + 2\varepsilon_m}\right|, \tag{5-32}$$

where k was the incident light wave vector, and the extinction cross section could be estimated with

$$C_{ext} = C_{abs} + C_{scat}. \tag{5-33}$$

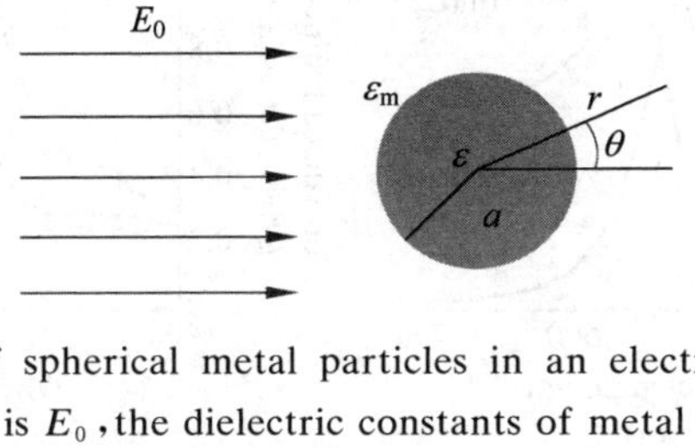

Fig. 5-7 Schematic of spherical metal particles in an electrostatic field. Incident electric field is E_0, the dielectric constants of metal particles and media are ε and ε_m, respectively

(For colored figure please scan the QR code on page 1)

Therefore, the scattering cross section was proportional to the sixth power about the radius, and the absorption cross section was proportional to the third power about the radius. For larger diameter particles, light scattering was a major contributor. In contrast, for particles with smaller diameters, the proportion of light absorption was larger. The above polarization rate derivation was based on a quasi-electrostatic model, treating metal NPs as a dipole, and ignoring the delay effect of the electric field in the particles and the radiation attenuation effect. Therefore, it could only be used to calculate particles with diameters much smaller than wavelength. When the particle was large, the particle diameter was comparable to the wavelength especially, since the skin depth of the electric field was smaller than the radius, the electric field in the whole particle was not uniform. Consequently, the effective electric field was weakened, and the particles could not be treated as dipoles, but the higher-order modes must be considered, such as quadrupole and octapole. Therefore, a more reasonable metal particle polarization rate was the modified long-wavelength approximation (MLWA), so that

$$\alpha_{\text{corr}} = \frac{\alpha}{1 - \frac{2}{3}ik^3 \frac{\alpha}{4\pi} - \frac{1}{a}k^2 \frac{\alpha}{4\pi}}. \qquad (5-34)$$

Fig. 5-8(a) shows the polarization rate curve of Au NPs with a diameter of 40 nm described in the MLWA model in water. Using this model, the red-

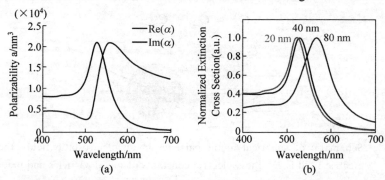

Fig. 5-8 (a) Polarity curve of gold NPs with a diameter of 40 nm in water (MLWA model), the solid and dashed lines represent the real and imaginary parts of the polarizability, respectively. (b) Extinction cross-section curves of gold NPs with diameters of 20 nm, 40 nm and 80 nm in water (normalized)
(For colored figure please scan the QR code on page 1)

shifted effect about the LSPR peak position of metal NPs with size increase could be perfectly explained, as shown in Fig. 5-8(b).

5.3.2 LSPs in coupled metallic NPs: parallel-polarized excitation

In this section, Photoluminescence (PL) and scattering phenomena[14] from metallic nanoparticles (MNP) have been explained and understood by several point of views. One of them is based on the classic harmonic oscillator model, which describes PL of single mode. In this study, we continue to expand this classic model to a coupling case, which involves two oscillators that interact with each other together with the excitation electric field. The new generated modes due to the coupling are carefully analyzed, including their behaviors varying with the coupling coefficients in different cases. Furthermore, for practical purpose, PL spectra and white light scattering spectra of two individual metallic nanostructures are calculated as examples employing the model to verify its validity.

When a MNP is illustrated by the incident light, the LSPs are excited, which influence the scattering and PL properties of the MNP. The optical properties of an individual MNP have been discussed in detail theoretically by one of the authors[15]. Here, we investigate the optical properties of the coupled MNPs in theory. Firstly, we employ the two-body system as the beginning before the many-body system is investigated.

Since there are plenty free electrons in the metallic nanostructure, and these electrons oscillate when excited by the external electric field, we treat the nanostructure as a resonator, the oscillators of which are the electrons. Due to the collectively oscillating, we can simplify the multiple electrons as only one electron. Consider that two metallic nanostructures treated as two oscillators are close to each other and are driven by the electric field of the excitation light. The schematic is shown in Fig. 5-9. In order to obtain the emission electric field from them, we need to find out the differential equations of them. Define $x_1(t)$ and $x_2(t)$ as the displacements relative to equilibrium positions of each oscillator, thus $\dot{x}_1(t)$ and $\dot{x}_2(t)$ the velocities, and $\ddot{x}_1(t)$ and $\ddot{x}_2(t)$ the accelerations. The equations should be in this form:

$$\begin{cases} \ddot{x}_1 + 2\beta_{01}\dot{x}_1 + \omega_{01}^2 x_1 - \dfrac{F_{21}}{m_e} = C_1 e^{-i\omega_{ex}t}, \\ \ddot{x}_2 + 2\beta_{02}\dot{x}_2 + \omega_{02}^2 x_2 - \dfrac{F_{12}}{m_e} = C_2 e^{-i\omega_{ex}t}. \end{cases} \quad (5-35)$$

Here, F_{21} and F_{12} are the interaction forces between oscillator 1 and oscillator 2, $C_1 = q_1 E_1/m_e$, $C_2 = q_2 E_2/m_e$, and m_e is the mass of electron. E_1 and E_2 are the amplitudes of the excitation electric field at the positions of the two oscillators, and usually $E_1 = E_2 = E_0$ is a good approximation; q_1 and q_2 are the charge of oscillator 1 and oscillator 2, respectively; β_{01} and β_{02} represent the damping coefficients, and ω_{01} and ω_{02} represent the inherent circular frequencies. The next step is to find out the interaction parts of the equations.

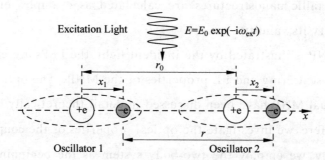

Fig. 5-9 Schematic of the coupling harmonic oscillator model. The electrons (grey, negative charged) oscillate collectively along x-axis near their equilibrium positions. The ions (white, positive charged) are at rest. r_0 is the distance between the two ions, while r is the distance between the two electrons. x_1 and x_2 are the displacements relative to equilibrium positions of each oscillator. The two oscillators both oscillate along x-axis when excited by the excitation light at the circular frequency of ω_{ex} which is x-polarized

The electric field introduced by a moving charged particle is given by[16]

$$E = \frac{q}{4\pi\varepsilon_0} \frac{\boldsymbol{r}}{(\boldsymbol{r} \cdot \boldsymbol{u})^3} [(c^2 - v^2)\boldsymbol{u} + \boldsymbol{r} \times (\boldsymbol{u} \times \boldsymbol{a})], \quad (5\text{-}36)$$

CHAPTER 5 Surface Plasmons

where q is the charge of the particle, ε_0 is the permittivity of vacuum, and $\boldsymbol{u} \equiv \dfrac{cr}{r} - \boldsymbol{v}$. Here, c is the velocity of light in vacuum, $\boldsymbol{v}$ and $\boldsymbol{a}$ are the velocity and the acceleration of the particle, respectively, and r is the displacement vector from the particle to field point. In our one-dimension case, when considering the electric field introduced by one oscillator acting on the other oscillator, the second part of Eq. (5-36) is zero due to the fact that $\boldsymbol{u}$ and $\boldsymbol{a}$ are parallel. Besides, we notice that the charged particle that moves is the electron while the positive ion is assumed to be at rest. Hence, the interacted electric field at one oscillator should be contributed to both positive charged ion and negative charged electron of the other oscillator. Therefore, the electric field can be written as

$$\begin{cases} E_{21} = -\dfrac{+e}{4\pi\varepsilon_0}\dfrac{1}{(r_0-x_1)^2} + \dfrac{-e}{4\pi\varepsilon_0}\dfrac{1}{r^2}\left(-\dfrac{c-\dot{x}_2}{c+\dot{x}_2}\right) \cong -\dfrac{e}{2\pi\varepsilon_0 r_0^2}\left(\dfrac{\dot{x}_2}{c}+\dfrac{x_2}{r_0}\right), \\ E_{12} = +\dfrac{+e}{4\pi\varepsilon_0}\dfrac{1}{(r_0+x_2)^2} + \dfrac{-e}{4\pi\varepsilon_0}\dfrac{1}{r^2}\left(+\dfrac{c+\dot{x}_1}{c-\dot{x}_1}\right) \cong -\dfrac{e}{2\pi\varepsilon_0 r_0^2}\left(\dfrac{\dot{x}_1}{c}+\dfrac{x_1}{r_0}\right). \end{cases}$$

(5-37)

Here, we use the conditions $v/c \ll 1$, $x/r_0 \ll 1$ for approximation, and $r = r_0 + x_2 - x_1$. Notice that E_{21} is the electric field in oscillator 1 introduced by one pair of the electrons and ions in oscillator 2, and E_{12} is the electric field in oscillator 2 introduced by one pair of the electrons and ions in oscillator 1. Hence, the interaction forces should be written as $F_{21} = -N_2 e E_{21}$ and $F_{12} = -N_1 e E_{12}$, where N_1 and N_2 are the effective numbers of free electrons in oscillator 1 and oscillator 2, respectively. We define the coupling coefficients as

$$\begin{cases} \gamma_{21} = \dfrac{N_2 e^2}{2\pi\varepsilon_0 m_e r_0^2 c}, & \gamma_{12} = \dfrac{N_1 e^2}{2\pi\varepsilon_0 m_e r_0^2 c}, \\ g_{21}^2 = \dfrac{N_2 e^2}{2\pi\varepsilon_0 m_e r_0^3}, & g_{12}^2 = \dfrac{N_1 e^2}{2\pi\varepsilon_0 m_e r_0^3}. \end{cases}$$

(5-38)

thus Eq. (5-35) can be written as

$$\begin{cases} \ddot{x}_1 + 2\beta_{01}\dot{x}_1 + \omega_{01}^2 x_1 - \gamma_{21}\dot{x}_2 - g_{21}^2 x_2 = C_1 e^{-i\omega_{ex} t}, \\ \ddot{x}_2 + 2\beta_{02}\dot{x}_2 + \omega_{02}^2 x_2 - \gamma_{12}\dot{x}_1 - g_{12}^2 x_1 = C_2 e^{-i\omega_{ex} t}. \end{cases} \quad (5\text{-}39)$$

For simplicity, we assume $N_1 = N_2 = N$, thus $\gamma_{21} = \gamma_{12} = \gamma$, $g_{21} = g_{12} = g$, and if we define $\dfrac{1}{\kappa} = \dfrac{\gamma^3}{g^4} = \dfrac{Ne^2}{2\pi\varepsilon_0 m_e c^3}$, it results in a simple form for γ and g:

$$\gamma = \frac{1}{\kappa}\left(\frac{c}{r_0}\right)^2, \quad g^2 = \frac{1}{\kappa}\left(\frac{c}{r_0}\right)^3. \quad (5\text{-}40)$$

Firstly, we consider the situation without coupling, i.e., $\gamma = g = 0$. In such conditions, Eq. (5-39) is degenerated into the simple form:

$$\begin{cases} \ddot{x}_1 + 2\beta_{01}\dot{x}_1 + \omega_{01}^2 x_1 = C_1 e^{-i\omega_{ex} t}, \\ \ddot{x}_2 + 2\beta_{02}\dot{x}_2 + \omega_{02}^2 x_2 = C_2 e^{-i\omega_{ex} t}. \end{cases} \quad (5\text{-}41)$$

The general solutions are

$$\begin{cases} x_1(t) = \exp(-\beta_{01} t \pm i\omega_{c1} t), \quad \exp(-i\omega_{ex} t), \\ x_2(t) = \exp(-\beta_{02} t \pm i\omega_{c2} t), \quad \exp(-i\omega_{ex} t). \end{cases} \quad (5\text{-}42)$$

Here, $\omega_{c1} = \sqrt{\omega_{01}^2 - \beta_{01}^2}$ and $\omega_{c2} = \sqrt{\omega_{02}^2 - \beta_{02}^2}$ represent the resonant circular frequencies, respectively, which are different from the inherent ones (ω_{01}, ω_{02}). The coefficients that represent amplitudes are omitted for the moment, which can be obtained with the initial conditions. The details of this kind of individual oscillator has been discussed carefully in our previous work[15].

Secondly, we start to consider the coupling situation without excitation light, i.e., $C_1 = C_2 = 0$. The equations are

$$\begin{cases} \ddot{x}_1 + 2\beta_{01}\dot{x}_1 + \omega_{01}^2 x_1 - \gamma\dot{x}_2 - g^2 x_2 = 0, \\ \ddot{x}_2 + 2\beta_{02}\dot{x}_2 + \omega_{02}^2 x_2 - \gamma\dot{x}_1 - g^2 x_1 = 0. \end{cases} \quad (5\text{-}43)$$

To solve Eq. (5-43), we can assume that $x_1(t) = A\exp(\alpha t)$ and $x_2(t) = B\exp(\alpha t)$, and substitute them back into Eq. (5-43), thus obtaining

$$\begin{cases} A(\alpha^2 + 2\beta_{01}\alpha + \omega_{01}^2) - B(\gamma\alpha + g^2) = 0, \\ B(\alpha^2 + 2\beta_{02}\alpha + \omega_{02}^2) - A(\gamma\alpha + g^2) = 0. \end{cases} \quad (5\text{-}44)$$

Obviously, to obtain non-zero solutions, α should satisfy:

$$(\alpha^2 + 2\beta_{01}\alpha + \omega_{01}^2)(\alpha^2 + 2\beta_{02}\alpha + \omega_{02}^2) = (\gamma\alpha + g^2)^2. \quad (5\text{-}45)$$

CHAPTER 5 Surface Plasmons

Notice that Eq. (5-45) has analytic solutions for α, marked as $\alpha_1, \alpha_2, \alpha_3$, and α_4. However, the expressions are so complex that we would not write in the text. Instead, to illustrate the physical significance of α, we rewrite it in this form:

$$\begin{cases} \alpha_1 = -\beta_1 + i\omega_1, & \alpha_2 = -\beta_1 - i\omega_1, \\ \alpha_3 = -\beta_2 + i\omega_2, & \alpha_4 = -\beta_2 - i\omega_2. \end{cases} \quad (5\text{-}46)$$

Here, ω_1 and ω_2 are the new generated resonant circular frequencies when the two oscillators couple. We can call them mode 1 and mode 2, respectively. In a more special case, i.e., $\beta_{01} = \beta_{02} = \beta_0$, $\omega_{01} = \omega_{02} = \omega_0$, the solutions of Eq. (5-45) are expressed easily:

$$\begin{cases} \omega_1 = \sqrt{\omega_0^2 + g^2 - (\beta_0 + \gamma/2)^2}, & \beta_1 = \beta_0 + \gamma/2, \\ \omega_2 = \sqrt{\omega_0^2 - g^2 - (\beta_0 - \gamma/2)^2}, & \beta_2 = \beta_0 - \gamma/2. \end{cases} \quad (5\text{-}47)$$

Thirdly, notice that the particular solutions for Eq. (5-39) are $x_1(t) = \exp(-i\omega_{ex}t)$ and $x_2(t) = \exp(-i\omega_{ex}t)$ (amplitudes are omitted). Therefore, combining these particular solutions and the general ones (Eq. (5-42)), we obtain the total solutions of Eq. (5-39) in a symmetric form:

$$\begin{cases} x_1(t) = A_1\exp(\Omega_1 t) + A_2\exp(\Omega_2 t) + A_3\exp(\Omega_3 t), \\ x_2(t) = B_1\exp(\Omega_1 t) + B_2\exp(\Omega_2 t) + B_3\exp(\Omega_3 t). \end{cases} \quad (5\text{-}48)$$

where $\Omega_1 = -\beta_1 + i\omega_1$, $\Omega_2 = -\beta_2 - i\omega_2$, and $\Omega_3 = -i\omega_{ex}$. We emphasize here that Eq. (5-47) is just a special case for ω_1 and ω_2, and the general case for them should satisfy Eq. (5-46). The initial conditions are $x_1(0) = x_2(0) = 0$, $\dot{x}_1(0) = \dot{x}_2(0) = 0$, $\ddot{x}_1(0) = \ddot{x}_2(0) = C_0$, where we assume that $C_1 = C_2 = C_0$ due to the subwavelength scale of the system. Hence, these coefficients are obtained as

$$\begin{cases} A_1 = B_1 = \dfrac{C_0}{(\Omega_1 - \Omega_2)(\Omega_1 - \Omega_3)}, \\[2mm] A_2 = B_2 = \dfrac{C_0}{(\Omega_2 - \Omega_3)(\Omega_2 - \Omega_1)}, \\[2mm] A_3 = B_3 = \dfrac{C_0}{(\Omega_3 - \Omega_1)(\Omega_3 - \Omega_2)}. \end{cases} \quad (5\text{-}49)$$

This results in the fact that $x_1(t) = x_2(t) = x(t)$.

At last, we deal with the far field radiation. For simplicity, we consider the electric field at the position $\boldsymbol{d}$, where $\boldsymbol{d}$ is perpendicular to x-axis, and $d = |\boldsymbol{d}|$ is the distance between field point and the center of the two oscillators. The assumption of $d \gg r_0$ is reasonable for far field radiation. Hence, the first part of Eq. (5-36) is ignored compared with the second part, thus giving the electric field introduced by oscillator 1 and oscillator 2 as

$$E_{\text{far}}(t) \cong \frac{Ne}{4\pi\varepsilon_0 c^2 d}[\ddot{x}_1(t) + \ddot{x}_2(t)] = D\ddot{x}(t), \tag{5-50}$$

where $D = \dfrac{Ne}{2\pi\varepsilon_0 c^2 d}$, and E_{far} is x-polarized. The emission intensity in the frequency domain, i.e., emission spectrum, can be evaluated by[15]

$$I(\omega) = \text{Re}\left\langle \int_0^\infty E_{\text{far}}^*(t) E_{\text{far}}(t + \tau) \exp(i\omega\tau)\, d\tau \right\rangle, \tag{5-51}$$

where $\text{Re}\langle Q \rangle$ is the real part of $\langle Q \rangle$, and $\langle Q \rangle = \dfrac{1}{t_0}\int_0^{t_0} Q\, dt$ is the time average of quantity Q. The calculated result is

$$I(\omega) = |A_1'|^2 \frac{1 - \exp(-2\beta_1 t_0)}{2\beta_1 t_0} \frac{\beta_1}{(\omega - \omega_1)^2 + \beta_1^2} +$$
$$|A_2'|^2 \frac{1 - \exp(-2\beta_2 t_0)}{2\beta_2 t_0} \frac{\beta_2}{(\omega - \omega_2)^2 + \beta_2^2} + |A_3'|^2 \sqrt{2\pi}\delta(\omega - \omega_{\text{ex}}),$$
$$\tag{5-52}$$

where $A_j' = A_j \Omega_j^2 D$ for $j = 1, 2, 3$. Here, we ignore the cross terms in the calculation because the time average is zero when $\omega_1 \neq \omega_2$.

As our previous work explains[15], the emission spectrum is separated into two parts, one is the inelastic part (I_{inela}) which corresponds to PL spectrum, and the other is the elastic part (I_{ela}) which corresponds to white light scattering spectrum. Rewrite Eq. (5-52) as

$$\begin{cases} I_{\text{inela}}(\omega) = |A_1'|^2 \dfrac{1 - \exp(-2\beta_1 t_0)}{2\beta_1 t_0} \dfrac{\beta_1}{(\omega - \omega_1)^2 + \beta_1^2} + \\ \qquad\qquad |A_2'|^2 \dfrac{1 - \exp(-2\beta_2 t_0)}{2\beta_2 t_0} \dfrac{\beta_2}{(\omega - \omega_2)^2 + \beta_2^2}, \\ I_{\text{ela}}(\omega) = |A_3'|^2 \sqrt{2\pi}\delta(\omega - \omega_{\text{ex}}). \end{cases} \tag{5-53}$$

Therefore, the PL spectrum is given by
$$I_{PL}(\omega) = I_{inela}(\omega), \tag{5-54}$$
while the white light scattering spectrum is given from Eq. (5-53) as long as ω_{ex} is substitute by ω:
$$I_{sca}(\omega) = I_{ela}(\omega_{ex} \to \omega) = \sqrt{2\pi} \mid A'_3(\omega_{ex} \to \omega)\mid^2. \tag{5-55}$$

To show the coupling modes for PL more clearly and to understand PL phenomenon more easily, we do not consider the electron distributions here as before[15], which contributes mostly to the anti-Stokes part of PL spectra, unless otherwise specified, though this model would be more accuracy for PL when assisted with the electron distributions.

Start from the coupling coefficients, g and γ.

Figure 5-10 (a) shows g and γ varying with the distance r_0, calculated from Eq. (5-40). It implicates that the coupling coefficients decrease with the increase of r_0, and γ is smaller than g. When r_0 is small enough, the coupling coefficients get large. Since these two coefficients are both related to r_0, we take one of them, i.e., g, as the coupling strength in the rest of this work.

Figure 5-10(b)~(d) show the new generated resonant circular frequencies (ω_1, ω_2) and damping coefficients (β_1, β_2) in different cases of the coupled oscillators as a function of the coupling strength g, calculated from Eqs. (5-45) and (5-46). The simplest one (Fig. 5-10(b)) is that the two oscillators are the same. The two new modes split when coupling, and the splitting increases with the increase of g. Here, we generally call the increasing ω "blue branch", and the decreasing ω "red branch". On the other hand, the two damping coefficients also splits, and one increases (corresponding blue branch), the other decreases (red branch). Notice that there is a cut-off coupling strength for the red branch at around $g_{cut} \approx \omega_0$. In Fig. 5-10(c), the situation is almost the same, i.e., the difference of ω and β between the two branches increase with the increase of g, and there also exists g_{cut}. The difference between Fig. 5-10(b) and (c) is that, to obtain the same level of splitting, the former needs a smaller g than

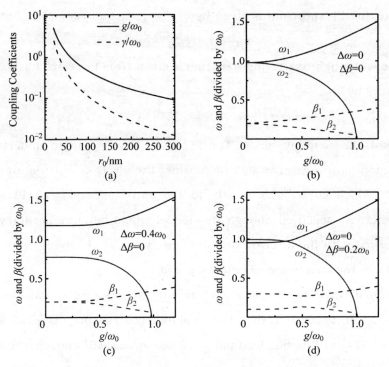

Fig. 5-10 (a) Coupling coefficients g (solid curve) and γ (dashed curve) as a function of r_0. (b)~(d) The new generated resonant circular frequencies (ω_1, ω_2) and damping coefficients (β_1, β_2) as a function of g. Here, $\omega_{01} = \omega_0 + \Delta\omega/2, \omega_{02} = \omega_0 - \Delta\omega/2$, $\beta_{01} = \beta_0 + \Delta\beta/2, \beta_{02} = \beta_0 - \Delta\beta/2$, and $\beta_0 = 0.2\omega_0$. The number of effective free electrons is estimated as $N = 10^6$

(For colored figure please scan the QR code on page 1)

the latter does. That is, the former gets a better coupling efficient than the latter does. In Fig. 5-10(d), due to the fact that $\omega_{c1} = \sqrt{\omega_{01}^2 - \beta_{01}^2} = 0.954\omega_0$ and $\omega_{c2} = \sqrt{\omega_{02}^2 - \beta_{02}^2} = 0.995\omega_0$, it results in $\omega_{c1} < \omega_{c2}$ with a small difference. The difference of ω between the two branches ($\omega_1 - \omega_2$) increases from negative value to zero and then increase to positive value as the increase of g. On the other hand, the difference of β between the two branches ($\beta_1 - \beta_2$) decreases and then increases as the increase of g. Also, g_{cut} exists in this case. The coupling efficient in Fig. 5-10 follows the relation: (b)>(d)>(c).

Furthermore, the effective free electrons number affects the splitting as shown in Fig. 5-11, giving three values of N as examples. Firstly, we find out the behaviors of ω_1 and ω_2 as g increases for each figure. In Fig. 5-11(a), ω_1 increases and then decreases, while ω_2 decreases, indicating that ω_1 has a maximal value. In Fig. 5-11(b), ω_1 decreases and then increases and finally decreases, while ω_2 increases and then decreases, indicating that both ω_1 and ω_2 have maximal values. In Fig. 5-11(c), ω_1 decreases, while ω_2 increases slightly and then decreases. In Fig. 5-11(d), ω_1 decreases, while ω_2 increases and then decreases, the curves of which almost coincide with each other at the range around $g = 0.3\omega_0$ to $g = 0.5\omega_0$. In Fig. 5-11(e), the behaviors are similar to the ones in Fig. 5-11(c). In Fig. 5-11(f), the behaviors are similar to the ones in Fig. 5-11(d), but the two curves cross rather than coincide. Secondly, we find out the similar behaviors for these parameters in a general view. In all cases, there are cut-off coupling strengths for both modes, writing as g_{cut1} and g_{cut2}, at which $\omega_1 = 0$ and $\omega_2 = 0$, respectively. The differences are, for smaller N (10^8 or 10^9), $g_{cut1} > g_{cut2}$, while for larger N (10^{10}), $g_{cut1} < g_{cut2}$. As g increases, the splitting of damping coefficients β_1 and β_2 gets larger. Furthermore, another interesting result is that there is a point g_0 (shown with black cross circle) at which $\beta_2(g_0) = 0$ for each case, and $g_0 < g_{cut2}$. This is different from the one in Fig. 5-10 where $g_0 > g_{cut2}$. When $g < g_0$, mode 2 behaves normally. However, when $g > g_0$, $\beta_2 < 0$ indicates that this is an exponentially increasing mode, which should be removed from the total solutions (Eq. (5-48)), resulting in the absence of mode 2. The most special case is when $g = g_0$ (or $g \to g_0^-$), which corresponds to a lossless (or low loss) mode. In frequency domain this mode would result in a narrow spectrum. However, the effective free electrons number that satisfy this condition is so large that it is almost impossible for a metallic nanostructure. Therefore, in the rest of this work, we only consider the number at the order of magnitudes of $N = 10^6$ unless otherwise specified.

In Eq. (5-48), A_1, A_2, and A_3 represent the amplitudes of the three corresponding modes of $x(t)$. When considering the far field, one should use the amplitudes of $\ddot{x}(t)$, i.e., A_1', A_2', and A_3'. Obviously, the frequency of the

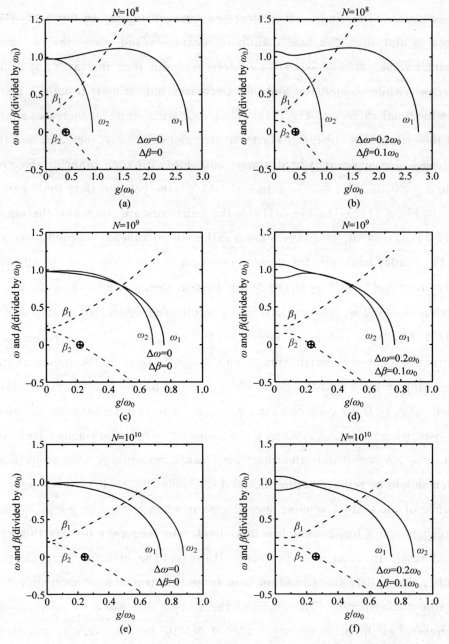

Fig. 5-11 The new generated resonant circular frequencies (ω_1, ω_2) and damping coefficients (β_1, β_2) as a function of g, varying with effective free electrons number N. (a), (c) and (e) represent the case of two same oscillators, where $\omega_{01} = \omega_{02} = \omega_0$ and $\beta_{01} = \beta_{02} = 0.2\omega_0$. (b), (d) and (f) Represent the case of two different oscillators, with $\Delta\omega = 0.2\omega_0$ and $\Delta\beta = 0.1\omega_0$. (a)~(f) Represent $N = 10^8$, $N = 10^9$, $N = 10^{10}$, respectively. The black cross circles represent the point at which $\beta_2 = 0$

(For colored figure please scan the QR code on page 1)

excitation light plays a significant role in the amplitudes. Figure 5-12 shows these amplitudes as a function of ω_{ex}. In the first case (Fig. 5-12(a)), i.e., two same oscillators, the coupled resonant circular frequencies (relative to ω_0) are calculated as $(\omega_1 - \omega_0)/\omega_0 = 0.11$ and $(\omega_2 - \omega_0)/\omega_0 = -0.13$. We find that to obtain the maximum intensities of mode 1 and mode 2 of the emission field, the circular frequency of the excitation light ω_{ex} should be close the corresponding circular resonant frequencies. For mode 3, there are two peaks when varying ω_{ex}, which correspond to around ω_1 and ω_2, respectively. In the second case (Fig. 5-12(b)), i.e., two different oscillators ($\omega_1 > \omega_2$), the coupled resonant circular frequencies (relative to ω_0) are calculated as 0.22 and -0.24, respectively, which, however, corresponds to a weak coupling due to the frequency splitting is small. This result has been identified in Fig. 5-10. Also, the intensities of mode 1 and mode 2 for far field reach their maximums when ω_{ex} is close to the resonant circular frequencies for each of them, and the two corresponding peaks appear for mode 3.

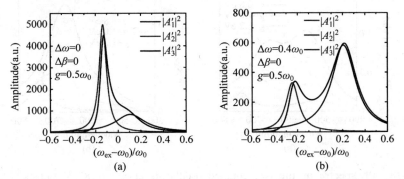

Fig. 5-12 Amplitudes of modes $\omega_1(|A_1'|^2)$, $\omega_2(|A_2'|^2)$, and $\omega_{ex}(|A_3'|^2)$ as a function of ω_{ex}. (a) $\Delta\omega = 0$, $\Delta\beta = 0$. (b) $\Delta\omega = 0.4\omega_0$, $\Delta\beta = 0$. The definitions of $\Delta\omega$ and $\Delta\beta$ are the same as Fig. 5-10 except for $\beta_0 = 0.1\omega_0$. Here, the coupling strength is $g = 0.5\omega_0$, and $C_0 = 1$ and $D = 1$ are used for normalization
(For colored figure please scan the QR code on page 1)

For practical purpose, we consider two metallic nanostructures, e.g., gold nanorods or nanospheres, as the two oscillators, each of which has an individual resonant mode. Figure 5-13 shows the coupling PL spectra for different coupling strengths at two different excitation wavelengths, calculated from Eq. (5-54).

With the increase of g, the splitting of the two modes of PL increases, and the total emission intensities decrease. The decrease of the intensities origin from Fig. 5-13(a) and (b). Take Eq. (5-13(a)) as an example to explain. The amplitudes depend not only on $|\omega_1 - \omega_{ex}|$ (this has been discussed in Fig. 5-12), but also on $|\omega_1 - \omega_2|$. When g increases, $|\omega_1 - \omega_2|$ increases, resulting in the decrease of the amplitude of mode 1. So does mode 2. Therefore, the PL intensities decrease as g increases. Besides, when excited by 532 nm laser, mode 1 is close to it, resulting in a larger intensity than the one of mode 2. While excited by 633 nm laser, mode 2 is close to it, resulting in a larger intensity than the one of mode 1. This is consistent with the results in Fig. 5-12. Here, unit "eV" and unit "Hz" for g satisfy the following relationship:

$$g(\text{eV}) = \frac{\hbar}{e} g \text{ (Hz)}, \tag{5-56}$$

where $\hbar$ is the reduced Planck constant. So does the damping coefficient β.

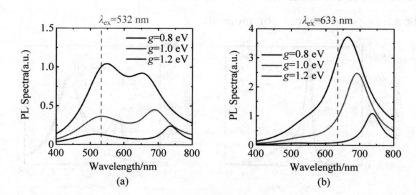

Fig. 5-13 PL spectra of the two coupled oscillators at $g = 0.8$ eV (black), 1.0 eV (orange) and 1.2 eV (blue), respectively, calculated from Eq. (5-54). The excitation light is at the wavelength of $\lambda_{ex} = 532$ nm (a) and $\lambda_{ex} = 633$ nm (b). Here, $\lambda_{c1} = 550$ nm and $\lambda_{c1} = 650$ nm represent the resonant wavelengths for each oscillator (before coupling), respectively; $\beta_{01} = \beta_{02} = 0.247$ eV. Vertical dashed lines stand for the position of 532 nm (green) and 633 nm (red), respectively

(For colored figure please scan the QR code on page 1)

Fig. 5-14 shows the coupling white light scattering spectra for different coupling strengths in different cases, calculated from Eq. (5-55). In Fig. 5-14 (a), i.e., two oscillators with different resonant wavelengths, with the increase of g, the splitting of the two modes increases, which behaves the same as PL does. However, the scattering intensities stay in the same level which is different from PL spectra. In Fig. 5-14 (b), i.e., two same oscillators, with the increase of g, mode 2 red-shifts, while mode 1 is hardly to be obtained. Also, the intensities stay in the same level. This behavior agrees well with the experiments[17-19]. For example, Fig. 5-15 shows the calculated scattering and PL spectra from our model and the experimental ones from Shen et al[19]. In Fig. 5-15(a), the scattering spectra of the experiment and the theory agree well with each other. The peak of the coupled one red-shifts from the single one. In Fig. 5-15(b), the PL spectra of the experiment and the theory agree a little well. Here, we use Fermi-Dirac distribution which is employed in a previous work to modulate the spectra[15]. The peaks do not agree very well, because more physical processes are not considered here, such as electron-phonon and electron-electron interactions[20]. These processes are useful for single mode of PL spectra. However, the coupled cases are more complex so that more

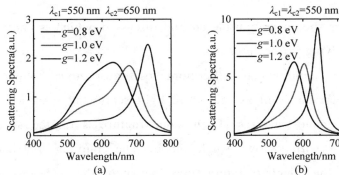

Fig. 5-14 White light scattering spectra of the two coupled oscillators at $g = 0.8$ eV (black), 1.0 eV (orange) and 1.2 eV (blue), respectively, calculated from Eq. (5-55). (a) The resonant wavelengths are different, $\lambda_{c1} = 550$ nm and $\lambda_{c2} = 650$ nm, respectively. (b) The resonant wavelengths are the same, $\lambda_{c1} = \lambda_{c2} = 550$ nm. Here, the damping coefficients are the same for all the oscillators, $\beta_0 = 0.247$ eV

(For colored figure please scan the QR code on page 1)

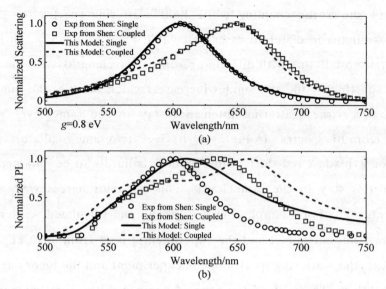

Fig. 5-15 (a) Normalized scattering spectra. (b) Normalized PL spectra. Black circles and red squares stand for the experimental data of single and coupled oscillators, respectively, copied from Shen et al[19]. Black solid curves and red dashed curves stand for the theory data of single and coupled oscillators, respectively, calculated from this model combined with Fermi-Dirac distribution. Here, $\lambda_{c1} = \lambda_{c2} = 610$ nm, $\beta_{01} = \beta_{02} = 0.187$ eV, $N = 3.7 \times 10^5$, $\lambda_{ex} = 532$ nm, $g = 0.8$ eV. The temperatures in PL spectra are $T = 400$ K for single case, and $T = 1400$ K for coupled case

(For colored figure please scan the QR code on page 1)

detailed processes should be considered and discussed, which is another topic and is worth studying in the future.

We develop a coupling classic harmonic oscillator model to explain the coupling PL spectra as well as the white light scattering spectra from two coupled metallic nanostructures. Each nanostructure is treated as a classic charged oscillator with its own single mode. The coupling coefficients are obtained from the electric interactions between the charges, and are proportional to the velocity and the acceleration of the oscillator, respectively. The behaviors of the two new generated modes due to the coupling are different under different conditions. In general, they split and the splitting gets large as the coupling strength g increases at the beginning. Meanwhile, tuning effective free electron number N, when g gets large enough, there exist cut-off coupling strengths for both modes, and a maximum frequency for one of the modes. Besides, PL spectra and white light scattering spectra are calculated from the model, and their behaviors varying with the coupling strength agree well with the experimental ones of other researchers' work. It is worth noting that this coupling model could be expanded to other wavebands dealing with two coupled single-mode resonators.

5.3.3 LSPs in coupled metallic NPs: vertical-polarized excitation

Optical properties of coupled MNPs have been widely reported due to their unique characteristics such as peak shift/splitting of the coupling spectra and electromagnetic enhancement at subwavelength scale, etc[21]. In Section 5.3.2, we have investigated the coupling spectra of two coupled MNPs with parallel polarized excitation. In this section, we investigate the vertical polarization case in detail. Different from the parallel one, the vertical one has its unique properties: (a) three coupling coefficients; (b) positive coupling terms in the coupling equations; (c) blue-shifts of the peaks with the increasing coupling strength for identical MNPs spectra, including scattering, absorption, and PL. The model shows that for the two resonant MNPs, the spectra reveal only one mode that blue shifts as the coupling strength increases; while for the two non-resonant MNPs, the spectra reveal two splitting modes, one of which blue shifts and the other of which red shifts as the coupling strength increases. The relative intensity of the two modes varies with the coupling strength. Furthermore, the PL efficiency (relative to the scattering) is about the order of 10^{-2}. Comparison with published experimental results shows the validity of this model. This work provides a deeper understanding on the optical properties of coupled MNPs and is beneficial to relevant applications.

Figure 5-16 shows the schematic of this model in x-y view. Two MNPs are treated as two oscillators with ions (positive charged) staying at rest and electrons (negative charged) oscillating around the ion. Here, the two oscillators are on the x-axis, and the distance between them is r_0. The y-polarized external electromagnetic field with angular frequency ω_{ex} and amplitude E_0 propagates along the z-axis and illustrates the coupled MNPs. Therefore, the polarization of the excitation light is perpendicular to the line between the two oscillators (vertical polarized excitation). We assume that the free electrons are forced to oscillate only along y-axis. Define $y_j(t)$, $\dot{y}_j(t)$, and $\ddot{y}_j(t)$ as the displacement, velocity, and acceleration in y-direction of the jth electron with $j=1,2$, respectively. The equations of the two electrons are written as

$$\begin{cases} \ddot{y}_1 + \beta_{01}\dot{y}_1 + \omega_{01}^2 y_1 - \dfrac{F_{21}}{m_e} = C_1 e^{-i\omega_{ex}t}, \\ \ddot{y}_2 + \beta_{02}\dot{y}_2 + \omega_{02}^2 y_2 - \dfrac{F_{12}}{m_e} = C_2 e^{-i\omega_{ex}t}. \end{cases} \quad (5\text{-}57)$$

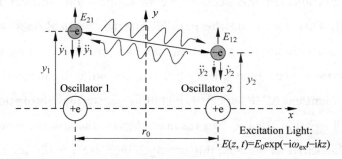

Fig. 5-16 Schematic of the coupled MNPs with vertical polarized excitation. The two MNPs are on the x-axis; circles filled with gray stand for electrons with negative charge with distance r between them, and circles filled with white stand for ions with positive charge with distance r_0 between them; y_1 and y_2 are the displacements from the equilibrium positions in y-direction of the two electrons, respectively; $\dot{y}_1$ and $\dot{y}_2$ are the corresponding velocities; $\ddot{y}_1$ and $\ddot{y}_2$ are the corresponding accelerations; E_{12} (or E_{21}) is the electric field introduced by Oscillator 1 (or 2) felt by electron 2 (or 1); the excitation electric field is y-polarized with angular frequency ω_{ex}

(For colored figure please scan the QR code on page 1)

Here, m_e is the electron mass, $C_1 = -eE_1/m_e$, $C_2 = -eE_2/m_e$, e is the elementary charge. Usually, $E_1 = E_2 = E_0$ is a good approximation due to the subwavelength distance between them; ω_{0j} and β_{0j} are the eigenfrequency (angular) and the damping coefficient of the jth oscillator, respectively; $F_{21} = -eE_{21}$ and $F_{12} = -eE_{12}$ are the interaction forces between the two oscillators, which can be derived from Eq. (5-36)[16]. Therefore, the electric field E_{21} and E_{12} can be written as

$$\begin{cases} E_{21} \cong \dfrac{e}{4\pi\varepsilon_0 r_0^2}\left(\dfrac{y_2}{r_0} + \dfrac{\dot{y}_2}{c} + \dfrac{r_0 \ddot{y}_2}{c^2}\right), \\ \\ E_{12} \cong \dfrac{e}{4\pi\varepsilon_0 r_0^2}\left(\dfrac{y_1}{r_0} + \dfrac{\dot{y}_1}{c} + \dfrac{r_0 \ddot{y}_1}{c^2}\right). \end{cases} \quad (5\text{-}58)$$

Here, we use the conditions $v/c \ll 1$, $y/r_0 \ll 1$, and $r_0 \dot{y}/c^2 \ll 1$, and ignore the higher-order infinitesimal for approximation. Due to the collective oscillation of large number of electrons in the MNPs, the interaction forces should be modified as $F_{21} = -N_2 eE_{21}$ and $F_{12} = -N_1 eE_{12}$, where N_1 and N_2 are the effective numbers of free electrons in Oscillator 1 and Oscillator 2, respectively. To make it clearer, we define the general coupling coefficients as

$$\frac{1}{\kappa} = \frac{e^2}{4\pi\varepsilon_0 m_e c^3}, \quad g_0^2 = \frac{1}{\kappa}\left(\frac{c}{r_0}\right)^3, \quad \gamma_0 = \frac{1}{\kappa}\left(\frac{c}{r_0}\right)^2, \quad \eta_0 = \frac{1}{\kappa}\left(\frac{c}{r_0}\right)^1, \quad (5\text{-}59)$$

then the coupling coefficients are

$$g_j^2 = N_j g_0^2, \quad \gamma_j = N_j \gamma_0, \quad \eta_j = N_j \eta_0, \quad j = 1, 2 \quad (5\text{-}60)$$

and Eq. (5-57) can be written as

$$\begin{cases} \ddot{y}_1 + \beta_{01}\dot{y}_1 + \omega_{01}^2 y_1 + \eta_2 \ddot{y}_2 + \gamma_2 \dot{y}_2 + g_2^2 y_2 = C_1 e^{-i\omega_{ex}t}, \\ \ddot{y}_2 + \beta_{02}\dot{y}_2 + \omega_{02}^2 y_2 + \eta_1 \ddot{y}_1 + \gamma_1 \dot{y}_1 + g_1^2 y_1 = C_2 e^{-i\omega_{ex}t}. \end{cases} \quad (5\text{-}61)$$

For simplicity, we define $\Omega_j(\alpha) = \omega_{0j}^2 + \beta_{0j}\alpha + \alpha^2$ and $G_j(\alpha) = g_j^2 + \gamma_j\alpha + \eta_j\alpha^2$ for $j = 1, 2$ which would be used in the following derivation.

To obtain the white light scattering spectra, α is substituted by $-i\omega_{ex}$, and we solve the following equations derived from Eq. (5-61) after assuming $y_j(t) = A_j e^{-i\omega_{ex}t}$ for $j = 1, 2$

$$\begin{pmatrix} \Omega_1(\alpha) & G_2(\alpha) \\ G_1(\alpha) & \Omega_2(\alpha) \end{pmatrix} \begin{pmatrix} A_1 \\ A_2 \end{pmatrix} = \begin{pmatrix} C_1 \\ C_2 \end{pmatrix}, \quad \text{with } \alpha = -i\omega_{\text{ex}}. \tag{5-62}$$

The solutions are

$$A_1(\omega_{\text{ex}}) = \frac{\Omega_2 C_1 - G_2 C_2}{\Omega_1 \Omega_2 - G_1 G_2}, \quad A_2(\omega_{\text{ex}}) = \frac{\Omega_1 C_2 - G_1 C_1}{\Omega_1 \Omega_2 - G_1 G_2}. \tag{5-63}$$

Therefore, the total scattering spectrum is derived after substituting ω_{ex} with ω:

$$I_{\text{sca}}(\omega) = \omega^4 \mid N_1 A_1(\omega) + N_2 A_2(\omega) \mid^2 \tag{5-64}$$

Here, the term ω^4 is due to the fact that the detected scattering field is usually the far field, which is proportional to $\ddot{y} \propto \omega^2$.

To obtain the absorption spectrum, we notice that the absorption is introduced by the term $\beta_{0j} \dot{y}_j$, thus the total absorption spectrum is written as

$$I_{\text{abs}}(\omega) = N_1 \mid \beta_{01} A_1(\omega) \mid^2 + N_2 \mid \beta_{02} A_2(\omega) \mid^2 \tag{5-65}$$

To obtain the PL spectrum, we should find the eigen solutions of the following equations:

$$\begin{cases} \ddot{y}_1 + \beta_{01} \dot{y}_1 + \omega_{01}^2 y_1 + \eta_2 \ddot{y}_2 + \gamma_2 \dot{y}_2 + g_2^2 y_2 = 0, \\ \ddot{y}_2 + \beta_{02} \dot{y}_2 + \omega_{02}^2 y_2 + \eta_1 \ddot{y}_1 + \gamma_1 \dot{y}_1 + g_1^2 y_1 = 0. \end{cases} \tag{5-66}$$

After assuming $y_j(t) = B_j e^{\alpha t}$ for $j = 1, 2$, the equations are written in matrix form:

$$\begin{pmatrix} \Omega_1(\alpha) & G_2(\alpha) \\ G_1(\alpha) & \Omega_2(\alpha) \end{pmatrix} \begin{pmatrix} B_1 \\ B_2 \end{pmatrix} = \begin{pmatrix} 0 \\ 0 \end{pmatrix}, \tag{5-67}$$

with initial conditions: $y_j(0) = A_j, \dot{y}_j(0) = 0$, for $j = 1, 2$. Obviously, to find a non-trivial solution, should satisfy

$$\begin{vmatrix} \Omega_1(\alpha) & G_2(\alpha) \\ G_1(\alpha) & \Omega_2(\alpha) \end{vmatrix} = 0. \tag{5-68}$$

A particular case is that the two oscillators are identical, i.e., $\omega_{01} = \omega_{02} = \omega_0$, $\beta_{01} = \beta_{02} = \beta_0$, $N_1 = N_2 = N$, $g_1 = g_2 = g$, $\gamma_1 = \gamma_2 = \gamma$, $\eta_1 = \eta_2 = \eta$, and $C_1 = C_2 = C_0$. In the rest of this section, we take this identical case as an example to illustrate the PL properties of the coupled system. The expressions derived from the general case (non-identical) are much more complicated than the

CHAPTER 5 Surface Plasmons

identical one, but the solving processes of them are similar. Back to the identical case, the solutions of α are

$$\begin{cases} \alpha_{p\pm} = \dfrac{-\beta_0 - \gamma \pm i\sqrt{4(1+\eta)(\omega_0^2 + g^2) - (\beta_0 + \gamma)^2}}{2(1+\eta)}, \\ \alpha_{m\pm} = \dfrac{-\beta_0 + \gamma \pm i\sqrt{4(1-\eta)(\omega_0^2 - g^2) - (\beta_0 - \gamma)^2}}{2(1-\eta)}. \end{cases} \quad (5\text{-}69)$$

Notice that these solutions should satisfy $\eta \neq 1$ (Case 1). If $\eta = 1$ and $\beta_0 \neq \gamma$ (Case 2), the solutions would be reduced to $\alpha_{p\pm}$ and α_m, with $\alpha_m = \dfrac{\omega_0^2 - g^2}{\gamma - \beta_0}$. In this case, α_m corresponds to the non-oscillation term with exponentially decreasing or increasing, depending on the sign of α_m, the latter of which should be removed from the solutions due to its divergence with time. If $\eta = 1$ and $\beta_0 = \gamma$ (Case 3), the solutions would only remain $\alpha_{p\pm}$. Incidentally, we can easily derive the relations of $g = \gamma = \beta_0$, $r_0 = c/\beta_0$, $N = \kappa/\beta_0$, and $Nc = \kappa r_0$ from Case 3. It is worth mentioning that Case 2 and Case 3 are not general cases, because they require a large number of electrons (large N) or small enough distance r_0 to achieve such strong coupling. Practically, Case 1 is a more general case when discussing coupled MNPs.

To make it clearer, we rewrite the solutions from Eq. (5-69) in a simple form:

$$\begin{cases} \alpha_1 = \alpha_{p-} = -\dfrac{\beta_1}{2} - i\omega_1, \quad \alpha_2 = \alpha_{p+} = -\dfrac{\beta_1}{2} + i\omega_1, \\ \alpha_3 = \alpha_{m-} = -\dfrac{\beta_2}{2} - i\omega_2, \quad \alpha_4 = \alpha_{m+} = -\dfrac{\beta_2}{2} + i\omega_2. \end{cases} \quad (5\text{-}70)$$

Here, there are two new eigenfrequencies (ω_1 and ω_2) and two new damping coefficients (β_1 and β_2). Obviously, they could be written as

$$\omega_1 = \sqrt{\dfrac{\omega_0^2 + g^2}{1+\eta} - \dfrac{\beta_1^2}{4}}, \quad \beta_1 = \dfrac{\beta_0 + \gamma}{1+\eta}, \quad \omega_2 = \sqrt{\dfrac{\omega_0^2 - g^2}{1-\eta} - \dfrac{\beta_2^2}{4}}, \quad \beta_2 = \dfrac{\beta_0 - \gamma}{1-\eta}. \quad (5\text{-}71)$$

Then, the solutions of Eq. (5-67) for $y_j(t)$ are written as

$$y_j(t) = \sum_{k=1}^{4} B_{jk} e^{\alpha_k t}, \quad \text{for } j = 1, 2. \tag{5-72}$$

There are 8 undetermined coefficients (B_{jk}), hence, we need 8 equations, i.e., 8 initial conditions. Below Eq. (5-67), we list 4 of them, and the rest 4 conditions are determined by $\ddot{y}_j(0)$ and $\dddot{y}_j(0)$ which can be derived from Eq. (5-66):

$$\begin{cases} A_1(\omega_{\text{ex}}) = A_2(\omega_{\text{ex}}) := A_0(\omega_{\text{ex}}), \\ \ddot{y}_1(0) = \ddot{y}_2(0) = -\dfrac{\omega_0^2 + g^2}{1+\eta} A_0 := \ddot{y}_0, \\ \dddot{y}_1(0) = \dddot{y}_2(0) = -\dfrac{\beta_0 + \gamma}{1+\eta} \ddot{y}_0 := \dddot{y}_0. \end{cases} \tag{5-73}$$

Here, A_0 could be simplified from Eq. (5-63):

$$A_0(\omega_{\text{ex}}) = \frac{C_0}{\Omega + G} = \frac{-C_0/(1+\eta)}{(\omega_{\text{ex}} + i\beta_1/2)^2 - \omega_1^2}. \tag{5-74}$$

The solutions of B_{jk} could be given by the matrix equation:

$$\begin{pmatrix} 1 & 1 & 1 & 1 \\ \alpha_1 & \alpha_2 & \alpha_3 & \alpha_4 \\ \alpha_1^2 & \alpha_2^2 & \alpha_3^2 & \alpha_4^2 \\ \alpha_1^3 & \alpha_2^3 & \alpha_3^3 & \alpha_4^3 \end{pmatrix} \begin{pmatrix} B_{j1} \\ B_{j2} \\ B_{j3} \\ B_{j4} \end{pmatrix} = \begin{pmatrix} A_j \\ 0 \\ \ddot{y}_j(0) \\ \dddot{y}_j(0) \end{pmatrix}, \quad \text{for } j = 1, 2. \tag{5-75}$$

Then, B_{jk} are listed below:

$$B_{jk} = \frac{\dddot{y}_j(0) - A_j \prod_{n \neq k}^{4} \alpha_n - \ddot{y}_j(0) \sum_{n \neq k}^{4} \alpha_n}{\prod_{n \neq k}^{4} (\alpha_k - \alpha_n)}$$

$$= \frac{\dfrac{(\omega_0^2 + g^2)(\beta_0 + \gamma)}{(1+\eta)^2} - \prod_{n \neq k}^{4} \alpha_n + \dfrac{\omega_0^2 + g^2}{1+\eta} \sum_{n \neq k}^{4} \alpha_n}{\prod_{n \neq k}^{4} (\alpha_k - \alpha_n)} A_0 := B_{0k}. \tag{5-76}$$

Therefore, the solutions of Eq. (5-66) are

$$y_1(t) = y_2(t) = \sum_{k=1}^{4} B_{0k} e^{\alpha_k t} := y_0(t), \tag{5-77}$$

After removing the insignificant coefficients, i.e., factor $f = \dfrac{e}{4\pi\varepsilon_0 c^2 d^2}$, [14]
where d is the distance between the system and the far field point, the far electric field is

$$E_{\text{far}}(t) = N_1 \ddot{y}_1(t) + N_2 \ddot{y}_2(t) = 2N\ddot{y}_0(t) := \sum_{k=1}^{4} B'_{0k} e^{\alpha_k t}, \qquad (5\text{-}78)$$

where $B'_{0k} = 2N\alpha_k^2 B_{0k}$. Hence, the emission spectrum can be evaluated by Eq. (5-51). Similarly, after removing the insignificant coefficients, i.e., factor $F_j = \dfrac{1 - \exp(-\beta_j t_0)}{\beta_j t_0}$ [15], where t_0 is the interaction time between light and the system, the total PL spectrum can be written as

$$I_{\text{PL}}^{\text{total}}(\omega) = \frac{\beta_1 |B'_{01}|^2}{(\omega - \omega_1)^2 + (\beta_1/2)^2} + \frac{\beta_1 |B'_{02}|^2}{(\omega + \omega_1)^2 + (\beta_1/2)^2} +$$

$$\frac{\beta_2 |B'_{03}|^2}{(\omega - \omega_2)^2 + (\beta_2/2)^2} + \frac{\beta_2 |B'_{04}|^2}{(\omega + \omega_2)^2 + (\beta_2/2)^2}. \qquad (5\text{-}79)$$

After ignoring the second and the fourth term due to the fact that they are far away from the resonance frequency, the PL spectrum can be evaluated as

$$I_{\text{PL}}(\omega) = \frac{\beta_1 |B'_{01}|^2}{(\omega - \omega_1)^2 + (\beta_1/2)^2} + \frac{\beta_2 |B'_{03}|^2}{(\omega - \omega_2)^2 + (\beta_2/2)^2}. \qquad (5\text{-}80)$$

Notice that for the identical case, $B'_{03} = B'_{04} = 0$ can be derived from Eq. (5-76). Hence, the identical PL spectrum is evaluated as

$$I_{\text{PL}}^{\text{identical}}(\omega) = \frac{\beta_1 |B'_{01}|^2}{(\omega - \omega_1)^2 + (\beta_1/2)^2}. \qquad (5\text{-}81)$$

Here, we emphasize that the whole derivation for the PL spectra is only applicable to the identical case, resulting in Eq. (5-81). In order to calculate the one of the non-identical case, i.e., Eq. (5-80), we should follow these steps: (1), (numerically) solve Eq. (5-68) to obtain the right modes for Eq. (5-70); (2), derive the initial conditions for Eq. (5-73) using Eq. (5-66) and $A_j(\omega_{\text{ex}})$ that are solved from Eq. (5-63); (3), solve Eq. (5-75) to obtain the amplitudes B_{jk}, also known as the first line of Eq. (5-76); (4), the rest is to follow Eq. (5-78), Eq. (5-79), and Eq. (5-80).

Now we evaluate the PL efficiency, which is a significant property of the MNP, by assuming that the far field emissions are from the oscillators that can be treated as dipoles. The total radiation power of a dipole with dipole moment p and angular frequency ω_p can be evaluated by[16]

$$\langle S \rangle = \frac{p^2 \omega_p^4}{12\pi\varepsilon_0 c^3}. \tag{5-82}$$

When calculating the power emitted by the scattering process, the dipole moment of the jth MNP can be evaluated by $p_{\text{sca}j} = eN_j |A_j|$, with the emission angular frequency ω_{ex}; while calculating the one of the PL process, the dipole moment of the jth MNP with the kth mode can be evaluated by $p_{\text{PL}jk} = eN_j |B_{jk}|$ (we only consider $k = 1, 3$ here), with the corresponding emission angular frequencies ω_1 and ω_2. Therefore, the total power ratio of PL to scattering can be evaluated by

$$\text{Ratio} = \frac{N_1^2 |B_{12}|^2 + N_2^2 |B_{21}|^2}{|N_1 A_1|^2 + |N_2 A_2|^2} \frac{\omega_1^4}{\omega_{\text{ex}}^4} F_1 + \frac{N_1^2 |B_{13}|^2 + N_2^2 |B_{23}|^2}{|N_1 A_1|^2 + |N_2 A_2|^2} \frac{\omega_2^4}{\omega_{\text{ex}}^4} F_2. \tag{5-83}$$

The factor F_j here is due to the integration of Eq. (5-51). This ratio represents the relative PL efficiency.

After preparing these formulas employing this coupling model, the optical properties of the coupled system could be obtained easily. First, we investigate the behavior of the coupling coefficients, i.e., g, γ, and η, in the identical case. Figure 5-17 shows these coupling coefficients varying with the electron number N and the distance r_0, calculated from Eq. (5-59) and Eq. (5-60). As N increases or/and r_0 decreases, they increase. The contour lines for g, γ, and η satisfy $N \propto r_0^{1.5}$, $N \propto r_0^2$, and $N \propto r_0$, respectively. The dashed lines are the examples ($g = 10^{16}$ Hz, $\gamma = 10^{16}$ Hz, and $\eta = 1$) to show the contour lines. We emphasize here that when r_0 is small enough, i.e., $r_0 < 1$ nm, the electron tunneling effect makes it hard to evaluate the interaction between the oscillators employing only classical electromagnetic method which is used in this paper. Therefore, this model fails to describe the behaviors of the oscillators at the sub-nanometer scale. A particular case is when $\eta \geqslant 1$, which corresponds to a so strong coupling (very large g and γ) that the practical MNPs cannot reach easily.

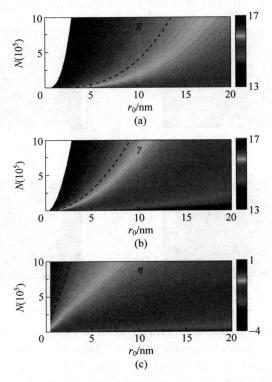

Fig. 5-17 The coupling coefficients g (a), γ (b), and η (c) varying with electron number N and distance r_0. The color bar of (a) and (b) is in unit of Hz with log scale, i.e., $\lg g$, and $\lg \gamma$, respectively. Dashed lines stand for the contour line of the values $g, \gamma = 10^{16}$ Hz. The color bar of (c) employs log scale, i.e., $\lg \eta$. Dashed line stands for the contour line of the value $\eta = 1$.
(For colored figure please scan the QR code on page 1)

Second, we investigate the eigen modes of the coupled system, i.e., ω_j and β_j for $j = 1, 2$, in the identical case.

Figure 5-18 shows the eigen modes varying with the electron number N and the distance r_0, calculated from Eq. (5-71). As the coupling coefficients increase, it is obvious that ω_1 and β_1 (Mode 1) increase; however, ω_2 and β_2 (Mode 2) decrease in the case of $\eta < 1$, but they increase in the case of $\eta > 1$. Due to the behaviors at $\eta < 1$ (the general case), we could naturally call Mode 1 the blue branch or blue mode, and call Mode 2 the red branch or red mode. Here, unit "Hz" and unit "eV" could be translated by $P(\text{eV}) = \dfrac{\hbar}{e} P(\text{Hz})$, where $\hbar$ is the reduced Planck constant and P is the parameter with unit "Hz" or "eV".

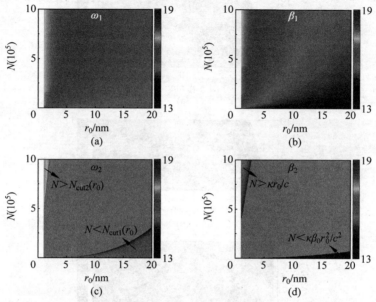

Fig. 5-18 Eigen modes of the coupled system. ω_1 (a), β_1 (b), ω_2 (c), and β_2 (d) varying with the electron number N and distance r_0. Here, the parameters of the individual MNP are $\beta_0 = 1.57 \times 10^{14}$ Hz (0.1034 eV), $\omega_0 = 3.14 \times 10^{15}$ Hz (2.0690 eV). The gray region in (c) and (d) stand for the vanishing of the red mode
(For colored figure please scan the QR code on page 1)

The blue mode could blue-shift along with the increasing damping coefficient as the coupling strength increases, which is only limited by the MNPs themselves, i.e., N is limited by the volume and r_0 is limited by the geometric structures or the electron tunneling effect. However, when it comes to the red mode, the behaviors are totally different. Both Fig. 5-18(c) and (d) show large areas filled with gray color, which indicates that the red mode vanishes in this area, i.e., only blue mode exists. The reason could be found in Eq. (5-71). Here, we take β_2 as an example to explain the reason. When $\eta < 1$, i.e., the denominator of β_2 is positive, as the coupling strength increases from 0, the numerator of β_2 decreases from β_0 until γ increases to β_0. After γ

continuous to increase, β_2 would be negative, indicating that the solutions $\alpha_{m\pm}$ should be abandoned due to its non-physical process (exponentially increasing of $y_j(t)$). This possible region is indicated by $(\beta_0>\gamma) \cap (\eta<1)$, i.e., $N<\kappa\beta_0 r_0^2/c^2$, as shown at the bottom right corner in Fig. 5-18(d). When $\eta>1$, i.e., the denominator of β_2 is negative, hence, Mode 2 has physical meaning only when γ is larger than β_0. As the coupling strength increases from a large value, $|\beta_0-\gamma|$ increases. This possible region is indicated by $(\beta_0<\gamma) \cap (\eta>1)$, i.e., $N>\kappa r_0/c$, as shown at the top left corner in Fig. 5-18(d). Therefore, Mode 2 in the view of β_2 is forbidden at the gray region in Fig. 5-18(d), or Region $FR_{\beta2}$: $\kappa\beta_0 r_0^2/c^2<N<\kappa r_0/c$. We analyze ω_2 similarly. When the value of ω_2 is an imaginary number, the value of $\alpha_{m\pm}$ would be a real number, indicating that Mode 2 is not an oscillation mode, thus abandoning Mode 2. Therefore, Mode 2 in the view of ω_2 is forbidden at the gray region in Fig. 5-18(c), or Region $FR_{\omega2}$: $N_{cut1}<N<N_{cut2}$, where $N_{cut1}(r_0)$ and $N_{cut2}(r_0)$ are the two roots of the equation $\omega_2 = \dfrac{\omega_0^2-g^2}{1-\eta}-\dfrac{\beta_2^2}{4}=0$, with N the uncertain number.

According to the above analysis and the parameters (ω_0 and β_0) we set in Fig. 5-18, the total forbidden region of Mode 2 should be $FR_{\omega2} \cup FR_{\beta2}$, i.e., $\kappa\beta_0 r_0^2/c^2<N<N_{cut2}$.

Third, we investigate the spectra of the coupled system, i.e., white light scattering spectra, absorption spectra, and PL spectra. Here, we choose g from (g,γ,η) to represent the coupling strength for clarify.

Figure 5-19 shows these spectra of resonant MNPs varying with the coupling strength g, calculated from Eq. (5-64), Eq. (5-65), and Eq. (5-81). Here, "resonant MNPs" stands for two identical MNPs, thus same resonant mode (both resonate at 600 nm). Two phenomena are illustrated. First, only one mode arises; second, the peaks blue shift and the linewidth increases as g increases. The first phenomenon is due to the identical condition. In scattering and absorption view, Eq. (5-74) shows no information of ω_2, indicating the vanishing of Mode 2; in PL view, the amplitude of Mode 2 is zero according to

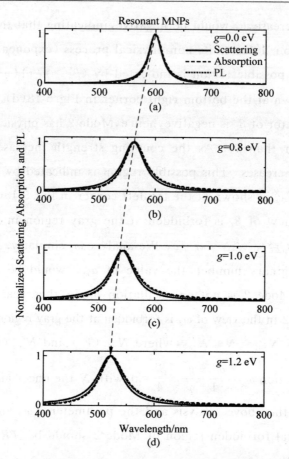

Fig. 5-19 Normalized scattering (black solid), absorption (black dashed), and PL spectra (red dotted) for resonant MNPs, with different coupling strengths. (a)~(d) stand for $g = 0.0, 0.8, 1.0,$ and 1.2 eV, respectively. Blue dashed arrow stands for the track of the resonant peaks. Here, the parameters are $N_1 = N_2 = 10^5$, $\beta_{01} = \beta_{02} = 0.1034$ eV, $\omega_{01} = \omega_{02} = 2.0690$ eV, and the excitation wavelength $\lambda_{ex} = 355$ nm
(For colored figure please scan the QR code on page 1)

Eq. (5-81), also indicating the vanishing of Mode 2, thus only Mode 1 existing. The second phenomenon could be explained by Eq. (5-71) together with Fig. 5-18 (a) and (b), where ω_1 and β_1 increase as g increases. It is worth mentioning that this blue-shift phenomenon is opposite to the case of parallel polarized excitation as Section 5.3.2 illustrates, where red-shift phenomenon is obtained. This is essentially caused by the fact that the coupling terms in the coupling equations (Eq. (5-61)) of them have opposite sign, i.e., positive sign for vertical polarization and negative sign for parallel polarization. Therefore, the primary peaks are different for these two cases. This phenomenon also agrees well with the results of Pasquale et al.[22], where they employed several structures consisted with different numbers of MNPs including dimers. The scattering spectra of the dimer show blue shift and red shift with vertical and parallel polarized excitations, respectively.

Figure 5-20 shows these spectra of non-resonant MNPs varying with the coupling strength g, calculated from Eq. (5-64) and Eq. (5-65). Here, "non-resonant MNPs" stands for two different MNPs, thus different resonant modes (resonate at 500 nm and 600 nm, respectively). The formula to calculate the PL spectra is not shown in this section because its amplitudes are pretty complicated as has been mentioned. Also, two phenomena are illustrated. First, there are two resonant peaks with different ratios of peak values for these three; second, as g increases, one peak blue shifts, the other red shifts. For the first phenomenon in detail, as g increases, for the scattering and PL spectra, the ratio of the amplitude of Mode 1 to the amplitude of Mode 2 increases; however for the absorption spectra, the ratio decreases. Furthermore, the line-width of Mode 1 increases and the one of Mode 2 decreases with the increasing g, which can be explained by Eq. (5-71) together with Fig. 5-18(b) and (d). Similar to the "resonant MNPs" case, the second phenomenon could also be explained by Eq. (4-71) and Fig. 5-18(a) and (c). In Fig. 5-20(d), we notice that Mode 2 of PL vanishes. Because when $g = 1.2$ eV, Mode 2 of PL reaches the forbidden region as discussed in Fig. 5-18, thus only Mode 1 remaining.

We use the power ratio of PL to scattering to represent the relative PL efficiency. As mentioned before, t_0 is the interaction time between light and the MNPs, therefore, we introduce a quantum explanation that t_0 can be evaluated by the PL lifetime of the MNPs, which is about the order of 10^{-14} s[23-25].

Figure 5-21 shows the ratio of the non-resonant MNPs as a function of g, the

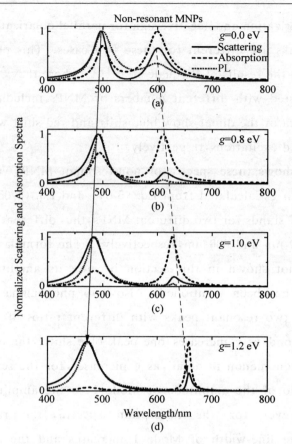

Fig. 5-20 Normalized scattering (black solid), absorption (black dashed), and PL spectra (red dotted) for non-resonant MNPs, with different coupling strengths. (a)~(d) stand for $g = 0.0, 0.8, 1.0,$ and 1.2 eV, respectively. Blue and red dashed arrows stand for the tracks of the blue-shifted (Mode 1) and red-shifted (Mode 2) resonant peaks, respectively. $N_1 = N_2 = 10^5, \beta_{01} = \beta_{02} = 0.1034$ eV, $\omega_{01} = 2.0718$ eV, $\omega_{02} = 2.0690$ eV, and the excitation wavelength $\lambda_{ex} = 355$ nm

(For colored figure please scan the QR code on page 1)

parameters of which are the same as the ones in Fig. 5-20. It indicates that as the coupling strength increases, the ratio first increases to its maximum and then decreases, with the values of the order of 2×10^{-2}. In Eq. (5-83), the first term plays a more important role than the second term does due to the facts that $\omega_1 > \omega_2$ and the peak of Mode 1 is closer to the excitation wavelength than the peak of Mode 2 does.

The behavior of the ratio varying with g is due to two competing

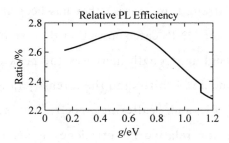

Fig. 5-21 Power ratio of PL to scattering as a function of coupling strength g, calculated from Eq. (5-83). The parameters of the MNPs and the excitation wavelength are the same as the ones in Fig. 5-20

parameters. The first is the factor F_1 that decreases with the increasing β_1 which is introduced by the increasing g; the second is the PL amplitude of Mode 1 increases as its peak gets closer to the excitation wavelength which is introduced by the increasing g.

We notice that there is a sudden dip at around $g_{cut} = 1.1$ eV. Because Mode 2 vanished when $g > g_{cut}$ as discussed in Fig. 5-18, the second term of Eq. (5-83) should be ignored when calculating the ratio, and the amplitudes B_{12} and B_{21} should be recalculated by resolving the equations (steps are below Eq. (5-81)) in the case of only one mode. Moreover, Eq. (5-83) is an approximation to evaluate the relative PL efficiency, because we only consider the emission of the dipoles with single frequency, rather than considering the integral of the PL spectra. As a result, the curve suddenly varies when g increases to g_{cut} because of the sudden vanishing of Mode 2.

In summary, in this section, we develop a coupling model based on classical electromagnetic method to explain the coupling properties of two MNPs with vertical polarized excitation. There are 3 coupling coefficients that influence the coupling and they vary with both the free electron number and the separation distance. The scattering, absorption, and PL properties of coupled MNPs are illustrated. Particularly, the modes of PL are analyzed in detail. The identical MNPs case shows that the mode blue shifts as the coupling strength increases, which is opposite to the parallel polarized excitation (the mode red shifts as the coupling strength increases) that has been investigated in Section 5.3.2. The non-resonant MNPs case shows that the two original modes behave differently. As the coupling strength increases, the peak of Mode 1 blue shifts while the peak of Mode 2 red shifts, and the intensity ratio of Mode 1 to Mode 2 becomes larger in PL and scattering spectra but smaller in absorption spectra. Furthermore, the relative PL efficiency is about the order of 10^{-2}. This work would be helpful to understanding the coupling properties of MNPs more deeply.

5.3.4 Plexciton model: coupling between plasmon and exciton

In this section, we talk about a coupling model for plexciton. Plexciton is the formation of new hybridized energy states originated from the coupling between a plasmon and an exciton. To reveal the optical properties of both the exciton and plexciton, we develop a classic oscillator model to describe their behavior[26]. Particularly, the coupling case, i.e., plexciton, is investigated theoretically in detail. In strong coupling, the phenomenon similar to electromagnetically induced transparency is achieved for the absorption spectra; the splitting behaviors of the modes are carefully analyzed, and the splitting largely depends on the effective number of electrons and the resonance coupling; the photoluminescence spectra show that the spectral shapes remain almost unchanged for weak coupling and change a lot for strong coupling; and the emission intensity of the exciton is strongly enhanced by the plasmon and can reach the order of 10^{10} for a general case. We also show the comparisons between our model and the published experiments to validate its validity.

We introduce a model that can evaluate the PL and absorption properties of individual and coupled nanoparticles (NPs, semiconductor or metal). First, we consider the individual NP, where we divide the interaction process into two steps: one is the absorption step, the other is the emission step. Second, we consider the coupling case between the two NPs, where the coupling process is also divided into two steps: one is the absorption coupling step, the other is the emission coupling step.

Actually, this model is applicable for three cases, i. e., plexciton, plasmon-plasmon, and exciton-exciton, because we classically treat both plasmon and exciton of the nanoparticles as the oscillation of electrons. We use this model to describe the plexciton here because this work focuses on the optical properties of the plexciton. In the model, the difference between the plasmon and the exciton is that the energy difference between the absorption peak and the emission peak of the exciton is larger than the one of the plasmon, the latter of which can be neglected. This determines the difference between the optical properties of the plasmon and the exciton. Therefore, the observed features would be general in some cases (e. g., energy splitting) but different in other cases (e. g., absorption and PL spectra).

Incidentally, the limits of this model are obvious. First, we treat each NP as an individual oscillator. However, the sizes of the NPs are not zero; that is, when the distance between them is comparable with their sizes (several or dozens nanometers), each NP should be treated as several oscillators coupled to each other rather than an individual one. Second, when the distance between them is small enough, e. g., with a surface distance less than 1 nm, the quantum effect allows the electron tunneling that this model cannot describe completely[27]. Therefore, the model in this work is applicable in the case where the distance between the NPs is larger than their sizes.

In the quantum mechanism, for a semiconductor, the electrons are mostly in the bound state and interact with the ions much more strongly than the ones of metals do. When the electrons are excited by the incident photons, they will stay in the excited state, and then the combinations of the electron-hole pairs result in the emission of photons. Referring to the above descriptions, for a SNP, we could treat the whole process in the classical view using the harmonic oscillator concept, in which we separate the process into two steps, i. e., absorption and emission, as shown in Fig. 5-22. The bound electron is treated as an oscillator with a certain resonance frequency due to the interaction with the ions, and there are two springs that provide the restoring. In the first step, when excited by a photon, the electron absorbs the energy and starts to oscillate through the interactions with the two springs; the first spring suddenly breaks as soon as the oscillator arrives at its maximum displacement, i.e., the amplitude. The "broken" refers to the thermal process (nonradiative process), in which part of the energy is converted into thermal energy. In the second step, the oscillator starts to oscillate through the interaction with the second spring, and photons are emitted simultaneously.

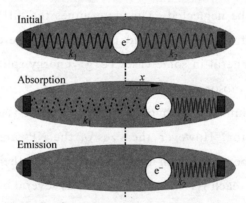

Fig. 5-22 Schematic of the two-step oscillator model for individual case. From above to bottom are initial state (no interaction with photons), absorption process, and emission process, respectively. The big gray ellipse stands for the individual exciton or plasmon. The blue and red curves stand for the springs with recovery factors of k1 and k2, respectively. The white circle with label e^- stands for the oscillator, which is the electron in this model. The vertical dash-dot line stands for the equilibrium position of the oscillators. The excitation light is polarized along the x-axis

(For colored figure please scan the QR code on page 1)

It is worth noting that, when describing the emission process of plasmon, the broken spring is not necessary because the energy difference between the absorption peak and the PL peak is small enough and can be neglected, which has been investigated in Section 5.3.2 and Section 5.3.3. However, in the case of the exciton, the blue shift of the absorption peak relative to the PL peak cannot be ignored, which can be easily understood from the perspective of energy transition in quantum mechanisms. In this classic model, the broken spring is necessary to explain this blue shift. Furthermore, although the model degenerates to the coupling between two oscillators after the broken spring, the amplitudes of the emission process are determined by the process before the broken spring (absorption process); i.e., the two processes of the model are related. Introducing the broken spring can enrich the physics of the plexciton. Therefore, the broken spring is necessary.

To start the deviation, define $x(t)$ as the displacement of an oscillator from its equilibrium position as a function of time t; thus, $\dot{x}(t)$ and $\ddot{x}(t)$ are the velocity and the acceleration. The excitation light is treated classically, with (circular) frequency ω_{ex} and electric field intensity E_0.

For the absorption step, the equation is in this form[15]:

$$\ddot{x} + 2\beta_a \dot{x} + \omega_a^2 x = C_0 e^{-i\omega_{ex}t}, \quad (5\text{-}84)$$

where $\omega_a^2 = (k_1 + k_2)/m_e$ is the absorption resonance frequency, β_a is the absorption damping coefficient, $C_0 = -eE_0/m_e$, m_e is the electron mass, and e is the elementary charge. The solution is

$$x(t) = A e^{-i\omega_{ex}t}, \quad (5\text{-}85)$$

with the amplitude $A = C_0/(\omega_a^2 - \omega_{ex}^2 - 2i\beta_a\omega_{ex})$. The energy stored in Spring 1 is

$$I_a(\omega_{ex}) = \frac{1}{2}k_1 |A|^2. \quad (5\text{-}86)$$

When Spring 1 is broken, this energy is converted into thermal energy, i.e., the absorption spectrum could be written as $I_a(\omega)$, in which ω_{ex} is replaced by ω.

For the emission step, the equation is in this form:

$$\ddot{x} + 2\beta_e \dot{x} + \omega_e^2 x = 0, \quad (5\text{-}87)$$

with initial conditions: $x(0) = A, \dot{x}(0) = 0$. Here, $\omega_e^2 = k_2/m_e$ is the emission resonance frequency and β_e is the emission damping coefficient. Obviously, $\omega_a > \omega_e$. The solution is

$$x(t) = \frac{A}{2i\omega_s}(\alpha_+ e^{\alpha_- t} - \alpha_- e^{\alpha_+ t}), \tag{5-88}$$

where $\omega_s = \sqrt{\omega_e^2 - \beta_e^2}$ is the resonance frequency of the emission and $\alpha_\pm = -\beta_e \pm i\omega_s$ are the conjugate eigenvalues of Eq. (5-87). The far field electric field produced by the electrons is evaluated by

$$E(t) \cong \frac{N_s e}{4\pi\varepsilon_0 c^2 d}\ddot{x}(t) = \frac{N_s e}{4\pi\varepsilon_0 c^2 d}\frac{A\omega_e^2}{2i\omega_s}(\alpha_- e^{\alpha_- t} - \alpha_+ e^{\alpha_+ t}), \tag{5-89}$$

where N_s is the effective electron number of the SNP, and d is the distance between the field point and the SNP. The far field electric field in frequency domain can be evaluated based on the optical cavity theory[15,28], i.e., Eq. (5-51). Therefore, the emission or the PL spectrum is

$$I_e = D\left[\frac{1}{(\omega - \omega_s)^2 + \beta_e^2} + \frac{1}{(\omega + \omega_s)^2 + \beta_e^2}\right] \cong \frac{D}{(\omega - \omega_s)^2 + \beta_e^2}, \tag{5-90}$$

where $D = |A|^2 \left(\frac{N_s e}{4\pi\varepsilon_0 c^2 d}\frac{\omega_e^2}{2\omega_s}\right)^2 \frac{\omega_e^2}{2t_0}(1 - e^{-2\beta_e t_0})$. Here, we omit the second term of Eq. (5-90), because the intensity of the first term is much larger than the intensity of the second term when ω is around ω_s, which corresponds to the general case of a practical PL spectrum. Define the PL excitation (PLE) as the integrated PL intensities varying with the incident frequency; thereby, it can be evaluated by

$$I_{\text{PLE}}(\omega_{\text{ex}}) = |A|^2 \int_{\omega_{\text{cut1}}}^{\omega_{\text{cut2}}} \frac{d\omega}{(\omega - \omega_s)^2 + \beta_e^2}, \tag{5-91}$$

where we retain A from D because the rest quantities in D are constant for a certain SNP and only A depends on ω_{ex}. For the practical purpose, we employ $\omega_{\text{cut1}} = \omega_s - 2\beta_e$ and $\omega_{\text{cut2}} = \omega_{\text{ex}}$ as the lower and upper limits of the integration term.

Actually, the PL progress of a MNP can also be treated in the same way as the one of a SNP. The difference is that, for MNP, $k_1 \ll k_2$, indicating that $\omega_a \approx \omega_e$ with ω_a a little larger than ω_e. This treatment is equivalent to the one in Section 5.3.2, where we considered only one spring rather than two.

CHAPTER 5 Surface Plasmons

Now, we consider the coupling between a SNP and an MNP. As mentioned above, the process is divided into absorption and emission. Here, we define $x_1(t)$ and $x_2(t)$ as the displacements of oscillators of SNP and MNP, respectively.

First, the absorption process is described as

$$\begin{cases} \ddot{x}_1 + 2\beta_{a1}\dot{x}_1 + \omega_{a1}^2 x_1 - \gamma_2 \dot{x}_2 - g_2^2 x_2 = C_1 e^{-i\omega_{ex}t}, \\ \ddot{x}_2 + 2\beta_{a2}\dot{x}_2 + \omega_{a2}^2 x_2 - \gamma_1 \dot{x}_1 - g_1^2 x_1 = C_2 e^{-i\omega_{ex}t}. \end{cases} \quad (5\text{-}92)$$

Here, generally $C_1 = C_2 = C_0$, β_{a1} and β_{a2} are the damping coefficients of the SNP and the MNP in absorption process, respectively, ω_{a1} and ω_{a2} are the absorption resonance frequencies of the SNP and the MNP before coupling, respectively, and γ_j and g_j ($j = 1, 2$) are the coupling coefficients, and they are derived from the interaction between the two electrons in motion based on electrodynamics where the high-order terms are ignored due to the approximation of small amplitudes of the oscillators, which have been obtained in Section 5.3.2. They are evaluated by

$$\gamma_j = \frac{N_j e^2}{2\pi\varepsilon_0 m_e r^2 c}, \quad g_j^2 = \frac{N_j e^2}{2\pi\varepsilon_0 m_e r^3}, \quad j = 1, 2, \quad (5\text{-}93)$$

where N_1 (or N_s) and N_2 (or N_m) are the effective electron numbers of the SNP and the MNP, respectively, and r is the distance between the centers of the SNP and the MNP. The solutions of Eq. (5-92) are

$$\begin{cases} x_j(t) = A_j e^{-i\omega_{ex}t}, \quad A_1 = \dfrac{B_2 + F_2}{B_1 B_2 - F_1 F_2} C_0, \quad A_2 = \dfrac{B_1 + F_1}{B_1 B_2 - F_1 F_2} C_0, \\ B_j = -\omega_{ex}^2 - 2i\beta_{aj}\omega_{ex} + \omega_{aj}^2, \quad F_j = g_j^2 - i\gamma_j \omega_{ex}, \quad j = 1, 2. \end{cases}$$

$$(5\text{-}94)$$

Second, the emission process is described as

$$\begin{cases} \ddot{x}_1 + 2\beta_{e1}\dot{x}_1 + \omega_{e1}^2 x_1 - \gamma_2 \dot{x}_2 - g_2^2 x_2 = 0, \\ \ddot{x}_2 + 2\beta_{e2}\dot{x}_2 + \omega_{e2}^2 x_2 - \gamma_1 \dot{x}_1 - g_1^2 x_1 = 0, \end{cases} \quad (5\text{-}95)$$

where the initial conditions are $x_j(0) = A_j$ and $\dot{x}_j(0) = 0$ ($j = 1, 2$). Here, β_{e1} and β_{e2} are the damping coefficients of the SNP and the MNP in emission process, respectively, and ω_{e1} and ω_{e2} are the emission resonance frequencies of the

SNP and the MNP before coupling, respectively. The solutions of Eq. (5-95) are similar to the ones in Section 5.3.2, but with different initial conditions. We assume $x_1(t) = Se^{\Omega t}$ and $x_1(t) = Me^{\Omega t}$, and substitute them into Eq. (5-95), thus obtaining the equation that Ω satisfies:

$$\prod_{j=1}^{2}(\Omega^2 + 2\beta_{ej}\Omega + \omega_{ej}^2) = \prod_{j=1}^{2}(\gamma_j\Omega + g_j^2). \tag{5-96}$$

Although Ω has analytic solutions, the expressions are too complex to be written here. Hence, we can rewrite the solutions of Ω in this form:

$$\Omega_1^{\pm} = -\beta_1 \pm i\omega_1, \quad \Omega_2^{\pm} = -\beta_2 \pm i\omega_2. \tag{5-97}$$

Thereby, combining with the initial conditions, the solutions of Eq. (5-95) can be written as

$$\begin{cases} x_1(t) = S_1 e^{\Omega_1^- t} + S_2 e^{\Omega_2^- t} + S_3 e^{\Omega_1^+ t} + S_4 e^{\Omega_2^+ t} \cong S_1 e^{\Omega_1 t} + S_2 e^{\Omega_2 t}, \\ x_2(t) = M_1 e^{\Omega_1^- t} + M_2 e^{\Omega_2^- t} + M_3 e^{\Omega_1^+ t} + M_4 e^{\Omega_2^+ t} \cong M_1 e^{\Omega_1 t} + M_2 e^{\Omega_2 t}. \end{cases}$$
$$\tag{5-98}$$

For the same reason as Eq. (5-90), we omit the solutions marked with "+", i.e., Ω_1^- and Ω_2^- are retained. To make it simple, we define $\Omega_1 = \Omega_1^-$ and $\Omega_2 = \Omega_2^-$. Therefore, the total emission far field in time domain can be written as

$$E(t) = E_1(t) + E_2(t) \cong \Omega_1^2 (K_s S_1 + K_m M_1) e^{\Omega_1 t} + \Omega_2^2 (K_s S_2 + K_m M_1) e^{\Omega_2 t}$$
$$:= A_1' e^{\Omega_1 t} + A_2' e^{\Omega_2 t}, \tag{5-99}$$

where $K_s = \dfrac{N_s e}{4\pi\varepsilon_0 c^2 d} \propto N_s$ and $K_m = \dfrac{N_m e}{4\pi\varepsilon_0 c^2 d} \propto N_m$. The total emission intensity in frequency domain can be evaluated by

$$I_{\text{tot}}(\omega) = \sum_{j=1}^{2} |A_j'|^2 \frac{1-e^{-2\beta_j t_0}}{2\beta_j t_0} \frac{\beta_j}{(\omega-\omega_j)^2 + \beta_j^2}. \tag{5-100}$$

Considering Fermi-Dirac distributions, the emission spectrum is tuned due to the electron temperature T. Therefore, the tuned emission intensity, i.e., the PL spectrum, can be evaluated by

$$I_{\text{PL}}(\omega) = I_{\text{tot}}(\omega) \frac{1}{1+e^{\hbar(\omega-\omega_{\text{ex}})/(k_B T)}}, \tag{5-101}$$

where $\hbar$ is the reduced Planck constant and k_B is the Boltzmann constant.

According to Eq. (5-97), there are two new modes, resulting in two peaks in the spectrum. One peak is related to the coupled SNP, the other peak is related to the coupled MNP; thereby, the PL spectrum can be divided into $I_{SNP}(\omega)$ and $I_{MNP}(\omega)$, where $I_{PL} = I_{SNP} + I_{MNP}$. Define the enhancement factor (EF) of SNP as

$$EF_{SNP} = \frac{I_{SNP}^{(CP)}}{I_{SNP}^{(ind)}}, \qquad (5\text{-}102)$$

where $I_{SNP}^{(CP)}$ and $I_{SNP}^{(ind)}$ stand for the peak intensities of the SNP mode of the coupled system and the individual SNP, respectively. We emphasize that, for the coupled system, the SNP mode ISNP is emitted not only by the SNP but also by the MNP, because both the SNP and the MNP emit the two modes with corresponding amplitudes, and the total electric field is the coherent superposition of the two emissions. Therefore, the intensity of the SNP mode can be highly enhanced due to the fact that $N_2 \gg N_1$; i.e., the MNPs help the SNPs emit photons through the channel of the MNPs with stronger intensity.

Notice that Eq. (5-100) is similar to Eq. (5-53). However, the latter was employed to calculate the PL spectra from two coupled MNPs with the same N, thus the same g and γ; the former will be employed to calculate the PL spectra from coupled SNPs and MNPs which have different parameters, e.g., N (N_s and N_m), etc. In this work, the symmetry is broken, and novelty phenomena may be obtained.

According to the model, the optical properties of the individual mode (SNP) and the coupling modes (SNP-MNP) are presented as following.

First, the properties of the individual SNP are obtained.

Figure 5-23 shows the calculated absorption, PLE, and PL of an individual SNP with single absorption mode and single emission mode. Here, ω_a, β_a, ω_e, and β_e are determined according to Ref. [29]. The absorption and PLE almost overlap for the larger frequencies, i.e., $\omega > \omega_a$, which is consistent with the experimental results of Ref. [29]. However, for smaller frequencies, i.e., $\omega < \omega_a$, the two curves do not overlap well. This is because, when we use Eq. (5-91) to

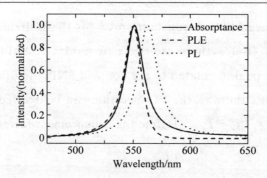

Fig. 5-23 Absorption (solid black line), PLE (dashed blue line), and PL (dot red line) calculated from Eqs. (5-86), (5-91), and (5-90), respectively. The parameters are $\omega_a = 2.251$ eV, $\beta_a = 0.0396$ eV, $\omega_e = 2.204$ eV, and $\beta_e = 0.0408$ eV

(For colored figure please scan the QR code on page 1)

calculate PLE for $\omega < \omega_a$, the anti-Stokes emission of PL is not integrated, the lack of which indicates the difference between PLE and the absorption. It is worth mentioning that a practical absorption of a SNP has multiple modes covering a wide range of wavelengths from ultraviolet to visible range. Here, for simplicity, we just employ one mode for absorption to demonstrate the optical properties.

Second, the coupled absorption process is investigated in detail according to Eq. (5-94).

Figure 5-24 shows the absorption intensities ($|A_j|^2$) of the coupled system varying with excitation wavelength at different distances r. We consider two situations, i. e., resonance coupling (Fig. 5-24(a), (b)) and nonresonance coupling (Fig. 5-24(c), (d)). In resonance coupling, the original absorption peaks of the two NPs are close. As the distance r decreases, i. e., the coupling strength increases, at first ($r > 10$ nm, weak coupling), the absorption amplitude of the SNP (Fig. 5-24(a), $|A_1|^2$) increases rapidly with almost no change in the line shape of the spectrum (one peak), while the one of the MNP (Fig. 5-24(b), $|A_2|^2$) remains almost unchanged in both intensity and line shape; then ($r \leqslant 10$ nm, strong coupling) both $|A_1|^2$ and $|A_2|^2$ appear as two splitting peaks with the blue one's intensity decreasing and the red one's intensity increasing, and the splitting increases with the increase of the

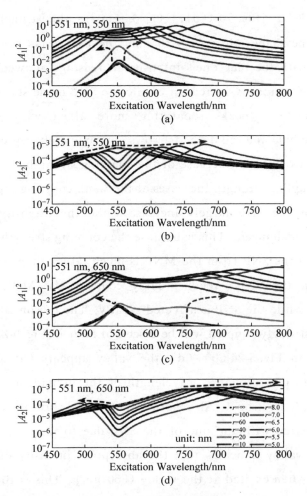

Fig. 5-24 Absorption amplitudes $|A_j|^2$ ($j=1,2$) of the coupled system with different distances r as a function of the excitation wavelength, calculated form Eq. (5-94). The dashed arrows stand for the trends of the blue and red shifts of the peaks as r decreases. The legend is shown in (d) and is suitable for (a)~(d). The parameters are (a)~(d) $N_1 = 10, N_2 = 10^5$; (a)~(d) $\omega_{a1} = 2.251$ eV, $\beta_{a1} = 0.0396$ eV (551 nm); (a),(b) $\omega_{a2} = 2.257$ eV, $\beta_{a2} = 0.1175$ eV (550 nm); (c),(d) $\omega_{a2} = 1.911$ eV, $\beta_{a2} = 0.1175$ eV (650 nm)

(For colored figure please scan the QR code on page 1)

coupling strength. In nonresonance coupling, the original absorption peaks of the two NPs are far apart. As the distance r decreases, at first ($r > 10$ nm), the amplitude of the SNP(Fig. 5-24(c), $|A_2|^2$) decreases followed by the increase with the appearance of the MNP mode (around 650 nm) when $r < 40$ nm, while the one of the MNP (Fig. 5-24(d), $|A_2|^2$) remains almost unchanged in both intensity and line shape; then ($r\leqslant10$ nm) the two peaks of $|A_1|^2$ start to separate more with the blue one's intensity increasing followed by decreasing and the red one's intensity increasing, while the SNP mode starts to appear in $|A_2|^2$ with the two peaks separating more along with the increasing intensities. The blue and red dashed arrows approximately represent the trends of the two modes with the increase of the coupling strength.

When the coupling strength increases at the weak coupling regime, why does the intensity of the SNP largely increase while the intensity of the MNP remains almost unchanged? This is because the coupling strength $g_1 \ll g_2$; i.e., the influence on the SNP from the MNP is much larger than the influence on the MNP from the SNP. When weakly coupled, the MNP is unaffected approximately, while the SNP is affected greatly. The reason why the splitting appears when strongly coupled will be discussed later in Fig. 5-25.

Notice that, in Fig. 5-24 (b), (d), the valley appears (at about 550 nm, corresponding to the resonance wavelength of the individual SNP) when the coupling strength is strong. Also notice that the absorption of the MNP is dominant compared with the one of the SNP due to the fact that $N_1 \ll N_2$. Therefore, the valley indicates that the absorption intensity of the system is extremely low when excited at the valley (550 nm). This is the phenomenon similar to EIT introduced by the classical mechanism. However, we should emphasize here that we only consider a single mode for the absorption of the SNP, while the actual case is that the SNP has multiple modes for the absorption, indicating the complicated coupling to achieve EIT.

Third, the coupled emission process is investigated in detail. Figure 5-25 shows the behaviors of the new generated modes of the coupled system for resonance coupling (Fig. 5-25(a), (b)) and nonresonance coupling (Fig. 5-25(c), (d)), respectively. The electron number of the MNP is kept as $N_2 = 10^5$. Hence, we use g_2 to represent one of the coupling strengths. For a certain value of N_1, as g_2 increases, the coupled resonance frequencies (ω_1, ω_2) split along with the

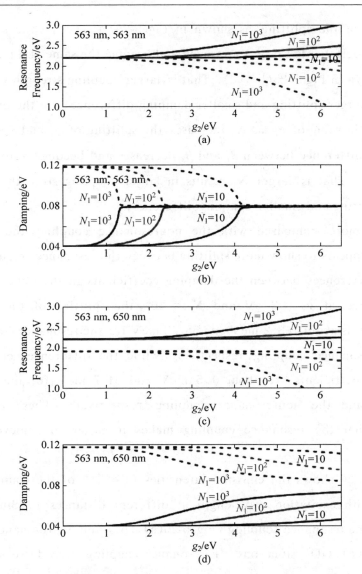

Fig. 5-25 Resonance frequencies ω_1 and ω_2 ((a), (c)) and damping coefficients β_1 and β_2 ((b), (d)) of the coupled system as a function of g_2 with different N_1, calculated from Eq. (5-97). Black, red, and blue lines stand for $N_1 = 10$, $N_1 = 10^2$, and $N_1 = 10^3$, respectively. Solid and dashed lines with the same color stand for the two splitting modes. Here, (a) ~ (d) $N_2 = 10^5$, $\omega_{e1} = 2.204$ eV, $\beta_{e1} = 0.0408$ eV (563 nm); (a), (b) $\omega_{e2} = 2.257$ eV, $\beta_{e2} = 0.1175$ eV (563 nm); (c), (d) $\omega_{e2} = 2.205$ eV, $\beta_{e2} = 0.1175$ eV (650 nm).

(For colored figure please scan the QR code on page 1)

increasing of the splitting, as shown in Fig. 5-25(a), (c), while the coupled damping coefficients (β_1, β_2) approach and become the same at a large enough g_2, as shown in Fig. 5-25(b), (d). That is, larger coupling strength (g_1 and g_2) results in larger splitting and smaller damping difference. On the other hand, for a certain value of g_2, as N_1 increases, the splitting of ω_1 and ω_2 increases, while the difference between β_1 and β_2 decreases and becomes zero at a large enough g_2. That is, larger N_1 results in larger coupling strength (g_1), thus larger splitting and smaller damping difference.

Furthermore, compared with the nonresonance coupling, the resonance coupling appears to show more splitting between the resonance frequencies and smaller differences between the damping coefficients in the same condition. For instance, at $g_2 = 2$ eV and $N_1 = 10^3$, the splitting of the resonance frequencies is about 163.2 meV and 55.42 meV for the resonance coupling and the nonresonance coupling, respectively, while the differences between the damping coefficients are about 0.5 meV and 64.7 meV for the resonance coupling and the nonresonance coupling, respectively. These phenomena indicate that the resonance coupling makes it easier to achieve stronger coupling.

Figure 5-26 shows the emission intensities ($|A'_j|^2$) of the coupled system varying with excitation wavelength at different distances r. Similarly, we consider the resonance coupling (Fig. 5-26(a), (b)) and nonresonance coupling (Fig. 5-26(c), (d)) situations. In resonance coupling, as r decreases, at first ($r>10$ nm), the emission amplitude of the SNP mode (Fig. 5-26(a), $|A'_1|^2$) increases rapidly with almost no change in the line shape of the spectrum, while the one of the MNP mode (Fig. 5-26(b), $|A'_2|^2$) remains almost unchanged in both intensity and line shape; then ($r \leqslant 10$ nm), both $|A'_1|^2$ and $|A'_2|^2$ appear as two splitting peaks with the blue one's intensity decreasing and the red one's intensity decreasing as well, and the splitting increases with the decrease of r. In nonresonance coupling, as r decreases, at first ($r >$

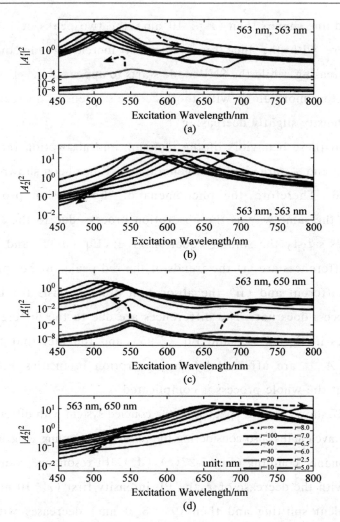

Fig. 5-26   Emission amplitudes $|A'_j|^2$ $(j=1,2)$ of the coupled system with different distances r as a function of the excitation wavelength, calculated from Eq. (5-99). The dashed arrows stand for the trends of the blue and red shifts of the peaks as r decreases. The legend is shown in (d) and is suitable for (a) ~ (d). The parameters are (a) ~ (d) $N_1 = 10, N_2 = 10^5, \omega_{a1} = 2.251$ eV, $\beta_{a1} = 0.0396$ eV (551 nm), $\omega_{e1} = 2.204$ eV, $\beta_{e1} = 0.0408$ eV (563 nm); (a),(b) $\omega_{e2} = 2.257$ eV, $\beta_{e2} = 0.1175$ eV (563 nm); (c),(d) $\omega_{e2} = 2.205$ eV, $\beta_{e2} = 0.1175$ eV (650 nm)

(For colored figure please scan the QR code on page 1)

10 nm), the amplitude of the SNP mode (Fig. 5-26(c), $|A'_1|^2$) decreases followed by the increase with the appearance of the MNP mode (around 650 nm), while the one of the MNP mode (Fig. 5-26(d), $|A'_2|^2$) remains almost unchanged in both

intensity and line shape; then ($r \leqslant 10$ nm), the two peaks of $|A'_1|^2$ start to separate more with the blue one's intensity increasing and the red one's intensity increasing, while the SNP mode starts to appear in $|A'_2|^2$ with the two peaks separating more along with the blue one's intensity decreasing and the red one's intensity slightly decreasing.

Notice that these behaviors of the emission and absorption intensities as a function of r and ω_{ex} are similar in general, which has been shown in Fig. 5-24 and Fig. 5-26. Therefore, the phenomena of the emission process can be explained in the same way as the absorption process, due to the fact that the two processes satisfy the similar equations, i.e., Eq. (5-92) and Eq. (5-95). The main differences are (i) the trends of the red peaks of the splitting peaks are usually different and (ii) the absorption process exists for EIT, but the emission process does not. The differences are due to the different values of the quantities in Eq. (5-92) and Eq. (5-95), and the fact that the emission intensities $|A'_j|^2$ are affected by the absorption intensities $|A_j|^2$, which indicates that the whole process is complicated.

Figure 5-27 shows the PL spectra of the coupled system with different r excited at different wavelengths, also considering the resonance coupling (Fig. 5-27(a),(b)) and nonresonance coupling (Fig. 5-27(c),(d)). In resonance coupling excited at 475 nm, with the decrease of r, the PL intensity first ($r \geqslant 10$ nm) increases with no evident splitting and then ($r = 8.0$ nm) decreases with splitting, followed by ($r = 6.0, 5.0$ nm) the increase of the blue peak due to the fact that the blue peak is closer to the excited wavelength with smaller r which corresponds to the resonance excitation. In resonance coupling excited at 532 nm, with the decrease of r, the PL intensity first ($r \geqslant 10$ nm) increases with no evident splitting and then ($r < 10$ nm) decreases with increasing splitting. Although the blue peak is close to 532 nm, the anti-Stokes emission of the PL spectra is restrained by the Fermi-Dirac distribution, resulting in the decrease of the intensity of the blue peak. In nonresonance coupling excited at 475 nm, with the decrease of r, the PL intensity first ($r \geqslant 10$ nm) remains

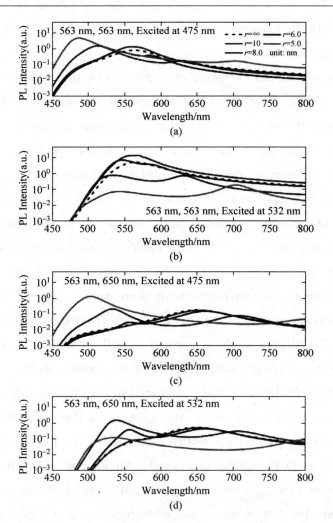

Fig. 5-27 PL spectra of the coupled system with different r and excited at 475 nm ((a), (c)) and 532 nm ((b), (d)), calculated from Eq. (5-101). Here, $T = 500$ K, $N_1 = 10$, and $N_2 = 10^5$, and ω_{e1}, β_{e1}, ω_{e2}, and β_{e2} are the same as in Fig. 5-26. The legend is shown in (a) and is suitable for (a)~(d)

(For colored figure please scan the QR code on page 1)

almost unchanged and then ($r < 10$ nm) increases in the blue peak with the increasing splitting due to the same reason as resonance coupling excited at 475 nm. In nonresonance coupling excited at 532 nm, with the decrease of r, the PL intensity first ($r \geqslant 10$ nm) remains almost unchanged and then ($r = 8.0$, 6.0 nm) increases in the blue peak with the increasing splitting, followed by

($r = 5.0$ nm) the decrease of the blue peak due to the same reason as resonance coupling excited at 532 nm.

In general, the coupled PL spectra are affected by the excitation wavelength because one of the new generated modes will be close to the excitation wavelength with the increase of the coupling strength, indicating the resonance excitation and resulting in the enhancement of the spectra, as well as the weakening of the spectra caused by Fermi-Dirac distribution.

To verify our model, a comparison with the experiments is necessary. Figure 5-28 shows the comparisons between the calculations of our model and the experimental data from Song[30]. The blue open squares show the experimental PL spectra of the individual SNP (CdSeTe/ZnS QD) with single resonance wavelength at about 800 nm, copied from their paper[30]; while the blue curve shows the corresponding PL spectra calculated from our model with proper parameters. The shapes of these two agree well with each other. Here, the peak positions, damping coefficients, and electron numbers of the two individual systems (MNP and SNP) are tuned to the observations of Ref. [30]. However, when calculating the coupling spectra, these parameters do not change, as shown below. The red open circles show the PL spectra of coupled system, i. e., the QD coupled to a single Au microplate with the distance between them of (18 ± 1.9) nm, while the red curve shows the corresponding PL spectra calculated from our model with the distance $r = 18.9$ nm. The two spectra agree well in both the enhancement and the shape in general. However, strictly speaking, the shape of the coupling spectrum does not agree well at $\lambda > 800$ nm, while the shape of the individual one behaves better. The spectra are affected by the electron-phonon and electron-electron interactions[20], which are not considered here; the coupling also affects the two interactions, which is more complicated than the case of the individual spectrum. We may discuss these effects of the coupling case in the future. Furthermore, we notice that the mode of the coupled system is almost the same as the mode of the individual SNP, indicating that the coupling is in the weak coupling regime, because there is no evident splitting in the coupling spectra. Although there is no splitting in weak coupling, there is an enhancement in the intensity of the SNP. The enhancement originates from the assistance of the MNP that emits the SNP modes through the channel of the MNP, which has been explained with Eq. (5-102).

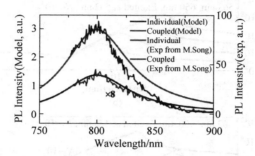

Fig. 5-28 PL spectra of the individual SNP (blue) and the coupled SNP-MNP (red). Blue and red curves stand for the individual and coupled systems, respectively, calculated from Eq. (5-101), with the following parameters: $r = 18.9$ nm, $T = 500$ K, $N_1 = 10$, $N_2 = 10^5$; $\omega_{a1} = 1.60$ eV, $\beta_{a1} = 0.06711$ eV (775 nm); $\omega_{e1} = 1.5505$ eV, $\beta_{e1} = 0.06711$ eV (800 nm); $\omega_{e2} = \omega_{a2} = 2.350$ eV, $\beta_{e2} = \beta_{a2} = 0.43481$ eV (528 nm). Dark green and violet curves stand for the experimental data of individual and coupled system, respectively, and are copied from Song[30]
(For colored figure please scan the QR code on page 1)

Fig. 5-29 shows the EF of the SNP as a function of r with different N_1 and N_2, and excited at different wavelengths, in the case of nonresonance coupling. In surface enhanced emission, nonresonance coupling is a more general case; moreover, resonance couplings have non-negligible background signals that are mostly from the MNP, which might submerge the weak signals from the SNP and hinder the detection. Therefore, here we only consider the general case, i. e., nonresonance coupling. For all the cases in Fig. 5-29, as r decreases, the EF first decreases a little and then increases rapidly to its maximum, followed by its decrease.

In Fig. 5-29(a), (b), N_2 is unchanged, and as N_1 increases, the maximum of the EF decreases from about 10^{10} ($N_1 = 1$) to 10^3 ($N_1 = 10^3$) excited at 475 nm and from about 10^8 ($N_1 = 1$) to 10^2 ($N_1 = 10^3$) excited at 532 nm. According to

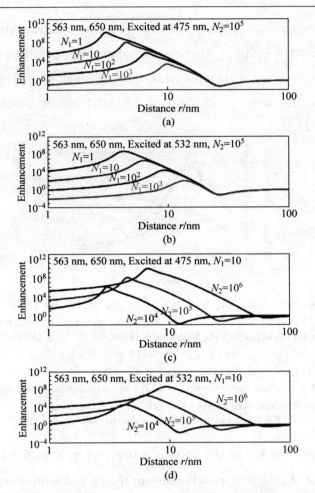

Fig. 5-29 EF of the SNP as a function of r with different N_1 ((a),(b)) and N_2 ((c),(d)), excited at 475 nm ((a),(c)) and 532 nm ((b),(d)), calculated from Eq. (5-102). Here, $T = 500$ K; $\omega_{a1} = 2.251$ eV, $\beta_{a1} = 0.0396$ eV (551 nm), $\omega_{e1} = 2.204$ eV, $\beta_{e1} = 0.0408$ eV (563 nm); $\omega_{e2} = 2.205$ eV, $\beta_{e2} = 0.1175$ eV (650 nm)

(For colored figure please scan the QR code on page 1)

Eq. (5-89) and Eq. (5-99), the emission intensity of the individual SNP is proportional to N_1^2, and the enhanced intensity of the SNP mode is proportional to $(N_1 S_1 + N_2 M_1)^2$. If the SNP and the MNP are strongly coupled and $N_2 \gg N_1$, the enhanced intensity of the SNP mode is approximately proportional to N_2; thus, $EF \propto (N_2/N_1)^2$. This is the reason why the EF decreases as N_1 increases. Furthermore, as just mentioned, $I_{SNP}^{(cp)} \propto N_2$, resulting

CHAPTER 5　Surface Plasmons

in the fact that the absolute intensity of the SNP mode of the coupled system is not that dependent on N_1. Therefore, for a certain MNP with certain N_2, although the EF can be very high when N_1 is low enough, the actually detected maximum signal of the SNP retains a certain order of magnitude. Moreover, the maximum related distance r increases as N_1 increases, indicating easier maximal coupling with larger N_1.

In Fig. 5-29(c),(d), N_1 is unchanged, and as N_2 increases, the maximum of the EF increases from about 10^6 ($N_2 = 10^4$) to 10^{10} ($N_2 = 10^6$) excited at 475 nm and from about 10^4 ($N_2 = 10^4$) to 10^8 ($N_2 = 10^6$) excited at 532 nm. It indicates that larger EF can be achieved by using MNP with larger N_2. This is because, in the coupling, the SNP mode is most emitted by the MNP, which is proportional to N_2. Moreover, the maximum related distance r increases as N_2 increases, indicating easier maximal coupling with larger N_2.

We notice that, in Fig. 5-29, the EF excited at 475 nm is generally larger than the EF excited at 532 nm, especially for the maximum, and there is a minimum for the EF before it reaches the maximum as r decreases. The corresponding case can be found in Fig. 5-26(c). When r decreases, especially in the strong coupling regime, the blue shift of the SNP mode slows down the increase of the intensity excited at 532 nm but speeds up the increase of the intensity excited at 475 nm. On the other hand, the individual intensity (black dotted line in Fig. 5-26(c)) excited at 532 nm is larger than the one excited at 475 nm. The above two reasons provide the results; i.e., the EF excited at 475 nm is larger than the one excited at 532 nm. Furthermore, in Fig. 5-26(c), as r decreases ($r > 40$ nm), the intensity decreases to a minimum, corresponding to the minimum of the EF in Fig. 5-29.

In summary, we develop a classic model to reveal the optical properties of the individual SNP and as well as the plexciton, i.e., the coupling between a SNP and a MNP. Good agreements between our model and the published experiments verify the validity of our model. The model divides the whole process into an absorption process and an emission process, both of which are analyzed and investigated carefully. In the coupled system, the absorption properties reveal the splitting and the EIT in the spectra; the emission properties reveal the splitting in the spectra and the enhancement of the SNP. Also, the PL spectra are illustrated and compared with the individual one; the spectral shapes are changed, i.e., modes split, and the intensities increase or decrease depending on the coupling strength. Moreover, the EF is analyzed in detail and varies with N_1, $\dot{N}_2$, ω_{ex}, and r. The maximum of the EF can reach the order of 10^{10} for a general case. The method based on this classic model could be of great convenience for the plexciton research. Furthermore, based on this classic model, the asymmetrical system, i.e., the two-body system with different individuals, could be easily analyzed, such as the SNP-MNP structure investigated here, a dimer with two different MNPs, coupling between the QDs and the MNPs, etc. Moreover, this model can be easily expanded to many-body system, such as the MNP-SNP-MNP structure, a long chain structure consisting of different MNPs, etc.

5.3.5 Fano Resonant Propagating Plexcitons and Rabi-splitting Local Plexcitons

In this section, we theocratically explore plexcitons of bilayer borophene synthesized on Ag(111) film in TERS system, where bilayer (BL) borophene located in the nanocavity between tip and substrate, stimulated by recent experimental synthesis Borophene synthesis beyond the single atomic-layer limit[31].

As has been mentioned above, the plasmon-exciton coupling interaction can be interpreted by a physical model consisting of two coupled driven oscillators, which is described by the following matrix equation:

$$\begin{pmatrix} \omega_p^2 - \omega^2 + i\omega\gamma_p & -g_e^2 \\ -g_p^2 & \omega_e^2 - \omega^2 + i\omega\gamma_e \end{pmatrix} \begin{pmatrix} A_p \\ A_e \end{pmatrix} = \begin{pmatrix} f_p \\ f_e \end{pmatrix} \quad (5\text{-}103)$$

Here A_p and A_e are the oscillator amplitudes, ω_p and ω_e are the resonant frequencies, γ_p and γ_e are the damping coefficients, and f_p and f_e are the amplitudes of the external forces with the driving frequency ω, with the relation $f_e = M f_p$, and M is a plasmonic-enhanced factor. The subscripts p and e in those quantities represent the plasmon and quantum emitter, respectively. The coupling strengths g_p and g_e describe the interaction between the plasmons and excitons, and $g_p^2 = M g_e^2$, which reveals that the influences of the coupling on each other are different. A non-Hermitian Hamiltonian H in Eq. (5-103) corresponds to a complex potential $V(x)$, i.e., to a complex refractive index distribution $n(x)$[32]. The imaginary part of the refractive index corresponds to the optical gain or loss of the dielectric medium. In the weak-coupling regime[33-34], the coupling strength is much less than one of the damping coefficients, $|g_e| \ll |\gamma_p|$ or $|g_p| \ll |\gamma_e|$. Fano resonance occurs when plasmon and exciton with different damping rates are weakly coupled, that is, by plasmon-exciton coupling with narrow (weakly damped) and broad (strongly damping) spectral lines[33-34]. The spectra show typical asymmetry with a sharp change between a dip and a peak. The phase of the undamped oscillator changes by π at the resonance, while the phase of the strongly damped oscillator varies slowly[33-34]. When $\omega_e = \omega_p$ in the Fano resonance, it is referred to as electromagnetically induced transparency (EIT)[35]. Very important are applications of Fano resonance in ultrasmall lasers based on the interference between a continuum of waveguide modes and a discrete mode of a nanocavity[36]. In the strong-coupling regime, the damping coefficients of both oscillators are weak[26], $|\gamma_e| \ll |g_e|$, and $|\gamma_p| \ll |g_p|$. Rabi splitting in the spectra can be observed, and the real parts of the split eigenfrequencies of the plexciton of two coupled plasmon-exciton systems are by $|\omega_+ - \omega_-| \approx 2\sqrt{g_e g_p}$, when the bare plasmonic and excitonic frequencies ω_p and ω_e are tuned to each other. The coupling strength determines both strong and weak couplings, and the transition from strong to weak coupling can be realized by altering the effective mode volume and the dielectric environment of a nanocavity[37]. Most observations of Fano resonance reveal the coupling interaction between the localized plasmons and excitons. Our findings demonstrate that the propagating plasmons can also couple with molecular excitons, and thus creating Fano-resonant propagating plexcitons.

The rich physics of Fano resonanceshas recently been explored in several different photonic and plasmonic systems[38-41]. Fano-resonant ultrathin film optical coatings have been reported in 2021[34]. Also, a self-pulsing photonic crystal Fano laser is reported in 2017[42]. Fano-resonant asymmetric metamaterials have been used for ultrasensitive spectroscopy and identification of molecular monolayers[38]. In tip-enhanced Raman spectroscopy (TERS), both Fano resonance and plexcitonic effect are observed due to the plasmon-exciton coupling interaction[43-44]. The plasmon-gradient effect in TERS has also been reported experimentally and theoretically, providing more information than dipole-Raman, in which multipole-Raman can be observed[45-48]. Rabi splitting plexcitons in the TERS system may take advantage of the on-the-coordinate anti-Stokes Raman scattering (CARS) by the splitting plexciton with higher energy, and the on the Raman and the fluorescence by the splitting plexciton with lower energy. The creation of Rabi splitting plexcitons not only requires a plasmonic nanocavity with a small effective mode volume but also is affected by the dielectric properties of a quantum emitter. Geohegan reported on the exchange of energy and electric charges in a hybrid composed of a two-dimensional tungsten disulfide (2D-WS_2) monolayer and an array of aluminum (Al) nanodisks[49]. Ray reported ultrafast real-time observation of double Fano resonances in discrete excitons and single plasmon-continuum[50]. Sun reported ultrafast investigation of individual bright exciton-plasmon polaritons in size-tunable metal-MoS_2 hybrid nanostructures[51]. We experimentally studied physical mechanism on exciton-plasmon coupling revealed by femtosecond pump-probe transient absorption spectroscopy[52].

Because of the unique optical, electric and thermoelectric effect properties, BL borophene is synthesized on Ag (111) film[31] and Cu (111) film[53], respectively, in high vacuum system, and is theoretically investigated in detail[53-54]. Because of the negative real part of the dielectric function in the wavelength region between 420 nm and 550 nm, BL borophene shows plasmonic properties[54-55]. Ultrasensitive TERS technology and BL borophene provide an ideal framework to deeply study the plexcitonic properties of plasmon-exciton, including Rabi-splitting plexcitons (strong coupling) and Fano-resonant plexcitons (weak coupling). Our studies demonstrate a two-dimensional Fano-resonant propagating plexciton along BL borophene.

In this paper, the Fano-resonant propagating plexcitons and Rabi-splitting local plexcitons in the TERS-BL borophene system are studied, respectively. Our results in the strong plasmon-exciton interaction regime manifest the strong Rabi splitting plexcitons, which can potentially enormously enhance CARS and fluorescence. The plasmon-gradient effect is also revealed theoretically. In the weak plasmon-exciton interaction regime, Fano resonance induces single-mode ultrasmall lasers at the nanoscale with an ultrahigh enhancement factor, reaching up to 10^8. The Fano resonance single-mode ultrasmall lasers with ultrahigh Q result in the propagation of plexciton along the surface of BL borophene, which can be considered the 2D plexcitonic emitter with different propagating modes of plexciton, which is very sensitive to the slight variety of wavelengths.

All electromagnetic calculations of TERS-BL borophene system were performed by using the finite element method (FEM) and COMSOL Multiphysics software. TERS-BL borophene system consists of an Ag tip above an Ag substrate, where the BL borophene locates in the tip-substrate nanocavity. In all simulations, the TM plane wave irradiates the nanocavity between the tip and substrate, where the incident angle and the electric field amplitude are 60° and 1.0 V/m, respectively. The size of the simulation region was set as 1000 nm × 1000 nm × 1000 nm, and such an extent is large enough for all calculations. Boundary reflection of the incident wave could be avoided by using the scattering and perfectly matched layers boundary conditions at all boundaries of the simulated domain. To improve the computational accuracy, we used the free tetrahedral mesh with a size of 0.5 nm in the nanocavity between the tip and substrate. The free tetrahedral mesh with the size of 2 nm was used for the tip apex. The mesh size of the free tetrahedral mesh was 5 nm in other regions of the TERS configuration. In the air domain, the mesh size was one-tenth of the wavelength. The relative dielectric function for Ag was taken from Ref. [56]. The dielectric function and absorption spectra of BL borophene are from Ref. [54]-[55].

Here we chose the scanning tunneling microscope-based TERS (STM-TERS) system to study the plasmon-exciton interaction. Figure 5-30(a) is the scheme of the STM-TERS system, where the BL borophene locates in the nanocavity between the Ag tip and Ag substrate, where the distance (d) is set as 1 nm. In all calculations, the cone angle of the tip is 10°, and the tip radius is 12 nm. Figure 5-30(b) shows the calculated electric field ($|E|^2$) in the STM-TERS system, which reveals two localized surface plasmon resonance (LSPR) peaks around 488 nm and 633 nm, and the electric fields can reach up to around 10^5. The result of Fig. 5-30(c) demonstrates two strong and weak absorption peaks around 488 nm and 633 nm in the calculated absorption spectrum of BL borophene. The wavelength matching between STM-TERS and BL borophene indicates a possible plasmon-exciton interaction around the two resonance wavelengths. The dielectric function of BL borophene in Fig. 5-30(d) shows the negative real part of dielectric functions perpendicular to the surface of BL borophene around 488 nm, which may result in strong coupling interaction between the plasmon and exciton. Figure 5-30(b)~(d) demonstrates that there may be a weak plasmon-exciton interaction around 633 nm.

In the weak-coupling regime for the dissipative (non-Hermitian) systems, the coupling strength is much less than one of the damping coefficients, $|g| \ll |\gamma_p|$ or $|g| \ll |\gamma_e|$, in Eq. (5-103). Fano resonance may occur from coupling

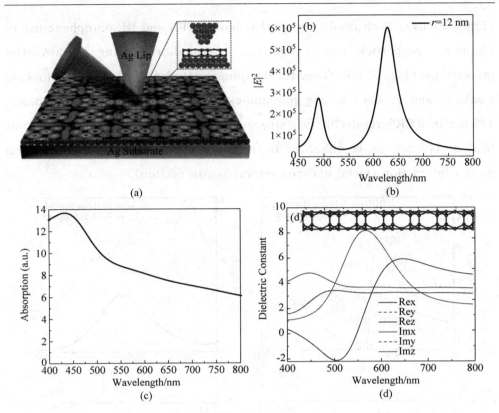

Fig. 5-30 (a) Schematic diagram of the TERS system with BL borophene located in the silver tip-substrate nanocavity. (b) Calculated electric field ($|E|^2$) plotted as a function of the wavelength. (c) The absorption spectrum of BL borophene. (d) Inhomogeneous dielectric function of BL borophene. Figure 5-30(c) and (d) are from Ref. [53]-[54]

(For colored figure please scan the QR code on page 1)

plasmon and exciton with strongly different damping rates producing narrow and broad spectral lines. Figure 5-31(a) shows a typical asymmetry spectral pattern with a sharp change between a dip and a peak in the TERS system. The interaction between the broad LSPR mode of TERS and the narrow resonance mode of BL borophene causes Fano resonance behavior in the wavelength range between 550 nm and 700 nm (Fig. 5-31(a)). Strong absorption around 488 nm can be observed in both plasmonic TERS system, and excitonic BL borophene (Fig. 5-30(b) and (c)), and the real part of dielectric functions perpendicular to the surface of BL borophene are negative

(Fig. 5-30(d)). Such results reveal that both TERS and BL borophene are of plasmonic properties around 500 nm, and indicates strong light-matter interactions (Fig. 5-31(b)), and the damping coefficients of both oscillators are weak, γ_p and $\gamma_e \ll g$. Strong plasmon-exciton coupling interaction around 488 nm in TERS results in two stronger plexcitonic peaks. The plexcitonic peak at 448 nm can be enhanced to the order of 10^7 due to the strong light-matter interaction (metal plasmon interacts with exciton).

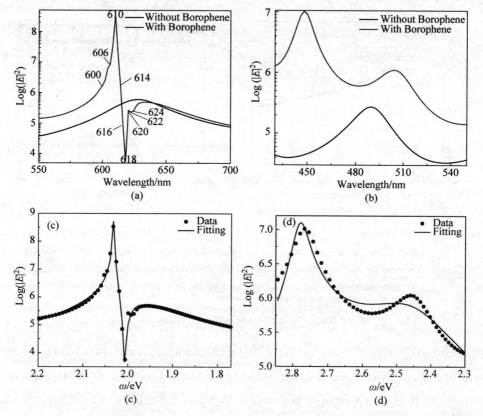

Fig. 5-31 Weak and strong interactions between plasmon and exciton. (a) Weak coupling, (b) strong coupling, (c) the fitting for Fano resonance, and (d) the fitting for Rabi splitting

(For colored figure please scan the QR code on page 1)

In general, the amplitude of Eq. (5-103) can be written as[26]

$$A_p = \frac{(\omega_e^2 - \omega^2 + i\gamma_e\omega)f_p + g_e^2 f_e}{(\omega_p^2 - \omega^2 + i\gamma_p\omega)(\omega_e^2 - \omega^2 + i\gamma_e\omega) - g_e^2 g_p^2}, \quad (5-104)$$

$$A_e = \frac{(\omega_p^2 - \omega^2 + i\gamma_p\omega)f_e + g_p^2 f_p}{(\omega_p^2 - \omega^2 + i\gamma_p\omega)(\omega_e^2 - \omega^2 + i\gamma_e\omega) - g_e^2 g_p^2}, \quad (5\text{-}105)$$

Fano resonance arises from the weak plasmon-exciton coupling with strongly different damping rates producing narrow and broad spectral lines. The phase shift of plasmon and exciton can be roughly estimated by

$$\theta_p = \arctan\left(\frac{\gamma_e\omega}{\omega_e^2 - \omega^2}\right), \quad (5\text{-}106)$$

$$\theta_e = \arctan\left(\frac{\gamma_p\omega}{\omega_p^2 - \omega^2}\right), \quad (5\text{-}107)$$

When $\omega = \omega_e$, and $f_e = 0$, Eqs. (5-104) and (5-105) can be simplified to

$$|A_p|^2 \approx 0, \quad (5\text{-}108)$$

$$|A_e|^2 \approx \frac{1}{g_e^2}|f_p|^2, \quad (5\text{-}109)$$

Eqs. (5-108) and (5-109) reveal the energy transfer from plasmon to exciton in plexciton mode at $\omega = \omega_e$.

Although Fano resonance is at the condition of $f_e = 0$, the actual electric field felt by the exciton is not 0. Notably, in TERS system, f_e could be very large, i.e., $f_e = Mf_p$ as mentioned above. Figure 5-31(c) is the fitted spectrum using Eq. (5-104), which agrees well with the Fano resonance line shape in Fig. 5-31(a). The fitting data are listed in the "Fano" line in Table 5-1. We further fit the result of Fig. 5-31(b) by using Eq. (5-104), which agrees with the Rabi splitting line shape caused by the strong plasmon-exciton coupling. The fitting data are listed in the "Rabi" line in Table 5-1. In Table 5-1, ω_p and ω_e stand for the resonant frequencies of non-coupled plasmon and exciton, respectively; γ_p and γ_e stand for the damping coefficients of non-coupled plasmon and exciton, respectively, which are related to the half-width of the spectra; g_e and $g_p = Mg_e$ stand for the coupling coefficients of exciton to plasmon and plasmon to exciton, respectively, which are related to the coupling strength between them; M stands for the factor (EF) of the electric field introduced by the plasmon, and we use the values of M at the respective resonance peaks of the plasmon spectra without borophene, i.e., about 488 nm and 633 nm, for the fitting of Fano resonance and Rabi resonance, with $f_e = Mf_p$ used here. Interestingly, Fig. 5-31(a) demonstrates an extreme lasering peak with a very narrow half-width around 610 nm. This high Q cavity Fano resonance for single mode laser can be strongly enhanced to $Q = 10^8$. Here, the enhancement factor is defined as the ratio of the maximum intensity to the full width at half maximum in the profile in Fig. 5-31(a). The intensity of exciton around the dip can be significantly enhanced by $\frac{1}{g_e^2}|f_p|^2$, which demonstrates the plasmon energy is significantly transferred to the exciton.

Table 5-1 Parameters of the fittings for Fano resonance and Rabi splitting

Quantity	Plasmon/eV		Exciton/eV		Coupling/eV	External Force	
Parameter	ω_p	γ_p	ω_e	γ_e	g_e	f_p	M
Fano	2.000	0.1070	2.032	0.002490	0.01130	291.1	−775
Rabi	2.533	0.3163	2.714	0.02795	0.1859	45.78	522

Since the exciton can obtain the ultrahigh by $\frac{1}{g_e^2}|f_{pe}|^2$ from the plasmon in the 616 nm to 622 nm region, as shown with the dip in Fig. 5-31(a), we further investigate the plexciton waveguide along the surface of BL borophene in this range. The side and top views of plexciton waveguides at different wavelengths are shown in Fig. 5-32. It can be found that plexcitonic waveguide can propagate along the surface of BL borophene at a specific wavelength. Furthermore, the waveguide modes are strongly manipulated by different wavelengths. The reason is that in the 616 nm to 622 nm region, plasmon energy strongly transfers to BL borophene, and the phase of plexciton waveguide is manipulated by Eq. (5-107). It is evidence that Fano resonant propagating plexcitons can be efficiently propagated longer distance with less loss. Moreover, propagating modes are very sensitive to the excited wavelength in Fano resonance.

From 624 nm to 630 nm, the profile of plexciton is gradually similar to that of a

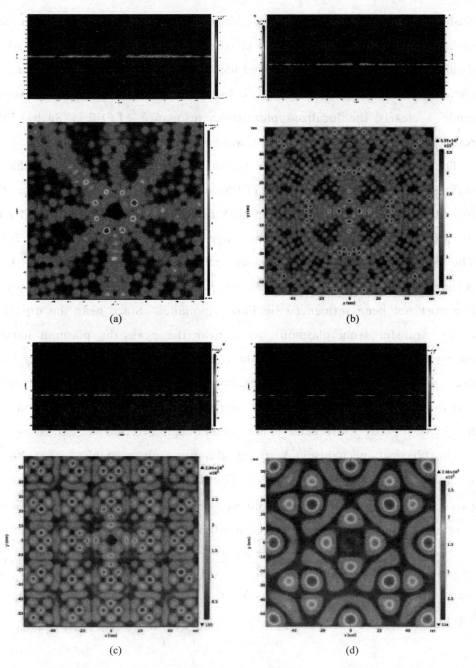

Fig. 5-32 Fano resonance propagating plexcitons at different wavelength. (a) at 616 nm, (b) at 618 nm, (c) at 620 nm, and (d) 622 nm

(For colored figure please scan the QR code on page 1)

plasmon in Fig. 5-31(a). Thus, the plexcitonic modes for those wavelengths are also studied, see Fig. 5-33. It is found that plexcitonic mode at 624 nm is unclear, which should be the edge of propagating and local plexcitonic modes. At 626 nm, the plexciton mode is gradually closed to localization. At 628 nm, the plexciton mode is clearly the localized plexcitonic resonance (LPER). Such LPER resonance at 630 nm is better than that at 628 nm.

The spectra in Fig. 5-31 shows typical asymmetry with a sharp change between a dip and a peak in TERS system. We also demonstrate the plexcitonic modes around the sharp peak in TERS. From 600 nm to 614 nm, the profile of plexciton is within the range of narrow sharp peak in Fig. 5-31. The plexcitonic modes for those wavelengths are also studied, see Fig. 5-34. It is found that plexcitonic mode at 600 nm is the LPER mode(Fig. 5-34), which is almost not been influenced by Fano resonance. Since near the dip, it is energy transfer from plasmon; while near the peak, the plasmon harvest energy from the exciton, which can be considered as the energy transfer from exciton to plasmon. In this case, plexciton on the BL borophene loses energy, and there is not enough energy to support plexciton waveguide. Then, the quality of plexciton waveguide is not good enough(Fig. 5-34(b)~(d)).

The physical mechanism of strong plasmon-exciton coupling can also be interpreted with perturbation theory. The real parts of the eigenfrequencies of two coupled plexciton peaks are split by $|\omega_+ - \omega_-| \approx 2\Delta = 2\sqrt{g_p g_e}$, when the bare oscillator frequencies ω_p and ω_e are tuned to each other.

$$\omega_\pm \text{(plexciton)} = \frac{\omega_p + \omega_e}{2} \pm \frac{1}{2}\sqrt{(\omega_p - \omega_e)^2 + 4\Delta^2}, \qquad (5\text{-}110)$$

According to data in Fig. 5-31(b), $\Delta \approx 0.155$ eV. The plasmon and exciton eigenfrequencies can be estimated by[43]

$$\omega_{p,e} = \frac{\omega_+ + \omega_-}{2} \pm \Delta \frac{I_+ - I_-}{I_+ + I_-}, \qquad (5\text{-}111)$$

where I_+ and I_- are the intensities of split plexcitonic peaks in Fig. 5-31(b). Note that $I_+ \gg I_-$ in Fig. 5-31(b), and $\frac{I_+ - I_-}{I_+ + I_-} \approx 0.80$. According to Eq. (5-110)

CHAPTER 5 Surface Plasmons

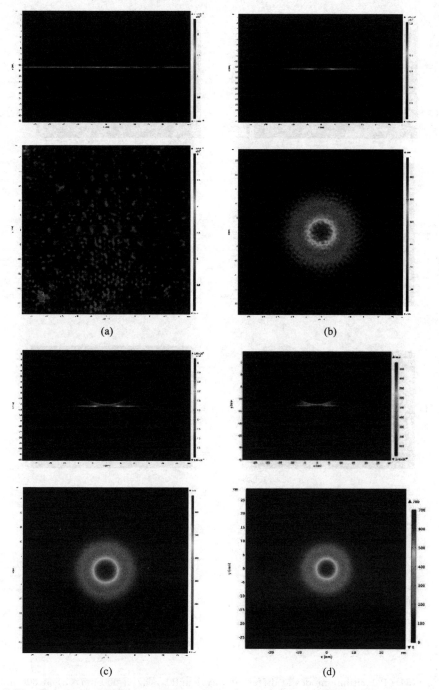

Fig. 5-33 Plexcitonic modes at a different wavelength. (a) at 624 nm, (b) at 626 nm, (c) at 628 nm, and (d) at 630 nm

(For colored figure please scan the QR code on page 1)

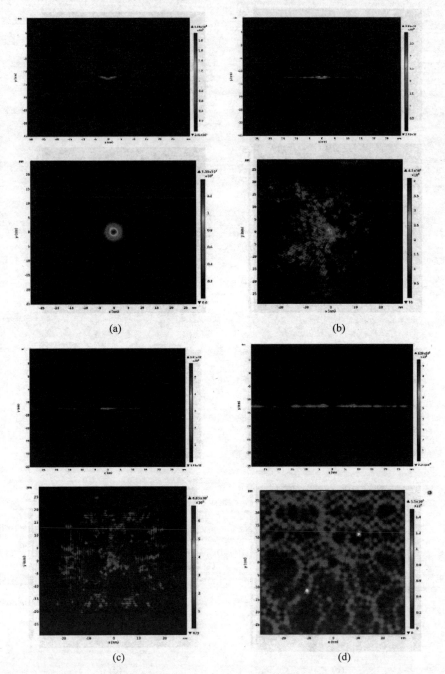

Fig. 5-34 Plexcitonic modes at different wavelengths. (a) at 600 nm, (b) at 606 nm, (c) at 610 nm, and (d) at 614 nm
(For colored figure please scan the QR code on page 1)

and (5-111), we can obtain that the plasmon around 488 nm in Fig. 5-30(b) is strongly coupled with the exciton around 453 nm in Fig. 5-30(c), and results in the strongly enhanced plexcitonic peaks around 448 nm and 504 nm in Fig. 5-31(b). Vacuum Rabi splitting, which is caused by strong exciton-plasmon coupling, arises from the quantum interaction between a molecule and a single photon. In the multi-molecule-plasmon system, strong optical response of molecules become very important, and enlarges the spectral splitting[57]. In this regard, both single-molecule quantum interaction and multiple-molecule optical interaction will result in the giant splitting in TERS, and the latter dominating the former[57]. The optical response of molecules will greatly affect the highly localized plasmon field in the tip-substrate nanocavity, and thus significantly change the exciton-plasmon interaction and the spectral splitting characteristic. In our studied multi-molecule-plasmon system, the dielectric function of the BL Borophene is highly dispersive, indicating sensitive optical response. The dielectric functions of the BL Borophene and Ag around 488 nm are of negative real part and manifest plasmonic properties, revealing strong exciton resonance and plasmon resonance. The optical match between the two types of resonance contributes to the strong exciton-plasmon coupling and giant Rabi splitting in TERS in spite of the large difference of $|\omega_p - \omega_e| = 0.248$ eV. The detailed physical mechanism can be seen in Fig. 5-35(a).

Strong plasmon-exciton interaction manifested by Rabi splitting can be used to improve plasmon-enhanced spectroscopy (i.e., Raman and fluorescence). We can improve the CARS s by using the Rabi splitting plexciton with higher energy, as is shown in Fig. 5-35(b). The calculated results demonstrate that the CARS signal can be enhanced by 10^{14} excited by 488 nm laser. The Raman peaks of BL borophene are below 700 cm^{-1}[54], so the CARS peaks of the target molecule enhanced by plexciton are background-free from BL borophene. The fluorescence enhancement by the splitting plexciton with lower energy can reach up to 10^5, which is 4.3 times higher than that by plasmon itself, see Fig. 5-35(c). And furthermore, plexciton enhanced Raman

($|E|^4$) can be reach up to 10^{12} (at around 470 nm), which is two more orders than that by plasmon alone. Strongly coupled emitter can also influence the mode of nanoresonators. The most apparent change in our studied system is splitting mode to plexciton modes with higher intensity once the structure is strongly coupled.

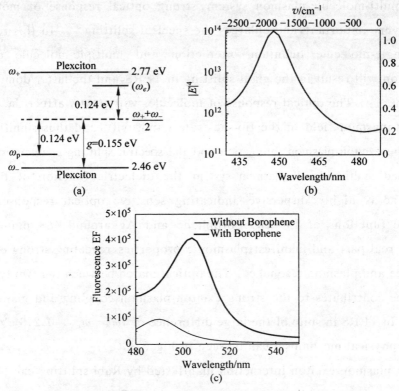

Fig. 5-35 Physical mechanism and applications of Rabi splitting plexciton. (a) The physical mechanism of Rabi splitting plexciton, (b) Rabi splitting plexcitonic peak at higher energy for the CARS spectrum, (c) Rabi splitting plexcitonic peak at lower energy for the of fluorescence
(For colored figure please scan the QR code on page 1)

We further studied the plexciton and plasmon fields and their gradient fields at the excitation wavelengths of 448 nm, 488 nm, and 504 nm. The results of Fig. 5-36 demonstrate that the electromagnetic field of plexcitonic peaks around 448 nm and 504 nm and plasmonic peaks at 488 nm are highly localized in the nanocavity between the tip and substrate, which can be considered as LPER.

CHAPTER 5 Surface Plasmons

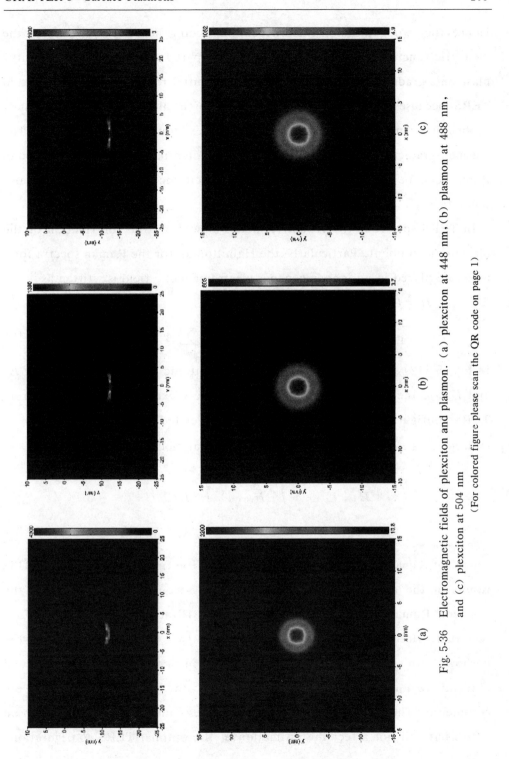

Fig. 5-36 Electromagnetic fields of plexciton and plasmon. (a) plexciton at 448 nm, (b) plasmon at 488 nm, and (c) plexciton at 504 nm

(For colored figure please scan the QR code on page 1)

Hence, the strong plasmon-exciton coupling, being negative real part of the dielectric function, can result in the LPER. Apart from plexcitonic intensity, plasmonic gradient intensity also plays an essential role along the substrate in TERS because it provides a plexcitonic force along the substrate and, probability, can promote the propagation of plexciton along the substrate. The plasmon-gradient effect in TERS provides potential opportunity multipole-Raman can be observed[45-48]. Thus, the plexcitonic gradient fields are also theoretically investigated.

In TERS spectrum, the IR-active modes arise from the contribution of the plexcitonic gradient. Particularly, the Hamiltonian for the Raman spectra for a molecule placed in an inhomogeneous electromagnetic field is written as[47],

$$H = H_0 + H_1 + H_2 + \cdots$$
$$= \alpha_{\alpha\beta}E_\beta E_\alpha + \frac{1}{3}\sum_{\alpha\beta\gamma} A_{\alpha,\beta\gamma} \nabla E_{\beta\gamma}E_\alpha + \frac{1}{3}\sum_{\alpha\beta\gamma} A_{\gamma,\alpha\beta} \nabla E_{\alpha\beta}E_\gamma + \cdots. \quad (5\text{-}112)$$

In Eq. (5-112), the $\nabla E_{\alpha\beta}$ and $\nabla E_{\beta\gamma}$ are the plexcitonic field gradient, and the E_α and E_β are the plexcitonic field, indicating the plexcitonic contribution from TERS configuration. The $\alpha_{\alpha\beta}$ and $A_{\alpha,\beta\gamma}$ are related to the polarizability of molecule. The intensity of molecular vibrational modes is given by[47]

$$I = I_1 + I_2 + \cdots = (H_0 + H_1 + H_2 + \cdots)^2 = (H_0 + 2H_1)^2$$
$$= \langle j | H_0 | i \rangle^2 + 4\langle j | H_0 | i \rangle\langle j | H_1 | i \rangle$$
$$= (\alpha_{\alpha\beta}E_\beta E_\alpha)^2 + 4 \times \alpha_{\alpha\beta}E_\beta E_\alpha \times \frac{1}{3}\sum_{\alpha\beta\gamma} A_{\alpha,\beta\gamma} \nabla E_{\beta\gamma}E_\alpha, \quad (5\text{-}113)$$

where the Raman shift is ignored, and $H_1 = H_2$. The first term in the Eq. (5-113) results in the dipole Raman modes, and the second term account for the gradient Raman modes (corresponding to IR-active modes) in TERS spectrum. Here, we only considered the first two terms, and neglected the unshown terms because of their small value even in the high plexcitonic field gradient regions. In our calculations, only the plexcitonic terms were considered. The polarizability term of molecules complicates the discussion on TERS but will not affect the estimation of the optimal TERS configuration, because the TERS spectrum is dominantly determined by the plexcitonic term.

Figure 5-37 reveals that the plexcitonic gradient is highly localized in the nanocavity. The reason why both plexciton and the plexcitonic gradient around 448 nm is of highly LPERs may be that the plasmon is strongly coupled with exciton (the real part of dielectric functions perpendicular to the surface of BL borophene are negative), which corresponds to the topical strong light-matter interaction. However, the plexcitonic gradient at the direct center under the tip is weakest, compared with the adjacent part of the center. The plexcitonic gradient depends upon the rapidity of the variation for the plexcitonic field. Our calculations reveal that the plexcitonic fields are highly localized at the direct center under the tip. In such area, the variation of plexcitonic field is very slow, and thus leads to the weakest gradient at the direct center under the tip.

In summary, in this section, Fano-resonant propagating plexcitons and Rabi-splitting local plexcitons have been observed in STM-TERS system, and the physical mechanism has been interpreted theoretically. Fano resonance occurs when the broad LSPR mode of TERS has coupled to the narrow resonance mode of BL borophene. The transition from propagating plexcitonic modes to local plexcitonic modes can be observed in the Fano resonance region by changing the excitation wavelength, demonstrating a strong dependence of plexcitonic mode on the excitation wavelength. The plasmonic properties of TERS and BL borophene around 488 nm arise from strong plasmon-exciton coupling interactions, resulting in two stronger plexcitonic peaks. We can selectively enhance CARS, TERS, and fluorescence signals using the Rabi splitting plexciton with higher and lower energy. In addition, both the plexcitonic and its gradient fields in the strong-coupling region nm are LPERs because of the topical strong light-matter interaction.

In future, twisted $MoSe_2$ Bilayers[58], and the twist angle superlattice of BL borophene[59] as substrate in TERS can be further investigated.

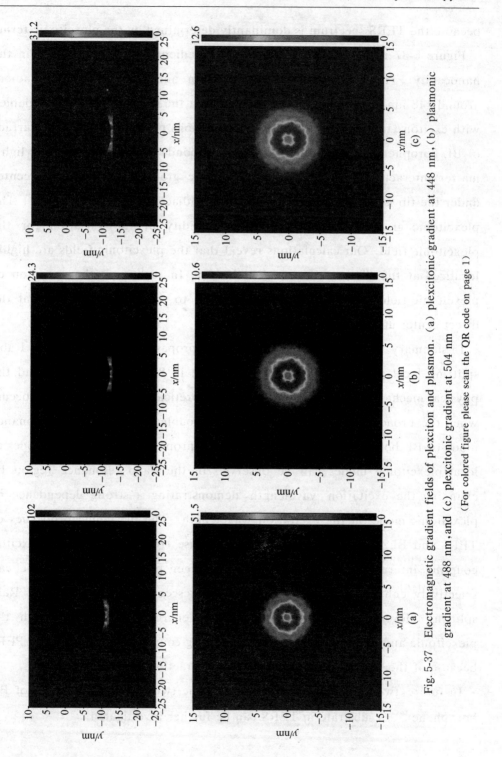

Fig. 5-37 Electromagnetic gradient fields of plexciton and plasmon. (a) plexcitonic gradient at 488 nm, (b) plasmonic gradient at 488 nm, and (c) plexcitonic gradient at 504 nm

(For colored figure please scan the QR code on page 1)

5.3.6 Plexciton revealed in experiment

In this section, we experimentally study the exciton-plasmon coupling of monolayer MoS_2-Ag nanoparticles hybrid system, using the transmission spectroscopy and femtosecond transient absorption spectroscopy[52]. The plasmon-exciton coupling can be well manipulated by the local surface plasmon resonance (LSPR), where the LSPR can be well controlled by monitoring the size of Ag nanoparticles. Optimization of coupling interaction through monitoring the overlap between LSPR peak and the excitonic states A and B, for example, has been demonstrate by the investigating the transmittance. Besides, the pump-probe femtosecond transient absorption reveals the dynamic nature of plasmon-exciton coupling interaction and especially the enlarged lifetime of plasmon-exciton coupling interaction. Our results can promote the deeper understanding and highlight the unique merits of plasmon-exciton coupling.

For this study, large-area monolayer MoS_2 was synthesized on sapphire substrates using chemical vapor deposition (CVD), see SEM image in Fig. 5-38(a). The high-resolution TEM (HR-TEM) image of as-fabricated monolayer MoS_2 is shown in Fig. 5-38(b), where the inset indicates that the MoS_2 is monocrystalline. From the UV-Vis transmission spectrum of MoS_2 on quartz (see Fig. 5-38(c)), we can observe that the transmittance of MoS_2 is 92.42% at 532 nm, which confirms the high quality of the as-fabricated monolayer MoS_2. The absorption peaks at 637.6 nm and 595.1 nm result from spin-orbit splitting of the valence band, both corresponding to excitonic transitions at the K-point of the Brillouin zone[60]. While the absorption peak at 425.3 nm corresponds to transitions involving bands located between the K- and G-points and has been attributed to van Hove singularities in the density of states[60]. The calculated UV-Vis transmission spectrum of MoS_2 on quartz in Fig. 5-38(d) strongly supports the experimental results in Fig. 5-38(c). The interaction

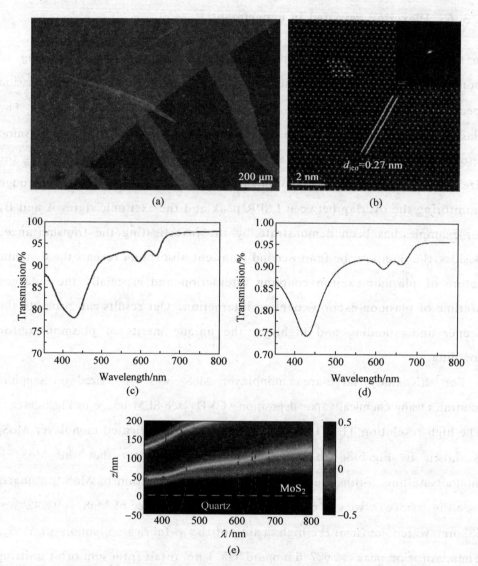

Fig. 5-38 (a) SEM image of synthesized MoS_2, (b) TEM image of synthesized MoS_2, (c) the transmission spectrum of synthesized monolayerMoS_2, (d) the calculated transmission spectrum of monolayer MoS_2, (e) the near electric field distribution of MoS_2 above the substrate, (f) the PL spectrum of monolayer MoS_2, and (g) the Raman spectrum of monolayer MoS_2

(For colored figure please scan the QR code on page 1)

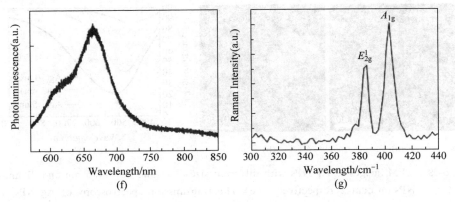

Fig. 5-38(Continued)

between monolayer MoS_2 and the substrate at three absorption peaks can be seen from Fig. 5-38(e), from which we can confirm that there is a very strong interaction around 425 nm, and the strongest electric field is about 100 nm above the substrate; while for the two weak absorption peaks as 637.6 nm and 595.1 nm, the interactions are weak, and the position with strongest electric field is about 150 nm above the substrate. Figure 5-38(f) is the PL spectrum of monolayer MoS_2, and the Raman spectrum is shown in Fig. 5-38(g) where the two Raman modes of E_{2g}^1 (385 cm^{-1}) and A_{1g} (406 cm^{-1}) confirm the 2-H structure of as-fabricated MoS_2[61].

The Ag nanoparticles with different sizes were synthesized on different quartzes for the measurements of femtosecond transient absorption spectroscopy, and their AFM images can be seen from Fig. 5-39(a) and (b) respectively. For Ag nanoparticles with smaller size cannot be easy be measured with AFM. The transmission spectra of Ag nanoparticles with different sizes are shown in Fig. 5-39(c), where we can find that with the size increase of Ag nanoparticles, the absorption intensities are significantly enhanced, and accompanied by the red shift of surface plasmon resonance (SPR) peak.

To reveal the strong exciton-plasmon coupling interaction, the UV-Vis transmission spectra of monolayer MoS_2/Ag NPs hybrids on quartz are measured (Fig. 5-40(a)~(c)), where the size of Ag nanoparticles are 6.1 nm,

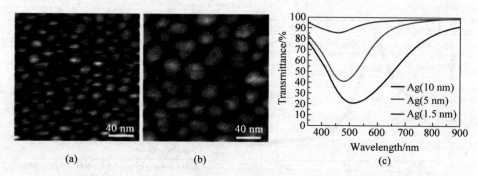

Fig. 5-39 AFM images of Ag NPs with different size: (a) and (b) 14.5 nm and 21-nm-Ag NPs on quartz, respectively. (c) The transmission spectroscopy of Ag NPs with different sizes on quartz, respectively

(For colored figure please scan the QR code on page 1)

14.5 nm and 21 nm, respectively. According to the transmission spectra, we can conclude that the LSPR peaks of Ag NPs with different size have a strong impact on the transmission peak of plasmon-exciton hybrids (plexciton), as the peak shifted and intensities enhanced. Compared with two exciton peaks of monolayer MoS_2, electromagnetically induced transparency dips can be found at two peaks of monolayer MoS_2-Ag nanoparticles hybrids, and this Fano interference is resulting from coherent dipole-dipole interaction between the discrete exciton states and broad plasmon resonance states. The electric field distribution of monolayer MoS_2/Ag NPs hybrid (the size of Ag nanoparticles is 21 nm) with incident laser of 490 nm, 617 nm, 658 nm (three strong absorption peaks) respectively can be seen from Fig. 5-40(d)~(f). The strongest electric fields are confined at the corners of Ag NPs, and in the meantime there is strong coupling electric field between Ag NPs, as well as at the interface in hybrid system. According to the differences among Fig. 5-40(d)~(f), we can conclude that the strength of electric fields strongly dependents on the match between the applied laser and LSPR peak of the hybrid system. To be more specific, when the wavelength of applied laser increases and away from the strongest absorption peak of hybrid system, the overall intensity of electric field is decreased, and the intensity of electric field which induced by coupling interaction is significantly dropped, especially between Ag nanoparticles.

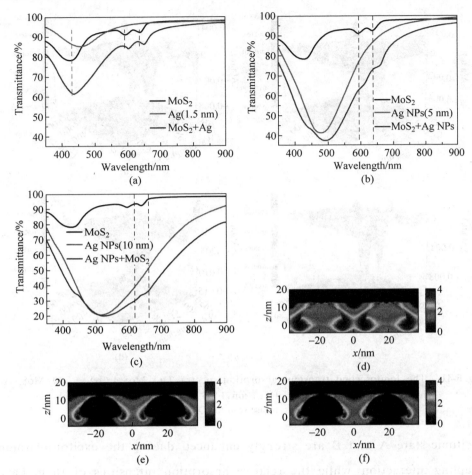

Fig. 5-40 (a)～(c) The transmission spectra of Ag NPs, monolayer MoS₂ and monolayer MoS₂-Ag NPs with different size of Ag NPs. (d)～(f) The electric field distribution of monolayer MoS₂-Ag NPs at 490 nm, 617 nm, 658 nm, where the size of Ag nanoparticles is 21 nm

(For colored figure please scan the QR code on page 1)

To further reveal the ultrafast dynamics of plasmon-exciton coupling interactions, the femtosecond transient absorption spectra of monolayer MoS_2-Ag NPs hybrid system have also been measured. According to Fig. 5-41, the absorption peaks that assigned to excitonic state A and B of MoS_2 are almost fixed at 637 nm and 595 nm. Since the plasmonic absorption of Ag nanoparticles with smallest size (6.1 nm) is far away from these two excitonic states, comparing Fig. 5-41(a) and (b), we can see that the overall intensities of these two

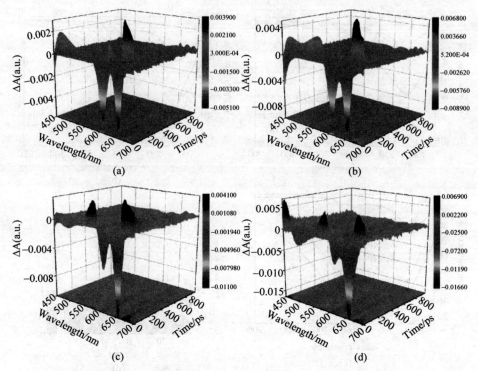

Fig. 5-41 The femtosecond transient absorption spectra, (a) MoS_2, (b) ~ (d) MoS_2-Ag hybrid with different sizes of 6.1 nm, 14.5 nm, and 21 nm, respectively
(For colored figure please scan the QR code on page 1)

excitonic state A and B are strongly enhanced due to the exciton-plasmon coupling interaction, while the relative absorption intensities of them have remained largely unchanged. With the size increase of Ag nanoparticles, the plasmonic absorption peaks are gradually red-shifted towards the two excitonic state A and B, and lead to a significant enhancement in the absorption intensity of excitonic state A at 637 nm, rather than the excitonic state B at 595 nm. The selective enhancement is apparent in Fig. 5-41(d), which can be interpreted with the relationship between plasmon-enhanced fluorescence. From Fig. 5-38 (f), the PL mainly is contributed from the absorption at excitonic state A. With the increase of size of Ag nanoparticles, the PL of MoS_2 can be strongly enhanced (Fig. 5-40 in Ref. [62]), due to strongly selectively enhanced absorption of excitonic state A. The plasmon-enhanced fluorescence (PEF) can be written as[63]

$$\sigma_{PEF}(\lambda_L,\lambda,d_{av}) = |M_1(\lambda_L,d_{av})|^2 |M_2(\lambda,d_{av})|^2 \times \frac{\sigma_{FL}(\lambda_L,\lambda)}{|M_d(\lambda,d_{av})|^2},$$

(5-114)

where $|M_1|^2$ and $|M_2|^2$ are enhanced factors related with the coupled effect between SPR with the incident wave and fluorescence emissions, respectively. $\sigma_{FL}(\lambda_L,\lambda)$ and $\sigma_{PEF}(\lambda_L,\lambda,d_{av})$ are the scattering cross-sections of MoS$_2$ in the free space and local field, respectively, where λ is the emission wavelengths, λ_L is the excitation wavelengths, d_{av} is defined as the average distance of fluorescent molecule from the metal substrate, $|M_{ad}|^2$ is the factor that shows the energy transfer. According to Fig. 5-38(c) and (f), the plasmon mainly enhanced the absorption and fluorescence at excitonic state A. Also supported by plasmon enhanced absorption in Fig. 5-40(c) and plasmon enhanced PL in Fig. 5-40 in Ref. [62], where the plasmon-enhanced PL are mainly contributed from excitonic state A.

To reveal the coupling interaction between plasmon and exciton for the monolayer MoS$_2$-Ag NPs hybrid system in detail, the 2D femtosecond ultrafast transient absorption spectra are presented, see Fig. 5-42. Figure 5-42(a) is the femtosecond transient absorption spectroscopy of monolayer MoS$_2$, where the pump laser is 400 nm with power of ~3 mW. The conditions of pump laser are designed to excite the indirect bandgap without the dissociation of excitons. To be more specific, according to Fig. 5-38(c), we can find out that the absorption rate of monolayer MoS$_2$ at 400 nm is 20%, and hence, the laser frequency at 400 nm is suitable for resonantly exciting the indirect bandgap (C excited state at 425 nm). Besides, we can get that the distance between excitons is about 3 nm, while previous researchers report that Bohr radius of excitons at C excited state is 0.60 nm, indicating that there is no spatial overlap between excitons, and the nearest distance between excitons is 1.8 nm. We can conclude that the dynamic process of exciton-exciton interaction is the intra-exciton interaction, which means the hole-electron excitations and recombination are within individual

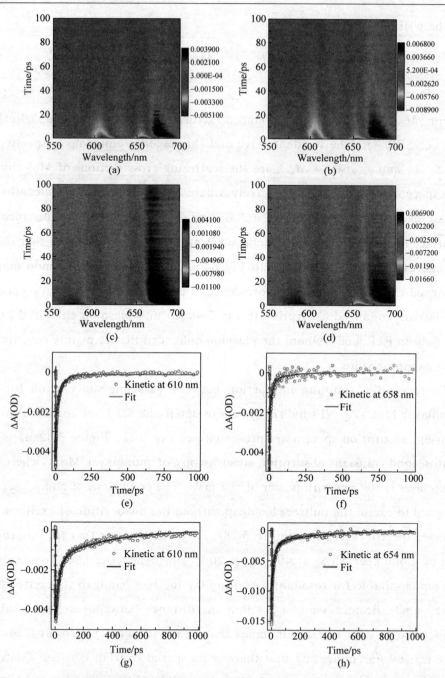

Fig. 5-42 The 2D femtosecond transient absorption spectra of (a) monolayer MoS_2, and (b)~(d) monolayer MoS_2-Ag NPs with different size of Ag NPs as 6.1 nm, 14.5 nm, and 210 nm, respectively. The fitted transient absorption spectra of (e)~(f) monolayer MoS_2, and (g)~(h) monolayer MoS_2-Ag NPs hybrid where the size of Ag NPs is 21 nm
(For colored figure please scan the QR code on page 1)

exciton, instead of inter-exciton interactions. In conclusion, we can confirm that there is no carrier at this condition for monolayer MoS_2-Ag NPs hybrid. Comparing Fig. 5-42(a) and (b)~(d), we can find out that the life times of these two excitonic states A and B, qualitatively, have been significantly influenced by the surface plasmon resonance of Ag nanoparticles.

To reveal the internal mechanism of plasmon-exciton interaction, the life times of these two excitonic states A and B for monolayer MoS_2 and the hybrid system are quantitatively fitted, respectively, as shown in Fig. 5-42(e)~(h). Here, for the hybrid system, we only fit the system when Ag nanoparticles are of largest size (size is 21 nm), due to the strongest LSPR effect, and the fitted data can be seen from Table 5-2. In the fitting procedure, three exponential functions were used, which has been confirmed that is a reasonable method[67-68]. For monolayer MoS_2 and monolayer MoS_2-Ag nanoparticles hybrid, there are three dynamics processes for A and B exciton states, which are the Auger scattering, electron-electron interaction, electron-phonon interaction. Among those, the fastest lifetimes (the Auger scattering) of B and A excitonic states are hardly influenced by the plasmon that induced by Ag NPs, see Fig. 5-43. The lifetimes of electron-electron interaction for monolayer MoS_2 are 2.46 ps and 0.85 ps for the B and A excitonic states, respectively; while in the case of hybrid system, the lifetimes are extended to 15 ps and 6 ps. The lifetimes of electron-photon interaction for monolayer MoS_2 are 30 ps and 21 ps for excitonic state B and A; while for the hybrid system, the lifetimes are significantly enlarged by plasmon up to 253 ps and 62 ps, respectively. The plasmon hot electrons with high kinetic energy transfer to monolayer MoS_2, and interaction between transferred plasmonic hot electrons and phonon of monolayer MoS_2 can result in the hot of phonon of monolayer MoS_2. The interaction between hot electrons and hot phonon can enhance the lifetime of electron-phonon interaction. Based on these experiments, the merit of exciton-plasmon coupling interaction in enlarging the lifetime of excitons is self-evident. Besides, as shown in Fig. 5-41, the intensity of steady absorption of A excitonic state is strongly enhanced by plasmon, so it's reasonable that the lifetime of electron-electron interaction and electron-photon interaction for the A excitonic state in Fig. 5-43 can be significantly enlarged with plasmon enhancement.

Table 5-2 The life times of excitonic states A and B of monolayer MoS_2 and monolayer MoS_2-Ag NPs hybrid system

Materials	PEAK/nm	Parameters	Lifetime/ps	Error
MoS_2	610	$T_0 = 0.08505$ ps	—	—
		$A_1 = 0.0264$	$t_1 = 0.15 \pm 0.035$	23%
		$A_2 = -0.0025$	$t_2 = 2.46 \pm 0.497$	20.2%
		$A_3 = -0.00275$	$t_3 = 29.99 \pm 2.92$	9.7%
	658	$T_0 = 0.1509$ ps	—	—
		$A_1 = 0.0104$	$t_1 = 0.24 \pm 0.0733$	30.5%
		$A_2 = -0.0066$	$t_2 = 0.8498 \pm 0.263$	30.9%
		$A_3 = -0.00241$	$t_3 = 21 \pm 3.1$	14.7%
Ag + MoS_2	610	$T_0 = 0.06718$ ps	—	—
		$A_1 = 0.023$	$t_1 = 0.1522 \pm 0.0349$	22.9%
		$A_2 = -0.00299$	$t_2 = 15.31 \pm 1.52$	9.93%
		$A_3 = -0.00118$	$t_3 = 252.6 \pm 63.6$	25.2%
	654	$T_0 = 0.3845$ ps	—	—
		$A_1 = -0.0127$	$t_1 = 0.27 \pm 0.0635$	23.5%
		$A_2 = -0.00474$	$t_2 = 6.02 \pm 1.04$	17.3%
		$A_3 = -0.00238$	$t_3 = 61.53 \pm 14.1$	22.9%

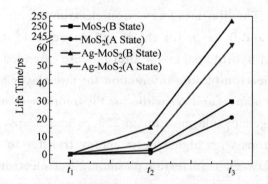

Fig. 5-43 The life times of three dynamics processes of monolayer MoS_2, and monolayer MoS_2-Ag NPs for excitonic states A and B respectively, where the size of Ag NPs is 21 nm. Note that the electron-coupled plasmon-phonon mode scattering[64] and the intrinsic electron-phonon scattering[65] of monolayer MoS_2 cannot been observed with ultrafast transient absorption spectroscopy, the reason may be that the scattering process is too weak to be fitted. To observe the electron-coupled plasmon-phonon mode scattering[64] and the intrinsic electron-phonon scattering[65] of monolayer MoS_2, ultrafast transient THZ spectroscopy may be a better method. With ultrafast transient THZ spectroscopy, in graphene/h-BN van der Waals heterostructures, new hybridized plasmon-phonon polariton mode has been reported[66]

(For colored figure please scan the QR code on page 1)

The mechanism of the plasmon-exciton coupling is that the incident photons first excite the local surface plasmons in Ag nanoparticle, which then couple to excitonic states in MoS_2, since the life time of collective electron oscillation (CEO) of plasmon is around 150 fs[67], but the lifetime of MoS_2 is in the picoseconds. Also, the lifetime of MoS_2 in the picoseconds is strongly enlarged by surface plasmon resonance. In our case, new modes, such as plexciton, may not be emerged due to hybridization of the surface plasmons and excitonic states, since the absorption peak does not split in Fig. 5-41, and only the intensity is strongly enhanced.

In summary, in this section, the complementary optical properties of TMDCs and plasmonic metal nanomaterials make them attractive components in developing probability and efficiency of catalytic reactions. When combined into hybrid system, the coupling interaction between plasmon and exciton results in advanced but unclear properties. Hence, in this paper, we focus on revealing the nature of plasmon-exciton coupling interaction, with transmission spectroscopy and pump-probe femtosecond transient absorption spectroscopy, where the degree of coupling interaction between plasmon and exciton are well manipulated. Among all these experiments, the femtosecond studies of exciton-plasmon interaction can vividly reveal the dynamic processes in MoS_2-Ag nanoparticles hybrid system, which is essential in advancing the whole field. In conclusion, our result is helpful to understand the nature of coupling interaction, and promote the applications of plasmon-exciton coupling in different fields.

5.3.7 LSPs in coupled metallic NPs: many-body

After the detailed discussions on two-body system, we would extend it to the many-body system, which is consisted with multiple MNPs arranged as a linear chain[69]. For simplicity, we assume that all the MNPs are the same with the same neighboring distance r_0. The schematic is shown in Fig. 5-44. We shall clarify here that some of the quantities are marked with letters different from Section 5.3.2.

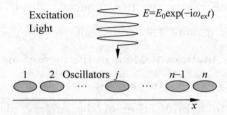

Fig. 5-44　Schematic of the coupling classic harmonic oscillator model, which is constituted by n oscillators arranged along x-axis. Each ellipse stands for an individual oscillator that oscillates along x-axis. The x-polarized excitation light with electric field amplitude E_0 and frequency ω_{ex} illuminates the system

The x-polarized incident light excites the system with frequency ω_{ex} and electric field amplitude E_0. Due to the fact that the coupling strength g decreases rapidly with the increase of r_0, we only consider the interaction between the neighboring particles. Therefore, the dynamical equations of these oscillators should be in this form:

$$\begin{cases} \ddot{x}_1 + 2\beta_0 \dot{x}_1 + \omega_0^2 x_1 - \gamma \dot{x}_2 - g^2 x_2 = K_1 \exp(-i\omega_{ex} t), \\ \vdots \\ \ddot{x}_j + 2\beta_0 \dot{x}_j + \omega_0^2 x_j - \gamma \dot{x}_{j-1} - g^2 x_{j-1} - \gamma \dot{x}_{j+1} - g^2 x_{j+1} = K_j \exp(-i\omega_{ex} t), \\ \quad\quad (\text{for } 2 \leqslant j \leqslant n-1) \\ \vdots \\ \ddot{x}_n + 2\beta_0 \dot{x}_n + \omega_0^2 x_n - \gamma \dot{x}_{n-1} - g^2 x_{n-1} = K_1 \exp(-i\omega_{ex} t). \end{cases}$$

(5-115)

Here, $x_j(t)$, $\dot{x}_j(t)$, and $\ddot{x}_j(t)$ are the displacement relative to equilibrium position, velocity, and acceleration of the jth oscillator, respectively, $K_j = -eE_j/m_e$, E_j is the external electric field felt by the jth oscillator, and e is the charge of electron.

Firstly, we deal with the scattering properties. Assume that $x_j(t) = A_j \exp(\alpha t)$ $(j = 1, 2, \cdots, n)$ are the solutions of Eq. (5-115). Define $B = -\gamma\alpha - g^2$, $C = \alpha^2 + 2\beta_0 \alpha + \omega_0^2$, and

CHAPTER 5　Surface Plasmons

$$D = \begin{pmatrix} C & B & C & & & & \\ B & C & B & \cdots & & & \\ & B & C & & & & \\ \vdots & & & \ddots & & & \vdots \\ & & & & C & B & \\ & & & \cdots & B & C & B \\ & & & & & B & C \end{pmatrix}_{n \times n}, \qquad (5\text{-}116)$$

where $D_{j_1 j_2} = C$ for $j_1 = j_2$, $D_{j_1 j_2} = B$ for $|j_1 - j_2| = 1$, and $D_{j_1 j_2} = 0$ for other cases. Also define:

$$\begin{cases} X = (A_1, \cdots, A_j, \cdots, A_n)^T, \\ K = (K_1, \cdots, K_j, \cdots, K_n)^T. \end{cases} \qquad (5\text{-}117)$$

After substituting $x_j(t)$ into Eq. (5-115), the amplitudes should satisfy the following equation:

$$DX = K, \qquad (5\text{-}118)$$

where the solution of α is $\alpha = -i\omega_{ex} t$. The solutions of the amplitudes are

$$X = D^{-1} K, \qquad (5\text{-}119)$$

where D^{-1} is the inverse of matrix D. The elements $D_{j_1 j_2}^{-1}$ can be written as[70-71]

$$D_{j_1 j_2}^{-1} = \frac{2}{n+1} \sum_{k=1}^{n} \frac{\sin\dfrac{j_1 k\pi}{n+1} \sin\dfrac{j_2 k\pi}{n+1}}{C + 2B\cos\dfrac{k\pi}{n+1}}. \qquad (5\text{-}120)$$

The total far field emission of the electric field is proportional to the accelerations of the oscillators[14-16]:

$$\ddot{x}(t) = \sum_{j=1}^{n} \ddot{x}_j(t) = (-i\omega_{ex})^2 \left(\sum_{j=1}^{n} A_j\right) \exp(-i\omega_{ex} t) = -\omega_{ex}^2 A^{(n)} \exp(-i\omega_{ex} t).$$

$$(5\text{-}121)$$

Actually, $A^{(n)}$ satisfy:

$$A^{(n)} = \sum_{j=1}^{n} A_j = \sum_{j=1}^{n} X_j = \sum_{j_1=1}^{n} \sum_{j_2=1}^{n} (D^{-1} K)_{j_1 j_2}. \qquad (5\text{-}122)$$

We give out the solutions of the amplitudes $A^{(n)}$ for $n = 1, 2, \cdots, 6$ as examples in the condition of $E_j = E_0$, thus $K_j = K_0 = -eE_0/m_e$ (for $j = 1, 2, \cdots, n$):

$$\begin{cases} A^{(1)} = \dfrac{1}{C} K_0, \quad A^{(2)} = \dfrac{2}{B+C} K_0, \quad A^{(3)} = \dfrac{4B - 3C}{2B^2 - C^2} K_0, \quad A^{(4)} = \dfrac{2B - 4C}{B^2 - BC - C^2} K_0, \\ A^{(5)} = \dfrac{B^2 + 8BC - 5C^2}{3B^2 C - C^3} K_0, \quad A^{(6)} = \dfrac{2(2B^2 + 2BC - 3C^2)}{B^3 + 2B^2 C - BC^2 - C^3} K_0. \end{cases}$$

(5-123)

Define $A'^{(n)} = -\omega_{ex}^2 A^{(n)}$, and employ the Fourier transform, thus obtaining the elastic emission spectrum:

$$I_{ela}(\omega) = |A'^{(n)}|^2 \sqrt{2\pi} \delta(\omega - \omega_{ex}). \tag{5-124}$$

Therefore, the white light scattering spectra should be given as:

$$I_{sca}(\omega) = I_{ela}(\omega_{ex} \to \omega) = \sqrt{2\pi} |A'^{(n)}|^2 (\omega_{ex} \to \omega). \tag{5-125}$$

Secondly, we deal with the PL properties. As our previous work demonstrates[15], PL term origins from the general solutions of the homogeneous linear equations:

$$DX = 0. \tag{5-126}$$

The necessary and sufficient conditions for the existence of nontrivial solutions of Eq. (5-126) is that the determinant of D is zero:

$$\det(D) = \frac{z_+^{n+1} - z_-^{n+1}}{z_+ - z_-} = 0, \tag{5-127}$$

where $z_\pm = \dfrac{1}{2}(C \pm \sqrt{C^2 - 4B^2})$. Eq. (5-49) determines the solutions of α.

Obviously, there are $2n$ solutions for Eq. (5-49). We can rewrite the solutions as $\alpha_{k\pm} = -\beta_k \pm i\omega_k$, $k = 1, 2, \cdots, n$. Due to the large difference from the excitation frequency, we omit the solutions of "α_{k+}" as our previous work does[14-15]. Therefore, the total solutions for Eq. (5-115) can be assumed and solved by

$$x_j(t) = S_j \exp(-i\omega_{ex} t) + \sum_{k=1}^{n} P_{jk} \exp(\alpha_{k-} t), \tag{5-128}$$

with initial conditions: $x_j(0) = 0, \dot{x}_j(0) = 0, \ddot{x}_j(0) = K_0, \cdots, \dfrac{d^n x_j}{dt^n}(0)$, for $j = 1$, $2, \cdots, n$. Here, S_j is the amplitude of the elastic term (scattering) of the jth oscillator; P_{jk} is the amplitude of the kth inelastic term (PL) of the jth oscillator. After considering the PL term for the solutions, the total far field emission of the electric field can be written as

$$\ddot{x}(t) = \sum_{j=1}^{n} \ddot{x}_j(t) = (-i\omega_{ex})^2 \sum_{j=1}^{n} S_j \exp(-i\omega_{ex} t) +$$

$$\sum_{j=1}^{n}\sum_{k=1}^{n} \alpha_{k-}^2 P_{jk} \exp(\alpha_{k-} t). \qquad (5\text{-}129)$$

Define $P'_k = \sum_{j=1}^{n}\sum_{k=1}^{n} \alpha_{k-}^2 P_{jk}$, employing Fourier transform and Fermi-Dirac distribution, the total PL spectrum of the system can be written as

$$I_{PL}(\omega) = \sum_{k=1}^{n} |P'_k|^2 \frac{1-\exp(-2\beta_k t_0)}{2\beta_k t_0} \frac{\beta_k}{(\omega-\omega_k)^2 + \beta_k^2} \times$$

$$\frac{1}{1+\exp[(\hbar\omega - \hbar\omega_f)/(k_B T)]}. \qquad (5\text{-}130)$$

Here, t_0 is the effective interaction time between the excitation light and the oscillators, $\hbar$ is the reduced Planck constant, ω_f is the so-called chemical potential, k_B is Boltzmann's constant, and T is the temperature. For $n=1$, the solution degenerates into the one of an individual particle, which has been discussed in Ref. [15]. For $n=2$, the solutions degenerate into the dimer case, which has been discussed in Ref. [14].

Figure 5-45(a)~(c) show the normalized scattering spectra of the chains with different particle number n varying with effective free electrons number N. The coupling strength is $g = 1.3$ eV. The primary LSPR peak redshifts as n increases. When $n \geq 3$, there exist other peaks at blue side of the primary peaks with small amplitudes for each case. Figure 5-45(d) shows the LSPR peak positions as a function of N for different cases. It indicates that for small n, the peaks almost stay unchanged as the increase of N. However, for large n, the peak positions (unit: nm) decrease at about $N = 10^6$. Notice that in Fig. 5-45, we keep g unchanged, and according to Eq. (5-40), we have the relation: $\gamma^3 = g^4/\kappa$. When N is small, γ is much less then g, thus influencing the interaction parts little. However, when N increases, κ decreases, resulting in the increase of γ, which influences the interaction parts greatly if γ is comparable with or even larger than g. Hence, the peak positions are influenced by N. Figure 5-45(e) shows the full width at half maximum (FWHM) of different cases as a function of N. For $n \geqslant 2$, the FWHM decreases first and then increases as N increases. The minimums of the FWHM occur at about $N = 10^5$, resulting in narrow shapes of the spectra as shown in Fig. 5-45(c). These minimums of the FWHM are explained later, together with the decrease of the FWHM in Fig. 5-45(d).

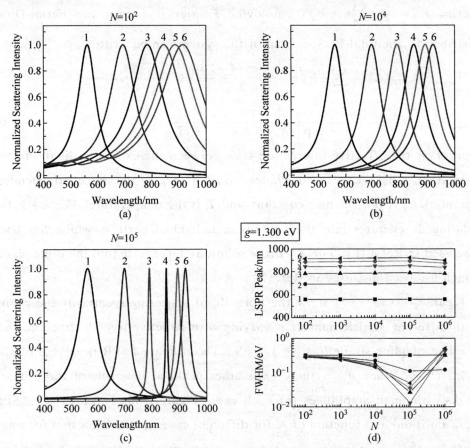

Fig. 5-45 (a)～(c) Normalized scattering spectra of the nanoparticle chains varying with effective free electron number N, i. e., $N = 10^2$, $N = 10^4$, and $N = 10^5$, respectively, calculated from Eq. (5-125). (d) LSPR peak position of the chains as a function of N. (e) The FWHM of the chains as a function of N. The spectra and curves are numerically labeled according to particle number n except for (e). Black, red, blue, green, cyan, and purple stand for n equaling from 1 to 6, respectively. Here, the coupling strength is kept as $g = 1.300$ eV

(For colored figure please scan the QR code on page 1)

To evaluate a practical value of N, we should consider two effects. Firstly, the effective free electron number N should be smaller than the total free electron number N_{tot} due to the screening effect of metal. For example, for a gold nanosphere with diameter of 64 nm, N_{tot} is evaluated as about 8.1×10^6 (assume that each gold atom has one free electron). Because of the screening effect, we assume that only the electrons at the edge of the particle participate the oscillating, and the effective depth is evaluated as about 0.2 nm, the half lattice parameter (about 0.4 nm) of gold, that is, the effective free electron number N is evaluated as 1.5×10^5, which is an order of magnitude smaller than N_{tot}. Secondly, for a practical system, the size of the nanoparticles should not be omitted. This indicates that when calculating the coupling strength g or γ, each nanoparticle has numerous oscillators distributing at the edge of the particle. Strictly, the coupling effective distance r_0 is not equal to the distance between the centers of the particles. If r_0 is still set as the distance between the centers, relatively, to gain a more reasonable coupling strength, the number N should be modified according to the shapes and sizes of the nanoparticles. However, this is not easy even if the coupled two particles are spheres, which are the most symmetric geometries. Hence, as an approximation, also for convenient and practical purpose, r_0 is evaluated to be the distance between the centers of the particles. Summarizing the above two reasons, i.e., screening effect and size effect, N is not the total free electron and should be substituted by the effective free electron number.

Figure 5-46(a) and (b) show the normalized scattering spectra of the chains with different n varying with coupling strength g. Here, $N = 1.5 \times 10^4$. The same as Fig. 5-45(a) ~ (c), the primary LSPR peak red-shifts as n increases. Obviously, the red-shift of $g = 1.3$ eV is larger than the one of $g = 1.0$ eV. In Fig. 5-46(c), LSPR wavelength increases as n increases (red-shift), and it also increases as g increases (red-shift). That is, the larger g or/and n is, the larger red-shift is. In Fig. 5-46(d), the FWHM decreases as n increases, and it also decreases as g increases, resulting in narrower shapes of the spectra with larger

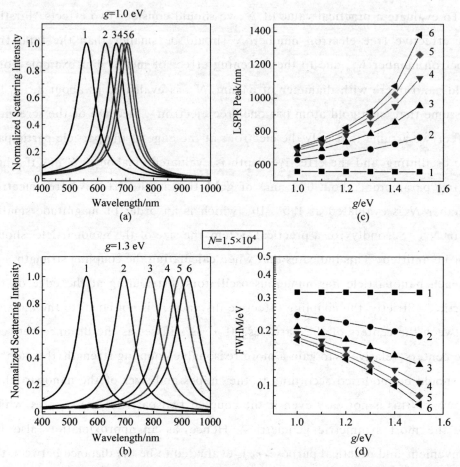

Fig. 5-46 (a), (b) Normalized scattering spectra of the nanoparticle chains varying with coupling strength g, i.e., $g = 1.0$ eV and $g = 1.3$ eV, respectively, calculated from Eq. (5-125). (c) LSPR peak position of the chains as a function of g. (d) The FWHM of the chains as a function of g. The spectra and curves are numerically labeled according to particle number n. Here, the effective free electron number is kept as $N = 1.5 \times 10^4$

(For colored figure please scan the QR code on page 1)

g or/and n in unit of "eV".

The behavior of the FWHM with the increase of N, g, and n can be explained as follow. Notice that the FWHM is positively related to the damping coefficient β_0 before the coupling ($g = 0$). For both the two cases, i.e., N increases with g unchanged (Fig. 5-45) and g increases with N unchanged (Fig. 5-46), γ increases simultaneously. In Eq. (5-115), both β_0 and γ are the

CHAPTER 5　Surface Plasmons

coefficients before the velocity terms, with opposite sign. The new damping coefficient β is affected by both of them while coupling. As a qualitative discussion, β is related to $|\beta_0 - \gamma|$. In Fig. 5-45, when N is small, e.g., $N = 10^2$, γ is much smaller than β_0. When N starts to increase, γ increases, resulting in β decreases, which indicates the decrease of the FWHM; when N increase to a large value, e.g., $N = 10^5$, γ is large enough compared with β_0, and β (FWHM) decreases to its minimum; when N continues to increase, e.g., $N = 10^6$, γ is very large and it results in the increase of β (FWHM). In Fig. 5-46, the situation is simple. Notice that $N = 1.5 \times 10^4$, indicating that γ is much smaller than β_0. When g increases, γ increases, resulting in the decrease of β (FWHM). These phenomena of the decreasing FWHM in scattering spectra are consistent with the ones in PL spectra, as shown in Fig. 5-47, which will be discussed in detail later.

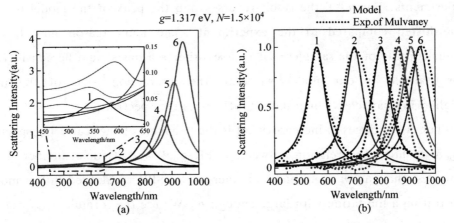

Fig. 5-47　(a) Scattering spectra of the nanoparticle chains calculated from Eq. (5-125). The inset of (a) is the zoom of the dashed box. (b) Normalized scattering spectra of the nanoparticle chains calculated from Eq. (5-125) (solid lines) and the experimental ones of P. Mulvaney[72] (dot lines). The spectra are numerically labeled according to particle number n. Here, $N = 1.5 \times 10^4$, $\omega_0 = 2.204$ eV, $\beta_0 = 0.156$ eV, and $g = 1.317$ eV

(For colored figure please scan the QR code on page 1)

In Fig. 5-47(a), we show the comparison between the scattering spectra of these chains calculated from this model and the experimental ones from P.

Mulvaney et al.[72]. Here, the intrinsic frequency ω_0 and damping coefficient β_0 of each identical nanoparticle are $\omega_0 = 2.204$ eV and $\beta_0 = 0.156$ eV, resulting in the resonant wavelength at 560 nm of an individual nanoparticle. The coupling strength and effective free electrons number are $g = 1.317$ eV and $N = 1.5 \times 10^4$. (a) shows the scattering spectra of these chains calculated from this model. As n increases, the scattering intensity increases and meanwhile the peak red-shifts. The inset of Fig. 5-47(a) illustrates the peaks with small intensities (for $n \geqslant 3$). The intensities of these small peaks increase and they also red-shift as n increases.

In Fig. 5-47(b), we compare the normalized scattering spectra of our model with the experiments of P. Mulvaney[72]. In general, they agree well with each other. Notice that the FWHMs of $n = 2$ and $n = 6$ agree not very well. For $n = 2$, the dimer case, the FWHM of our model is smaller than the one of the experiments. For $n = 6$ the result reverses, and the peak of the model is a bit blue-shifted compared to the experiment. The main reason may be the nonuniformity of the samples. Of course, other reasons should be considered. For example, in our model, we omit the interaction parts of the non-neighboring particles, which indeed influences the spectra.

When n continues to increase, what is the behavior of the scattering spectra? Fig. 5-48 shows the relation between LSPR peak and particle number n. It indicates that as n continues to increase, LSPR peak red-shifts more, illustrating a limitation with large enough n. We use the formula $\omega_{LSPR} = \omega_\infty + \sum_{j=1}^{3} a_j \exp(-n/\tau_j)$ to fit this relation, where a_j and τ_j are the fitted parameters, and ω_∞ is the predicted asymptotic frequency of an infinitely long nanoparticle chain. The fit gives $\omega_\infty = 1.1839$ eV, corresponding to the limitation in wavelength at $\lambda_\infty = 1047.3$ nm. Meanwhile, we show the relative deviation from the limitation as a function of n. It indicates that the numbers of the particles should be $n \geqslant 10, 15,$ and 20 when the expected relative deviations are less than 5%, 2%, and 1%, respectively.

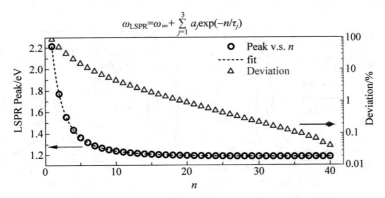

Fig. 5-48 LSPR peak position (unit: eV) as a function of particle number n (black open circles). Black dashed curve stands for the fit of the relation. Red open triangles stand for the corresponding relative deviations from the limitation of the peak position. Here, N, ω_0, β_0, and g are the same as the ones
(For colored figure please scan the QR code on page 1)

The above discussions are all based on the assumption that $K_j = K_0$ for $j = 1, 2, \cdots, n$, i.e., the external electric field felt by the particles is the same. What are the behaviors of the scattering spectra when these K_j are different? Here, we take $n = 6$ as an example to show the results with different K_j.

For the first case, we consider that the external electric field felt by the particles is the same in intensity but different in phases, i.e., $E_j = E_0 \exp(i \cdot j\varphi)$, where φ is the phase difference of the external electric field between the neighboring particles. Figure 5-49(a) shows the results varying with different phases φ. When φ increases from 0 to π, the intensity of LSPR peak (at around 941 nm) decreases rapidly, but the intensity of the subsidiary peak (at around 600 nm) changes little. Particularly, when $\varphi = \pi$, the intensity of the whole spectrum decreases almost to zero. This phenomenon is due to the interference among these particles. The emission intensity is tuned by the oscillation phases of the oscillators, which are affected by the excitation phases. The constructive and destructive interferences are corresponding to $\varphi = 0$ and $\varphi = \pi$, respectively.

For the second case, we consider that only one of the particles is excited by the incident light, i.e., $E_j = E_0$ for the jth particle but $E_{j'} = 0$ for $j' \neq j$. Figure 5-49(b) shows the results varying with different j. Due to the symmetry of the system,

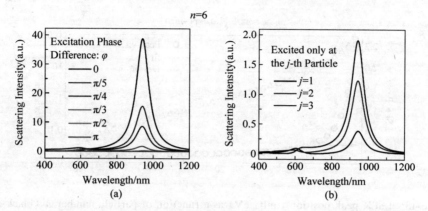

Fig. 5-49 Scattering spectra of 6 nanoparticles varying with different excitation conditions. (a) Varying with the excitation phase φ. (b) Only the jth particle is excited. Here, N, ω_0, β_0, and g are the same as the ones
(For colored figure please scan the QR code on page 1)

the situations of $j = 1, 2, 3$ are the same as $j = 6, 5, 4$, respectively. It indicates that when j increases, i.e., when the only excited particle is close to the middle of the chain, the intensity of LSPR peak increases. This is because the energy propagating to both sides of the excited particle attenuates along the chain due to the coupling only between neighboring particles, and when excited in the middle, it occurs the least attenuation. This can be explained employing a simple propagation model. For n particles, when excited only at the jth particle, we assume that the energy propagating from the jth particle to the j'th particle remains $P_0 \exp(-\eta |j - j'|)$, where P_0 is the energy of the jth particle, and $\eta > 0$ is a constant parameter related to the attenuation caused by the propagation. Therefore, the total energy P_{tot} of the system can be written as

$$P_{tot} = \sum_{j'=1}^{n} P_0 \exp(-\eta |j - j'|) = \frac{P_0}{e^{\eta} - 1} [(e^{\eta} + 1) - (e^{\eta} e^{-\eta j} + e^{-n\eta} e^{nj})].$$

(5-131)

Obviously, P_{tot} varies with j, and it reaches a maximum when $j = (n+1)/2$. Therefore, when the chain is excited in the middle, the emission intensity reaches a maximum.

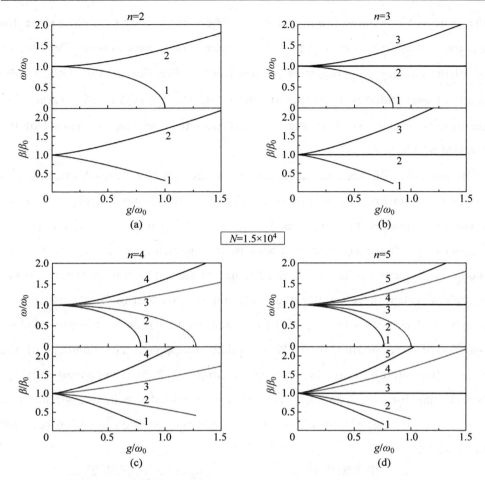

Fig. 5-50 The new generated resonant frequencies (ω_j) and damping coefficients (β_j) as a function of g in the cases of $n = 2$(a), $n = 3$(b), $n = 4$(c), and $n = 5$(d), respectively, calculated from Eq. (5-127). The curves are numerically labeled according to Mode j. Here, N, ω_0, and β_0 are the same

(For colored figure please scan the QR code on page 1)

The new generated PL modes are shown in Fig. 5-50, taking $n = 2, 3, 4, 5$ as examples. For a given particle number n, there are n modes of PL. Particularly, when n is even, the modes spit, half ($n/2$) of which blue shift, called the blue branches, the other half ($n/2$) of which red shift, called the red branches. When n is odd, however, there exists one mode with frequency and damping unchanged and equaling to ω_0 and β_0, respectively, and the rest modes split the same as even case. More special, Mode 1 and Mode 2 of $n = 2$ behave the same as Mode 2 and Mode 4 of $n = 5$. There exist cut-off coupling strengths g_{cut} for the red branches, at which the frequency decreases to zero. For Mode 1 of all the cases, g_{cut} decreases as n increases. Notice that in Eq. (5-130), the shape of each mode is a Lorentzian curve with the FWHM of $2\beta_j$. Hence, the increase (decrease) of β_j in Fig. 5-50 indicates the increase (decrease) of the FWHM of Mode j.

In Fig. 5-51, PL spectra calculated from Eq. (5-130) are shown for two different excitation wavelengths, i.e., 532 nm and 633 nm. In Fig. 5-51(a), it reveals the number of modes as 1, 2, 3, 4, and 3 for $n = 1, 2, 3, 4$, and 5, respectively. There are only 3 modes in the spectrum of $n = 5$, because one mode's wavelength is larger than 1000 nm so that it is not shown, the other mode's wavelength is smaller than 532 nm so that Fermi-Dirac distribution makes it disappear. In Fig. 5-51(b), due to the fact that some modes' wavelengths are smaller than 633 nm, they disappear. The numbers of the modes illustrated here are 1, 2, 2, 3, and 3 for $n = 1, 2, 3, 4$, and 5, respectively. Notice that, for $n = 2$ and $n = 5$, there are two modes behaving similarly at around 624 nm and 811 nm, respectively. This phenomenon has been revealed in Fig. 5-50.

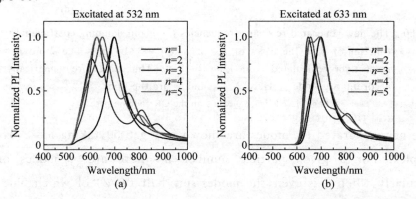

Fig. 5-51 Normalized PL spectra of nanoparticle chains calculated from Eq. (5-130), excited at 532 nm (a) and 633 nm (b), respectively. Black, red, blue, green, and cyan stand for n equaling from 1 to 5, respectively. Here, $N = 1.5 \times 10^4$, $g = 0.9$ eV, $\omega_0 = 1.774$ eV, and $\beta_0 = 0.101$ eV

(For colored figure please scan the QR code on page 1)

Furthermore, we should emphasize that, for most cases, the excitation electric field felt by these particles is treated as the same, i.e., the amplitudes and phases are the same. This approximation is reasonable when dealing with small n, e.g., $n \leqslant 6$. In Mulvaney's experiments[72], the diameter of one particle is 64 nm with gap of 1 nm, resulting in the length of the chain for $n = 6$ of about 390 nm (omitting the bend of the chain). For larger n, the length of the chain would be so large that it is larger than the excitation wavelength. In experiments, usually, the light is focused through the lens and the size of the facula is near the excitation wavelength due to the diffraction limit. In other words, a large n indicates that the electric field felt by the particles is not the same and it depends on the position of the facula.

In summary, in this section, we develop a multimer coupling classic harmonic oscillator model and employ it to explain the white light scattering and PL spectra of strongly coupled metallic nanoparticle chains. The model is suitable for $n \geqslant 1$, and we take $n = 1, 2, \cdots, 6$ as examples in this work. Comparisons with experiments of Mulvaney' work illustrate the accuracy and practicability of this model. Moreover, the scattering and PL properties are analyzed in detail, which depend on particle number n, coupling strength g, and effective free electron number N. For scattering spectra, larger n or/and larger g result in larger red-shift of LSPR peak and in smaller FWHM of the peak; small N, e.g., $N < 10^5$, influences the spectra little, while large N, e.g., $N \geqslant 10^5$, influences the spectra greatly. For PL spectra, the modes split due to the coupling and the splitting increases with the increase of g. The amplitudes of these modes are dependent on the excitation wavelength. This classic model is practical and accurate when dealing with the coupling of metallic nanostructures. Thereby, this work would be helpful to understanding the optical properties more deeply and gives a unified treatment for scattering and PL properties of strongly coupled multimer system. It is also useful for related applications utilizing strongly coupled system of nanophotonics.

5.4 Plasmonic Waveguide

5.4.1 The EM theory for calculating nanowires

For nonmagnetic media, the magnetic field and electric field in the frequency space satisfied the following relationship[73]

$$\nabla^2 \begin{Bmatrix} E(r) \\ H(r) \end{Bmatrix} + \frac{\omega^2}{c^2}\varepsilon(r) \begin{Bmatrix} E(r) \\ H(r) \end{Bmatrix} = 0, \tag{5-132}$$

among them, the E and H of Maxwell's equation were described as

$$E_i(r) = a_i v_i(r) + b_i w_i(r), \tag{5-133}$$

$$H_i(r) = -\frac{i}{\omega\mu_0} k_i [a_i w_i(r) + b_i v_i(r)], \tag{5-134}$$

where a_i and b_i were constant coefficients, and their extensions could be described as[73]

$$E_i(r) = \begin{cases} \left[\frac{im}{k_{i}\rho}a_i F_{i,m}(k_{i\perp}\rho) + \frac{ik_{\parallel} k_{i\perp}}{k_i^2}b_i F'_{i,m}(k_{i\perp}\rho)\right]\hat{\rho} + \\ \left[-\frac{k_{i\perp}}{k_i}a_i F'_{i,m}(k_{i\perp}\rho) - \frac{mk_{\parallel}}{k_i^2\rho}b_i F_{i,m}(k_{i\perp}\rho)\right]\hat{\phi} + \\ \frac{k_{i\perp}^2}{k_i^2}b_i F_{i,m}(k_{i\perp}\rho)\hat{z} \end{cases} e^{im\phi+ik_{\parallel}z}, \tag{5-135}$$

$$H_i(r) = -\frac{i}{\omega\mu_0}k_i \begin{cases} \left[\frac{im}{k_{i}\rho}b_i F_{i,m}(k_{i\perp}\rho) + \frac{ik_{\parallel} k_{i\perp}}{k_i^2}a_i F'_{i,m}(k_{i\perp}\rho)\right]\hat{\rho} - \\ \left[\frac{k_{i\perp}}{k_i}b_i F'_{i,m}(k_{i\perp}\rho) + \frac{mk_{\parallel}}{k_i^2\rho}a_i F_{i,m}(k_{i\perp}\rho)\right]\hat{\phi} + \\ \frac{k_{i\perp}^2}{k_i^2}a_i F_{i,m}(k_{i\perp}\rho)\hat{z} \end{cases} e^{im\phi+ik_{\parallel}z}.$$

$$\tag{5-136}$$

Among them, pay attention to the existence of the equations, $F_{1,m}(x) = H_m(x)$ and $F_{2,m}(x) = J_m(x)$. In general, the author considered the dipole along the radial ($P_0 \propto \hat{\rho}$), for the nanostructure, considering the quasi-static limit ($H \approx 0$)

CHAPTER 5 Surface Plasmons

to simplify, the following relationship exists

$$\nabla \cdot D = \rho_{ext}, \tag{5-137}$$

$$\nabla \times E = 0, \tag{5-138}$$

where $\rho_{ext}(r)$ was the density of external charge, in the interest system. The external source was a dipole positioned at r' outside the wire, and its radial coordinate was $\rho' = d$, so the following relationship exists

$$\rho_{ext}(r, r') = (P_0 \cdot \nabla')\delta(r - r'). \tag{5-139}$$

5.4.2 The decay rate in the plasmon mode

After calculation and derivation, the radiation decay rate and the non-radiation decay rate were

$$\frac{\Gamma_{rad}}{\Gamma_0} = \left| 1 + \frac{\epsilon - 1}{\epsilon + 1} \frac{R^2}{d^2} \right|^2 \quad (d \geqslant R), \tag{5-140}$$

$$\frac{\Gamma_{non\text{-}rad}}{\Gamma_0} \approx \frac{3}{16 k_0^3 (d - R)^3 \epsilon_1^{\frac{3}{2}}} \text{Im}\left(\frac{\epsilon - 1}{\epsilon + 1}\right), \tag{5-141}$$

respectively, where Γ_0 was the spontaneous emissivity about the emitter in a uniform dielectric, and ϵ was the dielectric constant, and note that, $|\epsilon| \gg 1$ and Im ϵ is small, $\text{Im}\left(\frac{\epsilon - 1}{\epsilon + 1}\right) \approx 2\text{Im}\frac{\epsilon}{(\text{Re } \epsilon)^2}$. When there was metal loss, it was not possible to clearly distinguish between $\Gamma_{non\text{-}rad}$ and Γ_{pl}, where Γ_{pl} was defined as the decay rate produced by the pole in the limit $\text{Im}\epsilon_2 = 0$.) For the dipole along the $\hat{\rho}$ orientation, the following relationship exists

$$\frac{\Gamma_{pl}}{\Gamma_0} = \frac{6\pi\epsilon_0}{k_0^3 \sqrt{\epsilon_1}} \left[\frac{\text{Im}\hat{\rho} \cdot E_r(r', r')}{p_0} \right]_{pole} = -\frac{6\pi\epsilon_0}{k_0^3 \sqrt{\epsilon_1}} \text{Im}\left[\hat{\rho} \cdot \nabla(\hat{\rho} \cdot \nabla') \Phi_r(r, r')|_{r=r'}\right]_{pole}$$

$$= -\frac{3}{\pi k_0^3 \sqrt{\epsilon_1}} \text{Im}\left[\int_0^\infty dh h^2 K_1^2(hd) \alpha_0(h)\right]_{pole}, \tag{5-142}$$

and Γ_{pl} could be described as

$$\Gamma_{pl} \propto \Gamma_0 \frac{K_1^2(\kappa_{1\perp} d)^2}{(k_0 R)^3}. \tag{5-143}$$

The "Purcell factor" defined by the ratio $P = \dfrac{\Gamma_{pl}}{\Gamma'}$ characterizes the efficiency coupled to the plasmon mode. In addition to the basic plasmon mode, the total emissivity of the other channels was $\Gamma' = \Gamma_{rad} + \Gamma_{nonrad}$. Figure 5-52 shows the numerical evaluation of the spontaneous emissivity and its optimization.

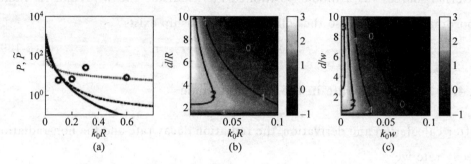

Fig. 5-52 (a) Solid line: Maximum Purcell factor $P = \dfrac{\Gamma_{pl}}{\Gamma'}$ for a NW, plotted as a function of R and optimized over the emitter position. Here, P is calculated in the quasistatic approximation. Dashed line: Same quantity, obtained by exact electrodynamic calculations. Dotted line: Effective Purcell factor $\widetilde{P}(R) = \dfrac{\widetilde{\Gamma}_{pl}(R)}{\widetilde{\Gamma}(R)}$ for a nanotip of final radius R, optimized over nanotip shape and emitter position. Solid points: Same quantity, calculated using numerical simulations (boundary element method). (b) Contour plot of lg P for a NW, as functions of R and $\dfrac{d}{R}$. (c) Contour plot of lg P for a nanotip, as functions of w and $\dfrac{d}{w}$.[73]

(For colored figure please scan the QR code on page 1)

5.4.3 The spontaneous emission near the nanotip

The spontaneous emissivity of the dipole positioned along the axis could be described as

$$\frac{\Gamma_{rad}}{\Gamma_0} = \left| 1 + \left[1 + \left(\frac{4d}{w}\right) \right]^{-1} \left(\frac{\varepsilon_2}{\varepsilon_1} - 1\right) \right|^2, \qquad (5\text{-}144)$$

$$\frac{\Gamma_{nonrad}}{\Gamma_0} = \frac{3}{8\varepsilon_1^{\frac{3}{2}}} \frac{1}{(k_0 d)^3} \text{Im}\left(\frac{\varepsilon_2 - \varepsilon_1}{\varepsilon_2 + \varepsilon_1}\right), \qquad (5\text{-}145)$$

CHAPTER 5 Surface Plasmons

$$\frac{\Gamma_{pl}}{\Gamma_0} = \tilde{\alpha}_{pl} \frac{1}{(k_0 w)^3 \left[1 + \left(\frac{4d}{w}\right)\right]} K_1^2 \left[C\sqrt{1 + \left(\frac{4d}{w}\right)}\right], \quad (5\text{-}146)$$

where $\tilde{\alpha}_{pl}$ was only determined by $e_{1,2}$ and derived from the following formula

$$\tilde{\alpha}_{pl} = \frac{24\pi}{\varepsilon_1^{\frac{3}{2}}} C^3 \frac{(\varepsilon_1 - \varepsilon_2) I_1(C) I_0(C)}{d\chi(C)/dx}. \quad (5\text{-}147)$$

In previous studies, coupling to a dielectric waveguide was able to produce a single photon. Firstly, the s total electric field of the system was expressed as

$$E_T(r) = \sum_{\mu = w,g} \sum_{i=1}^{N_\mu} C_{\mu,i}(z) E_{\mu,i}(r), \quad (5\text{-}148)$$

where w was the nanostructure system and g was the waveguide system, in addition, $i = 1, 2, \cdots, N_m$. The $N_w + N_g$-coupled mode equation could be described as

$$\sum_{\nu = w,g} \sum_{j=1}^{N_\nu} P_{\mu,i;\nu,j}(z) \frac{dC_{\nu,j}}{dz} = i\omega\varepsilon_0 \sum_{\nu = w,g} \sum_{j=1}^{N_\nu} K_{\mu,i;\nu,j}(z) C_{\nu,j}(z), \quad (5\text{-}149)$$

where the coefficients were derived from the following formula

$$P_{\mu,i;\nu,j}(z) = e^{i(k_{//\nu,j} - k^*_{//\mu,i})z} \int d\rho \left[E_{\nu,j}(\rho) \times H^*_{\mu,i}(\rho) + E^*_{\mu,i}(\rho) \times H_{\nu,j}(\rho)\right] \cdot \hat{z}$$
$$(5\text{-}150)$$

$$K_{\mu,i;\nu,j}(z) = e^{i(k_{//\nu,j} - k^*_{//\mu,i})z} \int d\rho E_{\nu,j}(\rho) \cdot E^*_{\mu,i}(\rho) \left[\varepsilon_T(\rho) - \varepsilon_\nu(\rho)\right], \quad (5\text{-}151)$$

among them, $e_T(\gamma)$ was the permittivity about the combined system[74].

5.4.4 SPP modes of Ag NW by One-End Excitation

In order to excite SPP in the NW, the simplest and most convenient way is directly focusing the light on one end of wire, due to the fact that local symmetry-breaking at a finite length NW terminus is sufficient to excite SPP. Assuming that the harmonic monochromatic laser beam with electric field $E_{inc} = E_0 e^{-i\omega t}$, illuminate on the end of wire perpendicular to the wire, the modes on the wire will be excited with the following form (ignoring the time factor):

$$E^j(r) = \sum_m \hat{A}^j_m F^j_m(k^j_{m\perp} \rho) \cos m\phi \cdot e^{ikm_{//}}, \quad (5\text{-}152)$$

where $m = 0, \pm 1$ corresponding to the modes which can be excited and $j = 1, 2$ refer to the medium around and wire.

As Fig. 5-53 shows, the wire excited by a paraxial Gaussian beam with an instantaneous electric field of the form $E_{inc} = E_0 e^{-i\varphi}$, where φ is the phase of incident light[75]. The fundamental $m = 0$ SPP mode or $m = -1$ is selectively excited when the input light phase is $\varphi = 0$ or $\pi/2$ with the polarization component along the wire. The $m = 1$ mode is generated for $\varphi = 0$ with polarization components perpendicular to the wire. For any polarization, the light can be divided into the two basic components with polarization along or perpendicular to the wire. The amplitude of the components excited is strongly depending on the polarization angle and of course, the shape of the illuminating terminal.

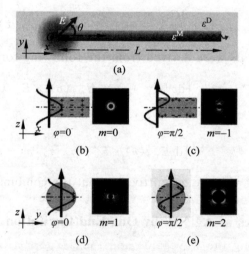

Fig. 5-53 (a) Schematic of excitation of plasmon modes in metallic NW in homogeneous environment. (b)~(e) Plasmon modes excited at different incident polarization angle θ and incident phase φ[75]
(For colored figure please scan the QR code on page 1)

5.4.5 Optical non-reciprocity with multiple modes based on a hybrid metallic NW

Non-reciprocity is an essential property or concept in physics and numerous applications[76]. For example, in the field of fluid engineering, passive flow control is a fundamental demand. In 1920, Nikola Tesla introduced the Tesla valve for the first time to passively promote the flow unidirectionally based on its unique design. In the field of biology, the aortic valve plays an important role in promoting unidirectional blood flow. In electronic engineering, the diode is a significant electronic component that conducts electricity primarily in one direction, i. e., high resistance from one direction and low resistance from the other direction. In optics, the unidirectional propagation property for photons, or optical non-reciprocity, is desirable.

In this section, in order to achieve optical non-reciprocity, we introduce a hybrid plasmonic nanostructure which consists of a metal (Ag) cavity waveguide together with two gratings with different parameters on both sides. The schematic of the hybrid metallic nanostructure is shown in Fig. 5-54. The waveguide is located through the Ag substrate along x direction with length L, width w and depth d. The gratings are located on both surfaces of Ag, and they are period in y direction, with period g_1, g_2, width a_1, a_2, depth h_1, h_2 and number N_1, N_2. The three-dimensional finite-difference time-domain (3D-FDTD) method is employed to investigate the optical properties of the hybrid structure. The total simulation size is 2000 nm × 2000 (6000) nm × 2000 nm and the step length is 2 nm × 2 nm × 5 nm. The linear polarized plane wave illuminates the Ag surface (y-z plane) perpendicularly, with the electric field intensity of $E_0 = 1$ V · m^{-1}. The electromagnetic field imports from one port, propagates along the waveguide and then exports from the other port. The permittivity of Ag is obtained from[56].

As a beginning, we investigate the properties of the waveguide without gratings. Here, the width and depth of the waveguide are constant, i. e, w = 50 nm and d = 200 nm. Figure 5-55 shows the reduced transmissivity (T) spectra of the structure varying with the length L. The reduced transmissivity represents the focus degree of the energy flux, that is, the waveguide helps to focus more energy to a small range (it is 50 nm × 200 nm in our work). It is

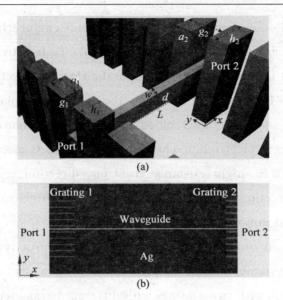

Fig. 5-54　Schematic of the hybrid nanostructure for 3D view (a) and x-y view (b). Silver is colored in gray. Yellow and blue rectangles stand for the waveguide and gratings, respectively, which are vacuum. To clearly illustrate the details of the structure, silver is not shown in (a). Ports 1 and 2 are two ports of the structure, and we define forward and backward propagations as from ports 1 to 2 and from ports 2 to 1, respectively

(For colored figure please scan the QR code on page 1)

obvious that with y-polarized illumination, multiple peaks are observed. Besides, the larger L is, the more peaks are supported. This phenomenon is the same as the resonant spectra of nanowire, as our previous works have reported[77-78]. This is because SPPs are excited on the interface of metal and dielectric (or vacuum) and can propagate along the interface. In our case, SPPs propagate along the waveguide, meanwhile the waveguide can be treated as a nanocavity which also supports LSPR modes. In contrast, with x-polarized illumination, the field cannot propagate through the waveguide. This property can be used as a switch whose key is the polarization of the incident light, i.e, y-polarized illumination corresponds to "on" and z-polarized illumination corresponds to "off". Therefore, to investigate the properties of the "on" mode, we only consider the y-polarized illumination in the rest of this work. Additionally, as a fundamental demonstration, we select $L = 1000$ nm to continue our design for the structure. Certainly, other sizes should also be suitable, depending on the resonant modes that one requires.

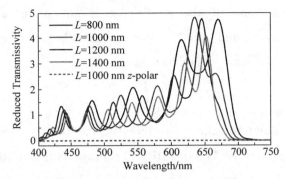

Fig. 5-55 Reduced transmissivity of the waveguide without gratings. Black, red, blue and cyan solid lines stand for the length of 800 nm, 1000 nm, 1200 nm and 1400 nm, respectively, with y-polarized incident light. Red dashed line stands for the length of 1000 nm with z-polarized incident light. The spectra intensities are divided by 0.0025 which origins from the ratio of cross section area (50 nm × 200 nm) of the waveguide to the total cross section area (2000 nm × 2000 nm) of the incident light

(For colored figure please scan the QR code on page 1)

We now achieve multiple modes by only employing a metallic cavity waveguide. However, due to the symmetry of the structure, the forward and backward propagation properties would be the same, which hinders the optical nonreciprocity of the system. Therefore, we should break the symmetry to obtain excellent non-reciprocal properties.

Next, we employ gratings on both side of the waveguide, and the parameters of these two gratings are different, which ensures the asymmetry of the system. All the parameters are shown in Table 5-3. When the electric field propagates from ports 1 to 2 (forward), as shown by the black solid line in Fig. 5-56 (a), the transmission spectrum still illustrates multiple peaks. Compared with the one without gratings (red solid line in Fig. 5-55), not only the number of the resonant modes increases, but also the FWHM of each mode decreases. On the other hand, when the electric field propagates from ports 2 to 1 (backward), as shown by the black dot line in Fig. 5-56 (a), the transmission spectrum also illustrates multiple peaks. The most critical thing is that, the peaks of the forward and the peaks of the backward are interlaced. In other words, when the electric field at the wavelength of λ is permitted to propagate forward (one of the peaks), it is not permitted to propagate backward (the associated dip). We define the contrast of the two transmissivities at the wavelength of λ as:

$$C(\lambda) = \frac{T_f(\lambda) - T_b(\lambda)}{T_f(\lambda) + T_b(\lambda)}. \tag{5-153}$$

Here, T_f and T_b stand for the transmissivities of the forward and backward propagations, respectively. The quantity C represents the ability or strength of the non-reciprocal properties of the system. The larger $|C|$ is, the better non-reciprocity is. Due to the positive values of T_f and T_b, the range of C is from -1 to 1. There are two special conditions. One is at $|C|=0$, which indicates that there is no non-reciprocity in the system, such as the waveguide without gratings in this work. The other is at $|C|=1$, which indicates the perfect non-reciprocity of the system. In this case the electric field only propagate unidirectionally, behaving as a "optical diode". Meanwhile, the isolation ratio in unit of dB can be defined as:

$$\mathrm{IR}(\lambda) = 10\lg\frac{T_f(\lambda)}{T_b(\lambda)}. \tag{5-154}$$

Obviously, IR and C are equivalent in describing the optical non-reciprocity.

Table 5-3 The parameters of the hybrid metallic nanostructure

Length unit/nm	L 1000	w 50	d 200
N_1 10	h_1 100	a_1 100	g_1 300
N_2 10	h_2 50	a_2 200	g_2 300

For our hybrid structure, the isolation ratio and the contrast are shown in Fig. 5-56(b). This illustrates that there are multiple peaks and dips in the spectra, which indicate the forward and backward non-reciprocities, respectively. For the purpose of availability, we use IR>10 as a criterion to indicate effective non-reciprocity. Our structure shows seven excellent modes for forward propagation, i. e., 410 nm, 440 nm, 476 nm, 514 nm, 548 nm, 608 nm and 654 nm. Meanwhile, for backward propagation, there are three excellent modes, i. e., 424 nm, 454 nm and 488 nm. Notice that at 678 nm there is a large dip, but there is no resonant mode here, hence we do not consider this mode. Particularly, at the wavelength of 548 nm, IR reaches a maximum of 29.1 dB, the FWHM of which is 16.2 nm.

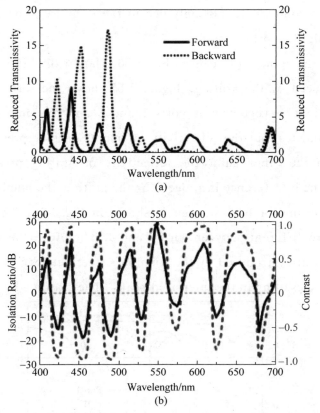

Fig. 5-56 (a) Reduced transmissivity of the hybrid metallic structure. Black solid and dot lines stand for T of forward and backward propagations, respectively. The spectra intensities are divided by 8.3×10^{-4} which origins from the ratio of cross section area (50 nm × 200 nm) of the waveguide to the total cross section area (6000 nm × 2000 nm) of the incident light. (b) Isolation ratio (black solid curve, left y-axis) and contrast (red dashed curve, right y-axis) of the structure. The illuminating light is y-polarized

(For colored figure please scan the QR code on page 1)

Here, to demonstrate the good performance of our structure, we make some comparisons with other works. Fan et al present an all-silicon passive optical diode and achieve the isolation ratio at least 18 dB at the wavelength of about 1630 nm[79]. Liang et al employ collision effect of thermal atoms to achieve broadband optical non-reciprocal with maximum isolation ratio close to 40 dB and bandwidth over 1.2 GHz[80]. Bino et al utilize Kerr effect in an optical microresonator to achieve optical non-reciprocity with isolation ratio of larger than 20 dB at the wavelength of 1550 nm[81]. In general, the performance of our structure is competitive, not only due to the large isolation ratio, but also due to the multiple working modes.

Notice that the number of peaks in Fig. 5-56 is larger than those in Fig. 5-55. This is introduced by the grating. Figure 5-57 shows the reflectivity of bare gratings (without waveguide) of ports 1 and 2 and the one of smooth Ag surface. It indicates that the reflectivity of smooth surface presents a smooth spectrum. On the other hand, the reflectivity of grating presents multiple peaks due to the interference introduced by the grating, the number of which is larger than the one of bare waveguide shown in Fig. 5-55. The positions of these peaks are different between ports 1 and 2. Therefore, the multiple peaks in Fig. 5-56(a) are introduced by the coupling effect between the gratings and the waveguide, resulting in the interlaced peaks. The reason why we choose the

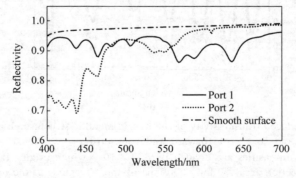

Fig. 5-57 Reflectivity of bare grating and smooth Ag surface. Solid and dot curves stand for the gratings of ports 1 and 2, respectively. Dash dot curve stands for the smooth surface without grating. The illuminating light is y-polarized

CHAPTER 5 Surface Plasmons

reflectivity of bare grating rather than the transmissivity to analyze the modes is that, without the waveguide the transmissivity of bare grating is almost zero due to the bulk metal reflecting most energy of the incident light.

To clearly demonstrate the optical non-reciprocity of the structure, we analyze the electric field modes in details. Taking $\lambda = 440$ nm and $\lambda = 488$ nm as examples, we show the electric field distributions of different cases in Fig. 5-58. Comparing Fig. 5-58 (a) and (b), it is obvious that at the wavelength of 440 nm, the forward propagation is enhanced, the electric field is localized inside the waveguide and the intensity is highly enhanced, the maximum enhancement factor (EF) of which is about EF = 15.6. Here, we define EF as:

$$\text{EF} = E_{\max}/E_0, \tag{5-155}$$

where $E_{\max}$ is the maximum of the electric field intensity inside the waveguide. In contrast, the backward propagation at 440 nm is impeded, or the electric field intensity is too weak (EF = 0.86) compared with the forward one. Therefore, the optical non-reciprocity is achieved at 440 nm for forward propagation. On the other hand, comparing Fig. 5-58 (c) and (d), at the wavelength of 488 nm, the forward propagation is impeded, with weak electric field intensity (EF = 3.91), while the backward propagation is enhanced, with localized electric field inside the waveguide and highly enhanced intensity (EF = 13.5). Therefore, the optical non-reciprocity is achieved at 488 nm for backward propagation. Other modes result in the same conclusion after further analysis (not shown).

From Fig. 5-56, we have noticed that at the wavelength of 548 nm, although the forward transmissivity is not as large as that of others, the value of IR is close to 30 dB, which indicates almost perfect optical non-reciprocity. We take the 548 nm wavelength as an example to analyze each component of the electric field of forward propagation. Figure 5-59 shows the electric field distributions of E_x(a) and E_y(b), illuminated with y-polarized light. E_z is not shown here because its value is zero, indicating there is no z component of electric field in this figuration. E_x is localized near the edges of the waveguide

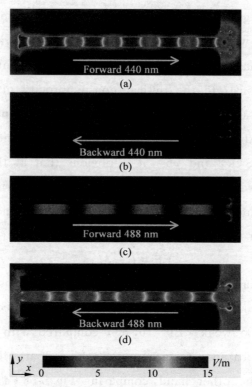

Fig. 5-58 Electric field distributions of x-y view. (a) and (b) forward and backward propagations at 440 nm. (c) and (d) forward and backward propagations at 488 nm. The color bar stands for the intensity of the electric field. The illuminating light is y-polarized

(For colored figure please scan the QR code on page 1)

with weak intensity and rapidly depletes along y axis within few nanometers. Only at both ports is there strong intensity of E_x. Meanwhile, E_y is mostly localized inside the waveguide with strongly enhanced intensity. Therefore, the main component of the electric field propagating inside the waveguide is the y component.

In summary, in this section, we theoretically employing a hybrid metallic nano waveguide to realize optical non-reciprocity with multiple modes in the visible range. The isolation ratios are as high as around 10~30 dB, resulting in unidirectional propagations. Some of the modes support non-reciprocity of forward propagations, while the others support non-reciprocity of backward

CHAPTER 5 Surface Plasmons

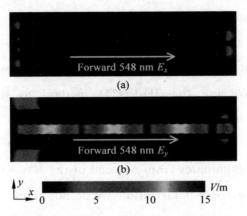

Fig. 5-59 Electric field distributions of x-y view for forward propagation at the wavelength of 550 nm. (a) Electric field intensity of x component E_x. (b) Electric field intensity of y component E_y. Black dashed lines represent the edges of the waveguide. The color bar stands for the intensity of the electric field. The illuminating light is y-polarized

(For colored figure please scan the QR code on page 1)

propagations. Particularly, at the wavelength of 548 nm, the isolation ratio reaches to a maximum value of close to 30 dB with FWHM of 16 nm. Furthermore, the electric field distribution is investigated and it is localized inside the waveguide, the intensity of which is strongly enhanced by about 1 order of magnitude for these resonant modes. Hence, this work paves a way to realize multiple modes of optical nonreciprocity and is essential to potential applications.

5.4.6 Strongly enhanced propagation and non-reciprocal properties of CdSe NW

In this section, we theoretically investigate the propagation and nonreciprocal properties of CdSe NW[82]. Combining the NW with nanostructures, i.e., MgF_2 and Ag substrates with grooves, we obtain highly improved propagation performance at communication wavelength of 1550 nm when illuminated with different polarizations. Furthermore, high contrast of the non-reciprocal property without time modulations is also obtained at 1550 nm by breaking the symmetry of the system.

The schematic of the hybrid structure is shown in Fig. 5-60, where CdSe NW lies on MgF_2 substrate, followed by Ag substrate. The grooves are etched on Ag when necessary. We use the 3D-FDTD method to simulate the optical properties of the structure. The length of the NW is $L = 20$ μm unless otherwise specified. The electric field intensity of the incident plane wave is $E_0 = 1$ V/m, which illuminates from one end of the NW with y-polarization or x-polarization. Here, for y-polarization, we do not employ the grooves; while it is for x-polarization, the grooves are necessary to enhance the optical properties of the system; moreover, the asymmetric grooves are employed to achieve non-reciprocal performance of the system for x-polarization. The permittivities of CdSe, MgF_2 and Ag are obtained from Ref. [83], Ref. [84]

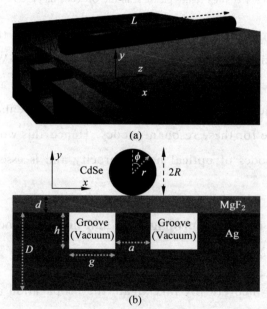

Fig. 5-60 Schematic of the hybrid structure. (a) The three-dimensional view. CdSe NW (blue) is on the substrates, which consist of MgF_2 (green) and Ag (gray). The two grooves are symmetrically etched inside Ag. The incident light illuminates from one end of the NW. L: length of the NW. (b) x-y view. r and ϕ: radius and degree in cylindrical coordinate; R: radius of CdSe NW; d: thickness of MgF_2; D: thickness of Ag; a: distance between the two grooves; h: depth of the grooves; g: width of each groove

(For colored figure please scan the QR code on page 1)

and Ref.[85], respectively.

Firstly, we implement our design with y-polarization illuminating employing the simplest structure, i.e., the CdSe NW is placed on the Ag substrate (without MgF_2). Figure 5-61 shows the mode distributions of the structures varying with R. The resonance wavelengths increase (from around 1100 nm to 1700 nm) with the increasing R. For R = 70 nm, 90 nm, and 110 nm, electric fields are mostly localized at the incident side of the NW within about 5 μm. Particularly, for R = 130 nm, the electric field is localized at the other side (the output side) at the wavelength of around λ = 1700 nm, which indicates that the PSPP propagates at least for 20 μm at this case, with a field enhancement factor of EF = 15.1. Here, EF is defined as EF = $|E_{max}/E_0|^2$, where E_{max} is the peak intensity of electric field at the output side of the NW. However, this good behavior of the propagation property is at the wavelength of 1700 nm

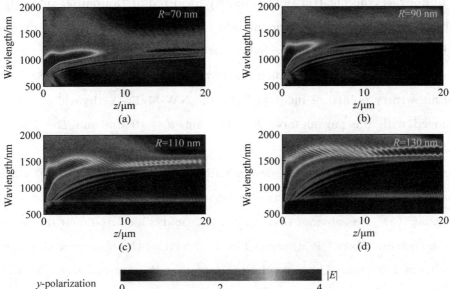

Fig. 5-61 Mode distributions of the structures varying with the radii R of NWs without MgF_2. The incident light is y-polarized. The thickness of Ag substrate is D = 150 nm. The position of line along z direction is at 2 nm above the intersection line between CdSe and Ag. (a)~(d) stand for the radii of 70 nm, 90 nm, 110 nm and 130 nm, respectively. Color bar stands for the intensity of electric field $|E|$, the unit of which is V/m

(For colored figure please scan the QR code on page 1)

rather than 1550 nm. Therefore, we would implement the simulation for further investigation.

An additional substrate of MgF_2 between the NW and Ag is a good idea to enhance the propagation property, due to the low value of dielectric constant of MgF_2 and the coupling among them when a gap is constructed. Since the resonance wavelength of the structure would be red-shifted with the existence of MgF_2, we choose the $R = 110$ nm one, the resonance wavelength of which is a bit less than 1550 nm, to investigate the behavior of the propagation property. Figure 5-62(a)~(c) show the mode distributions of the structures varying with the thickness d of MgF_2. Furthermore, Fig. 5-62(d) shows the electric field at $z = 20$ μm and $\lambda = 1550$ nm, as a function of d, including the one without MgF_2 ($d = 0$) shown in Fig. 5-61(c). It indicates that the best one is at $d = 10$ nm with EF = 13.2, while others behave not well. Furthermore, for $d = 10$ nm case, the electric field is mostly localized at the output side (right) of the NW rather than the incident side (left), which results in the excellent propagation performance of the hybrid structure.

As a brief summary, to enhance the propagation properties of CdSe NW at 1550 nm with y-polarized incident light, the NW-MgF_2-Ag hybrid structure is employed, with the parameters: $R = 110$ nm, $d = 10$ nm and $D = 150$ nm, resulting in EF = 13.2.

Secondly, we implement our design with x-polarized illuminating based on the above structure (Fig. 5-62(a)), the mode distributions of which are shown in Fig. 5-63(a). It is obvious that the electric field is localized near the incident side, the intensity is not as strong as Fig. 5-62(a), and the resonance wavelength is far from 1550 nm. These results indicate that the optimal structure for y-polarization is not suitable for x-polarization. Therefore, it should be redesigned to achieve more effective properties for x-polarized incident light. Here, we employ two grooves which are symmetrically etched inside Ag substrate. With the initial parameters ($a = 120$ nm, $h = 3600$ nm and $g = 4000$ nm) of the grooves, the mode distributions are shown in Fig. 5-63(b). Compared with

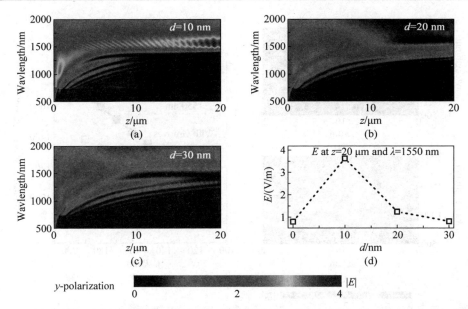

Fig. 5-62 (a)～(c) Mode distributions of the structures varying with the thickness d of MgF_2, which are 10 nm, 20 nm and 30 nm, respectively. The incident light is y-polarized. Other parameters: $D = 150$ nm and $R = 110$ nm. The position of line along z direction is at 2 nm above the intersection line between CdSe and MgF_2. (d) The intensity of the electric field as a function of d at $z = 20$ μm and $\lambda = 1550$ nm. Color bar stands for the intensity of electric field $|E|$
(For colored figure please scan the QR code on page 1)

Fig. 5-63(a), (b) reveals a better performance, i.e., the electric field is more likely localized near the output side with the enhanced intensities, and the resonance wavelength is red-shifted. However, the resonance wavelength is still far away from 1550 nm, and it should be red-shifted more. As indicated in Fig. 5-61, the resonance wavelength increases when R increases. Hence, we could shift it to 1550 nm by increasing R. Figure 5-63(c) shows the R dependent resonance wavelength without and with grooves, respectively. In addition to the fact just mentioned, i.e., the resonance wavelength is directly proportional to R approximately, it also reveals that the resonance wavelength with grooves is larger than that without grooves. For the purpose of strong field enhancement at 1550 nm, we will set $R = 175$ nm for the following design.

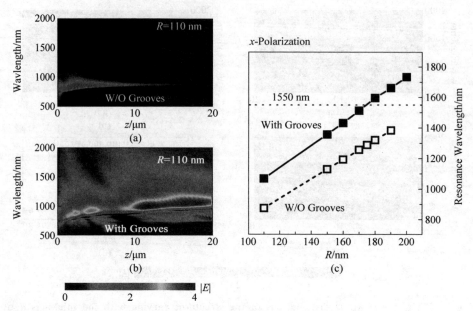

Fig. 5-63 Mode distributions of the structures without (a) and with (b) grooves, respectively. Here, $R = 110$ nm for both cases. The position of line along z direction is at $r = 108$ nm and $\phi = \pi/2$. (c) Resonance wavelength as a function of R, with (solid curve filled squares) and without (dashed curve open squares) grooves. The incident light is x-polarized. The horizontal dashed line stands for 1550 nm label. Here, $D = 18000$ nm and $d = 10$ nm. The parameters of the grooves: $a = 120$ nm, $h = 3600$ nm and $g = 4000$ nm

(For colored figure please scan the QR code on page 1)

We start from an initial set of parameters of the grooves, and change one of them in the order of a, h and g respectively while remaining the other two unchanged. Figure 5-64(a) shows the electric field intensity as a function of a, remaining $h = 3600$ nm and $g = 4000$ nm unchanged. The intensities increase first and then decrease when a increases. The largest one corresponds to $a = 100$ nm. Figure 5-64(b) shows the one as a function of h, remaining $a = 100$ nm and $g = 4000$ nm unchanged. The intensities increase first and then decrease when h increases. When h increases to 6000 nm, the intensity increases again. The largest one corresponds to $h = 4500$ nm. Figure 5-64(c) shows the one as a function of g, remaining $a = 100$ nm and $h = 4500$ nm unchanged. The intensities fluctuate when g increases. The largest one corresponds to $g = $

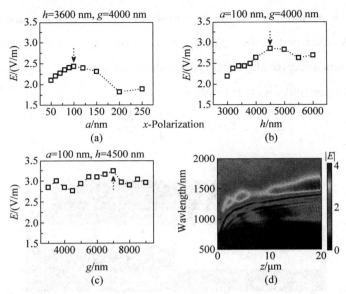

Fig. 5-64 Optimizing the structure by changing a, h and g. The incident light is x-polarized. Here, $R = 175$ nm, $D = 18000$ nm and $d = 10$ nm. (a) $h = 3600$ nm, $g = 4000$ nm, E varying with a. (b) $a = 100$ nm, $g = 4000$ nm, E varying with h. (c) $a = 100$ nm, $h = 4500$ nm, E varying with g. The intensities of E of (a)~(c) are all at $z = 20$ μm and $\lambda = 1550$ nm. The inset red arrows point out the locally optimal parameters. (d) Mode distributions of the optimal structure with the parameters of $a = 100$ nm, $h = 4500$ nm and $g = 7000$ nm. The position of line along z direction is at $r = 173$ nm and $\phi = \pi/2$. Color bar stands for $|E|$
(For colored figure please scan the QR code on page 1)

7000 nm. After the three-step optimizing, we achieve the optimal structure, the mode distributions of which are shown in Fig. 5-64(d). The electric field is mostly localized near the output side of the NW, the resonance wavelength is at 1550 nm, and the corresponding field enhancement factor is EF = 10.5.

As a brief summary, to enhance the propagation properties of CdSe NW at 1550 nm with x-polarized incident light, the NW-MgF$_2$-Ag-groove hybrid structure is employed, with the parameters: $R = 175$ nm, $d = 10$ nm, $D = 18000$ nm, $a = 100$ nm, $h = 4500$ nm and $g = 7000$ nm, resulting in EF = 10.5.

Thirdly, in order to obtain non-reciprocal property with x-polarized illuminating, the symmetry of the system should be broken. In our case, it is the symmetry about xy plane of the grooves that must be broken, as shown in

Fig. 5-65. The parameters of the left grooves (a_l, h_l and g_l) are kept as the optimal ones described above. While the ones of the right part are changed to achieve the non-reciprocal property. To describe quantitatively, we define the contrast of the output electric fields as:

$$\text{Contrast} = \frac{I_l - I_r}{I_l + I_r}. \tag{5-156}$$

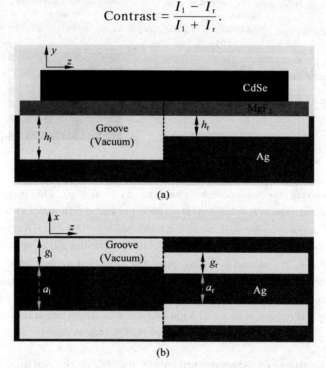

(a)

(b)

Fig. 5-65 Schematic of the asymmetric system. (a) Side view (yz plane) and (b) top view (xz plane, the NW and MgF$_2$ are not drawn for clarify). The grooves are separated into two parts by xy plane in the middle of the NW. The parameters of the grooves, i. e., a, h and g, are marked with subscripts "l" and "r" referring to the ones of left and right parts, respectively. Here, $R = 175$ nm, $D = 18000$ nm and $d = 10$ nm

(For colored figure please scan the QR code on page 1)

Here, I_l and I_r stand for the output power of the electric fields when illustrated from left and right ends, i. e., propagating along forward ($+z$) direction and backward ($-z$) direction, respectively. Notice that the power is directly proportional to the square of the electric field intensity in the corresponding end of the NW, which can be described as: $I_j = E_j^2$, $j = l, r$. Here, E_l and E_r

stand for the electric intensity at the output end of the NW when illustrated from left and right ends, respectively.

We start from an initial set of parameters of the grooves, i.e., $h_r = 4500$ nm, $a_r = 100$ nm and $g_r = 7000$ nm, and we will change one of them in the order of h_r, a_r and g_r respectively while remaining the other two unchanged. Figure 5-66(a) shows the contrast and the electric field intensity as a function of h_r, remaining $a_r = 100$ nm and $g_r = 7000$ nm unchanged. As h_r decreases, E_1 decreases; while the contrasts show fluctuation. The value of contrasts reaches a maximum of 0.32 at $h_r = 800$ nm, with $E_1 = 1.9$ V/m. Figure 5-66(b) shows the ones as a function of a_r, remaining $g_r = 7000$ nm and $h_r = 800$ nm unchanged. As a_r increases, E_1 slightly decreases; while the contrast reaches its maximum of 0.43 at $a_r = 20$ nm, with $E_1 = 2.0$ V/m. Figure 5-66(c) shows the ones as a function of g_r, remaining $h_r = 800$ nm and $a_r = 20$ nm unchanged. As g_r decreases, E_1 firstly remains almost unchanged, and then decreases when $g_r < 2000$ nm; while the contrast behaves similarly, with a sudden increase when $g_r < 2000$ nm, reaching its maximum of 0.91 at $g_r = 200$ nm, with $E_1 = 1.0$ V/m. However, at $g_r = 600$ nm, the contrast and E_1 of which are 0.89 V/m and 1.44 V/m, respectively. The difference of the contrasts between them is little, while the difference of E_1 between them is considerable. Therefore, after the three-step optimizing, we achieve the optimal non-reciprocal structure, the parameters of which are $h_r = 800$ nm, $a_r = 20$ nm and $g_r = 600$ nm.

To better understand the results, the mode distributions of the optimal structure are shown in Fig. 5-66(d) and (e) for propagating along $+z$ and $-z$ directions, respectively. For $+z$ direction, the modes of the left part show standing waves with strong amplitudes at the wavelengths from about 1000 nm to about 1700 nm. This is due to the interference between the propagating and reflecting waves. The latter is introduced by the interface (xy plane) at the middle of the grooves, which origins from the asymmetry of the system. When the propagating wave travels to the interface, reflection and transmission occur simultaneously. The transmitted wave continues to travel along the right part

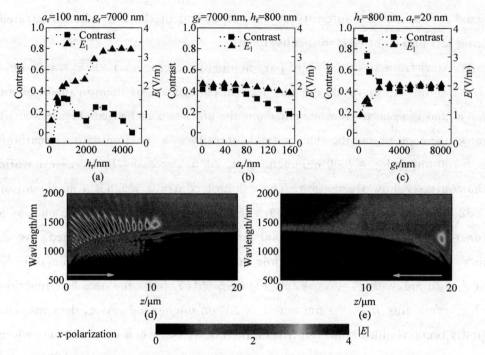

Fig. 5-66 Optimizing the non-reciprocal structure by changing h, a and g. The incident light is x-polarized. Here, $R = 175$ nm, $D = 18000$ nm, $d = 10$ nm, $h_1 = 4500$ nm. $a_1 = 100$ nm and $g_1 = 7000$ nm. (a) $a_r = 100$ nm, $g_r = 7000$ nm, the contrast and E_1 varying with h_1. (b) $g_r = 7000$ nm, $h_r = 800$ nm, the contrast and E_1 varying with a_r. (c) $h_r = 800$ nm, $a_r = 20$ nm, the contrast and E_1 varying with g_r. The intensities of E_1 of (a)~(c) are all at $z = 20$ μm and $\lambda = 1550$ nm. (d) and (e), mode distributions of the optimal non-reciprocal structure illustrated from left and right ends, respectively (propagating directions marked with yellow arrows), with the parameters of $h_r = 800$ nm, $a_r = 20$ nm and $g_r = 600$ nm. The position of line along z direction is at $r = 173$ nm and $\phi = \pi/2$. Color bar stands for $|E|$
(For colored figure please scan the QR code on page 1)

of the system. This results in the effective propagation from left end to right end at 1550 nm with $+z$ direction. For $-z$ direction, the modes of the right part also show standing waves but with weak amplitudes. Similar analysis as above, when the wave travels to the interface, the transmitted wave with weak amplitude would hardly travel along the left part of the system. This results in the weak electric field at the output end at 1550 nm with $-z$ direction. Therefore, the contrast of the system is high enough.

As a brief summary, to achieve high contrast of non-reciprocal property of CdSe NW at 1550 nm with x-polarized incident light, NW-MgF$_2$-Ag-groove hybrid structure is employed, with the parameters: $R = 175$ nm, $d = 10$ nm, $D = 18000$ nm, $a_1 = 100$ nm, $h_1 = 4500$ nm, $g_1 = 7000$ nm, $a_r = 20$ nm, $h_r = 800$ nm and $g_r = 600$ nm, resulting in the high contrast of 0.89, with EF = 2.07 for the forward propagation.

So far, we gain the above performances by tuning the parameters of the nanostructures. How do these structures control the behaviors of the photons?

Firstly, the combination between the NW and Ag substrate allows the SPPs supported by the latter to enhance the electric field intensity in the NW, thus being beneficial to the propagation property of the NW. By tuning the radius of the NW, the resonance wavelength can be located at the one we need. Secondly, MgF$_2$ layer is employed to separate the NW and Ag substrate, thus forming a nanogap between them. This gap results in the cavity effect that mostly supports y-polarized modes of SPPs. Furthermore, the SPPs supported on Ag surface helps to enhance the electric field intensity in the gap, as well as in the NW. Thirdly, the two nanogaps are formed due to the grooves. There gaps can support x-polarized modes of SPPs. Also, the SPPs supported in these gaps help to enhance the electric field intensity in both the gaps and the NW, shown in Fig. 5-66(b). By tuning the radius of the NW, the resonance wavelength can be located at the one we need. Fourthly, the non-reciprocity is achieved by breaking the symmetry of the structures. Different parameters of the grooves support different modes. The contact surface ($z = 10$ μm) of the two kinds of grooves behaves as a "mirror". For $+z$ propagation (Fig. 5-66(d)), the electromagnetic wave is reflected by this "mirror", thus forming standing wave and resonance at the wavelength of 1550 nm. This resonance enhances the field intensity and helps propagate to the right end of the NW. However, for $-z$ propagation (Fig. 5-66(e)), although weak standing wave is obtained, the right grooves could not support a strong resonance, thus little field enhancement. Hence, the wave seldom propagates to the left end of the NW.

That is the reason why the non-reciprocity is achieved.

In summary, in this section, we design the hybrid structures for CdSe NW to strongly enhance its propagation properties at communication wavelength of 1550 nm with y-polarization and x-polarization, resulting in the localized electric field near the output side of the NW with the field enhancement factor of EF = 13.2 and EF = 10.5, respectively. The former consists of NW, MgF_2 substrate and Ag substrate, while the latter consists of NW, MgF_2 substrate, Ag substrate and two grooves. Moreover, the non-reciprocal property is obtained by breaking the symmetry of the system at λ = 1550 nm. Details of the parameters for each situation are summarized in Table 5-4. This kind of nanostructure could be easily expanded to other wave bands such as visible and infrared regions by tuning the parameters of the structure. We hope our results are helpful to the applications in photonic integrated circuit and optical communications.

Table 5-4 The optimal parameters for each situation, i. e., propagation property of y- and x-polarized illuminations, and non-reciprocal property of x-polarized illumination

Parameters (unit: nm) Polarization	Propagation property		Non-reciprocal property
	y	x	x
R	110	175	175
d	10	10	10
D	150	18000	18000
a (or a_1)		100	100
h (or h_1)		4500	4500
g (or g_1)		7000	7000
a_r			20
h_r			800
g_r			600
EF			2.07
Contrast			0.89

5.5 Unified treatments for LSPs and PSPs

In this section, we theoretically investigated the SPP modes supported in a metal cylinder with hemispheres on both ends. When the length of the cylinder is small (i.e., dozens of nm), it can be considered as a metal nanorod, which supports LSPR modes[77]. When the length of the cylinder is large enough (i.e., several μm), it can be considered as a metal NW, which support PSPP modes. We employ Maxwell's Equations to approximately solve out the electric fields on the intersurface between the metal and the surrounding dielectric, employing both cylindrical and spherical coordinates. Analyzing the propagation length of metal NWs and the transmitted spectra of both metal nanorods and NWs, we obtain a unified formula to describe both the modes of LSPR and PSPP. Therefore, we conclude that LSPR and PSPP result from the same mechanism by combining the propagating modes together with the constructive interference effect. Firstly, we develop the model which can describe the behavior of the electromagnetic fields. Secondly, we employ the obtained formula to calculate the propagation length and the transmitted spectra of metal nanorods and NWs.

The schematic of the metal NW is shown in Fig. 5-67(a). Here, the metal NW with the relative dielectric constant of ε_m is surrounded by the dielectric with the relative dielectric constant of ε_d. The length (cylinder part) and the radius of the NW are L and R, respectively (total length $L_0 = L + 2R$). Specially, the two ends of the NW are smoothly consisted of hemispheres with the radius of R, indicating a real structure compared with the samples in the experiments. When excited by an incident light, the electromagnetic field propagates as a surface wave through the interface between the metal and the dielectric.

First of all, for simplicity, we only consider a NW with infinite length and

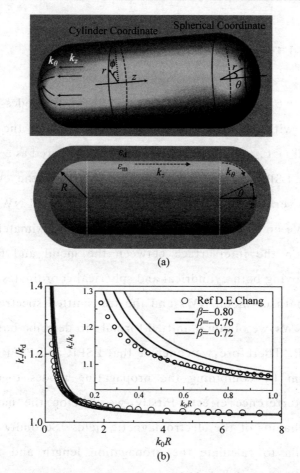

Fig. 5-67 Schematic of a metal NW surrounded by a dielectric for 3-dimensional (upper) and sectional (under) views, respectively. The length (cylinder part) and the radius of the NW are L and R, respectively. k_z and k_θ stand for the wave vectors of z direction and θ direction in cylindrical and spherical coordinates, respectively. ε_m and ε_d stand for the relative dielectric constants of metal and dielectric, respectively. (b) The ratios of k_z to k_d as a function of $k_0 R$. Solid lines stand for our results of different parameters β. Open circles are the results of Ref.[73]. The conditions here are the same: $\varepsilon_d = 2$, $\varepsilon_m = -50$ and the vacuum wavelength $\lambda_0 = 1000$ nm. The inset is the zoom of $k_0 R$ from 0.2 to 1.0

(For colored figure please scan the QR code on page 1)

solve the Maxwell Equations in cylindrical coordinate (r, ϕ, z) approximatively. For the fundamental mode, the electric and magnetic fields can be described approximatively as follows:

CHAPTER 5 Surface Plasmons

$$\boldsymbol{E} = r^{\alpha} \begin{pmatrix} a_{r,j} \\ 0 \\ a_{z,j} \end{pmatrix} \exp\left[\pm ik_{r,j}(r-R) + ik_{z,j} - i\omega t\right], \qquad (5\text{-}157)$$

$$\boldsymbol{H} = r^{\beta} \begin{pmatrix} 0 \\ b_{\phi,j} \\ 0 \end{pmatrix} \exp\left[\pm ik_{r,j}(r-R) + ik_{z,j} - i\omega t\right]. \qquad (5\text{-}158)$$

Here, α and β are the introduced parameters describing the damping of the fields propagating through r direction, a_r, a_z and b_ϕ are the amplitudes of electric and magnetic fields. The subscript $j = d$ and m, correspond to the situations of dielectric and metal, respectively. The wave vectors k_r and k_z stand for the directions along r and z. The signs "+" and "−" before $ik_{r,j}$ correspond to the fields in dielectric and metal, respectively. ω is the circular frequency of the wave. Although the imaginary part of k_r has already describe the damping of the field, it is not accurate enough. Therefore, we should introduce α and β to additionally describe the damping together with k_r, to develop our approximate model more accurately. We employ Maxwell's equations as follows:

$$\nabla \times \boldsymbol{H} = \varepsilon_0 \varepsilon_j \frac{\partial \boldsymbol{E}}{\partial t} = -i\varepsilon_0 \varepsilon_j \omega \boldsymbol{E}, \quad j = d \text{ and } m, \qquad (5\text{-}159)$$

together with the boundary conditions at $r = R$:

$$\varepsilon_d a_{r,d} = \varepsilon_m a_{r,m}, \quad a_{z,d} = a_{z,m}. \qquad (5\text{-}160)$$

The relation between $k_{r,d}$ and $k_{r,m}$ is obtained:

$$\frac{\beta + 1 + ik_{r,d}R}{\beta + 1 - ik_{r,m}R} = \frac{\varepsilon_d}{\varepsilon_m}, \quad k_{z,d} = k_{z,m} = k_z \qquad (5\text{-}161)$$

Besides, the relations among the wave vectors are obviously satisfied as follows:

$$k_j^2 = k_0^2 \varepsilon_j = k_{r,j}^2 + k_{z,j}^2, \quad j = d \text{ and } m, \qquad (5\text{-}162)$$

where $k_0 = \omega/c$ is the wave vector in vacuum. Combining Eqs. (5-161) and (5-162) we can solve out the wave vectors in dielectric and metal as follows:

$$k_{r,d} = \frac{i}{R} \frac{(1+\beta)\varepsilon_m - \varepsilon_d \Delta}{\varepsilon_d + \varepsilon_m}, \qquad (5\text{-}163)$$

$$k_{r,m} = -\frac{i}{R}\frac{(1+\beta)\varepsilon_d - \varepsilon_m \Delta}{\varepsilon_d + \varepsilon_m}, \tag{5-164}$$

$$k_z^2 = k_0^2 \frac{\varepsilon_d \varepsilon_m}{\varepsilon_d + \varepsilon_m} + \frac{(\varepsilon_m^2 + \varepsilon_d^2)(1+\beta)^2 - 2(1+\beta)\varepsilon_d \varepsilon_m \Delta}{R^2(\varepsilon_d + \varepsilon_m)^2}, \tag{5-165}$$

where $\Delta = \sqrt{(1+\beta)^2 - k_0^2 R^2 (\varepsilon_d + \varepsilon_m)}$.

We notice that Eq. (5-165) reduces to $k_z = k_0 \sqrt{\frac{\varepsilon_d \varepsilon_m}{\varepsilon_d + \varepsilon_m}}$ when the radius $R \to \infty$, corresponding to the PSPP wave vector along a planar interface, which confirms the correctness of our approximation. Furthermore, we also compare our results with the ones of Chang et al.[73] in Fig. 5-67(b). Here, $k_z = \sqrt{\varepsilon_d} k_0$. We notice that k_z decreases as the radius of the cylinder R increases. Actually, according to Eq. (5-165), when R is small enough, this relationship should be reciprocal, i.e., $k_z \propto 1/R$. Also notice that at the condition of Fig. 5-67(b), according to Eq. (5-163), the perpendicular component of the wave vector $k_{r,d}$ is pure imaginary, resulting in the confined and nonradiative modes near the interface. Moreover, different values of parameter β are illustrated. It indicates that larger $|\beta|$ agrees better with the smaller R part, and smaller $|\beta|$ agrees better with the larger R part. In this work, we mainly consider the radii at nanoscale, which correspond to the small R part. Here, $\beta = -0.80$ is the optimal one for $k_0 R < 1$ with the result that agrees well with the one of Chang. Therefore, in the rest of this work, we utilize $\beta = -0.80$ as a good approximation.

Employing Eq. (5-165), we can calculate the propagation length of a metal NW emerged in a dielectric. The propagation function of the electric field can be written as

$$E = E_0 \exp(ik_z z - i\omega t). \tag{5-166}$$

Here, E_0 is the initial intensity of the electric field, and $k_z = k_1 + ik_2$ is a complex parameter because ε_m is a complex number, thus obtaining

$$E = E_0 \exp(-k_2 z)\exp(ik_1 z - i\omega t). \tag{5-167}$$

Hence, considering the energy propagation, which is proportional to E^2, the propagation length of SPPs can be described as

$$L_{\text{SPP}} = 1/(2k_2). \quad (5\text{-}168)$$

Here, k_2 is the imaginary part of k_z from Eq. (5-165). This formula will be used to analyze the propagation length of a metal NW surrounded by a dielectric.

Secondly, due to the shapes on both sides of the NW, when the wave propagates to one end, the spherical coordinate (r,θ,ϕ) should be employed. The derivative process is almost the same. In spherical coordinate, electromagnetic fields of the fundamental mode are written as

$$\boldsymbol{E} = r^{\alpha} \begin{pmatrix} a_{r,j} \\ a_{\theta,j} \\ 0 \end{pmatrix} \exp\left[\pm ik_{r,j}(r-R) + ik_{\theta,j}r(\pi/2 - \theta) - i\omega t\right], \quad (5\text{-}169)$$

$$\boldsymbol{H} = r^{\beta} \begin{pmatrix} 0 \\ 0 \\ b_{\phi,j} \end{pmatrix} \exp\left[\pm ik_{r,j}(r-R) + ik_{\theta,j}r(\pi/2 - \theta) - i\omega t\right]. \quad (5\text{-}170)$$

Here, the fundamental mode stands for the zero component of E_ϕ, H_r, and H_θ. We employ Eq. (5-159) and boundary conditions at $r = R$:

$$\varepsilon_d a_{r,d} = \varepsilon_m a_{r,m}, \quad a_{\theta,d} = a_{\theta,m} \quad (5\text{-}171)$$

The relation between $k_{r,d}$ and $k_{r,m}$ is obtained:

$$\frac{\beta + 1 + ik_{r,d}R}{\beta + 1 - ik_{r,m}R} = \frac{\varepsilon_d}{\varepsilon_m}, \quad k_{\theta,d} = k_{\theta,m} = k_\theta. \quad (5\text{-}172)$$

Besides, the relations among the wave vectors are obviously satisfied as follows:

$$k_j^2 = k_0^2 \varepsilon_j = k_{r,j}^2 + k_{\theta,j}^2, \quad j = d \text{ and } m. \quad (5\text{-}173)$$

Notice that Eqs. (5-161) and (5-162) are similar to Eqs. (5-172) and (5-173). Hence, the wave vectors $k_{r,j}$ are the same as Eqs. (5-163) and (5-164), and $k_\theta = k_z$.

Now, we assume that the wave propagates circularly along the surfaces, i.e., surfaces of the cylinder and the sphere. Due to the continuity of the wave vector, β for sphere is identical to that for cylinder. When the wave propagates along the sphere surface, it would concentrate to the "tip" of the sphere. In other words, the energy line density (ELD) varying with θ is approximately described as (neglecting the loss of the energy):

$$\text{ELD} = \frac{E_0^2}{2\pi R \sin\theta}. \tag{5-174}$$

This indicates that the "tip" is at $\theta = 0$, resulting in an infinity energy, which is impossible. In other words, when employing this classical view (propagation of wave) to describe the electric field, it will cause the "tip catastrophe", i.e., the divergency at small θ. This may be concerned with the quantum theory. Here, in the classical view, we assume that the energy should be transmitted or reflected through the interface in order to avoid the divergency at $\theta = 0$. That is, the wave would simultaneously reflect and transmit at one end ($\theta = 0$) of the NW, every time it travels to the end. The former continues to propagate along the surface, while the latter propagates in the free space. To obtain the latter, we define the incident (E_I), reflected (E_R) and transmitted (E_T) electric fields as

$$\begin{cases} E_I = E_{I0} \exp\left[-ik_{r,m}(r-R) + ik_\theta r(\pi/2 - \theta) - i\omega t\right], \\ E_R = E_{R0} \exp\left[ik_{r,m}(r-R) + ik_\theta r(\pi/2 - \theta) - i\omega t\right], \\ E_T = E_{T0} \exp\left[ik_d(r-R) - i\omega t\right]. \end{cases} \tag{5-175}$$

The boundary conditions at $r = R$ and $\theta = 0$ are

$$\begin{cases} E_I + E_R = E_T \\ \frac{\partial}{\partial r}(E_I + E_R) = \frac{\partial}{\partial r}E_T. \end{cases} \tag{5-176}$$

Combining Eqs. (5-175) and (5-176), we obtain the reflectivity (Refl) and transmissivity (Trans) of one circular of the wave propagating on the NW:

$$\begin{cases} \text{Refl} = \dfrac{E_{R0}}{E_{I0}} = \dfrac{k_{r,m} + k_d - k_\theta \pi/2}{k_{r,m} - k_d + k_\theta \pi/2}, \\ \text{Trans} = \dfrac{E_{T0}}{E_{I0}} = \dfrac{2k_{r,m}\exp(ik_\theta R\pi/2)}{k_{r,m} - k_d + k_\theta \pi/2}. \end{cases} \tag{5-177}$$

To calculate the final emission of the wave, all the circulars should be considered. Besides, when the NW is surrounded by the dielectric, the transmissivities $\frac{2k_0}{k_0+k_d}$ and $\frac{2k_d}{k_0+k_d}$ should be added to the incident and transmit wave, respectively. Because in experiments, the light is always incident from vacuum (or air) in the beginning and detected in vacuum (or air) in the end. These two coefficients are deduced from a simple model with a plane boundary between vacuum and dielectric, and the plane wave transfers from one side to the other side. The details can be found in Ref.[16]. A coefficient of geometric, $q = \text{Refl}^2 \exp[2i(k_z L + k_\theta R\pi/2)]$, which results from the phase difference in the propagation, is added to the propagation function through every circular. Here, Refl^2 stands for the two reflections at the two ends, $\exp(2ik_z L)$ stands for the phase difference through a round trip in the cylinder, and $\exp(2ik_\theta R\pi/2)$ stands for the addition phase when the wave propagates from $\theta = 0$ to $\theta = \pi/2$ that occurs at both hemispheres. Then, the final transmissivity (in energy) of the wave is obtained as

$$I_T = \left| \frac{2k_0}{k_0+k_d} \frac{\text{Trans}}{1-q} \frac{2k_d}{k_0+k_d} \right|^2. \quad (5\text{-}178)$$

This formula will be used to analyze the modes of a metal NW in different dielectrics. The propagation length is a significant property of NW, Fig. 5-68 shows the propagation length of Ag NW with different radii and dielectric constants. The permittivity of Ag is from Ref.[85]. Solid lines are calculated from our model, while open squares and circles are the experimental data from Ref. [86]. It indicates that LSPP increases with the increase of incident wavelength. Particularly, Fig. 5-68(a) shows that LSPP increases with the increase of the radius, while Fig. 5-68(b) shows that LSPP decreases with the increase of ε_d. The behaviors of these results are similar to other researchers' qualitatively. However, a quantitative comparison between our model and experiments is impossible. Because the exact radii of the measured NW are not always reported, and the experiments on the NW are always done on a slide rather than surrounded by the dielectric. Here, we compare our results with the experimental ones from Shegai et al.[86], where they measured the Ag NWs with the radii of about 25~100 nm on a glass slide. Notice that ε_d of glass is about 2.1. This indicates that our results agree well with their experiments.

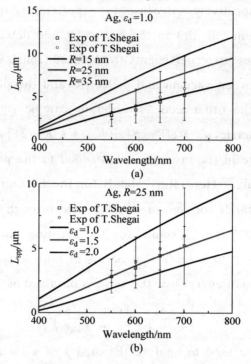

Fig. 5-68　The propagation length of Ag NW varying with the incident wavelength. The solid lines are calculated from our model. The open squares and circles with error bar are the experimental data from Ref. [86], the typical radii of which are 25~100 nm. (a) LSPP of different radii of NW is calculated at the dielectric constant of $\varepsilon_d = 1.0$. (b) LSPP of different dielectric constants is calculated at the radius of $R = 25$ nm

(For colored figure please scan the QR code on page 1)

The transmitted spectrum is another important property of NW. Figure 5-69 shows the transmitted spectra of typical Ag NWs with lengths at several μm and radii at dozens of nm. The spectra show multiple modes which are caused by the constructive interference of the electric field in NW. This phenomenon has been illustrated in Ref. [87] and our previous study[78]. Particularly, Fig. 5-69(a) indicates that the longer the NW is, the more modes it supports. The relative intensities of the spectra decrease as the length increases. Because the loss is larger when the electric wave propagates for a longer distance. Figure 5-69(b)

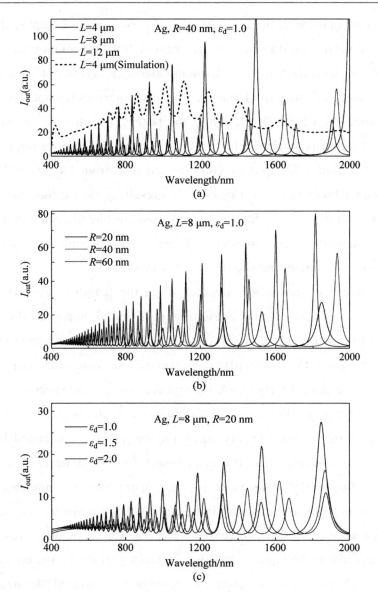

Fig. 5-69 The transmitted spectra of Ag NWs with different conditions. (a) $R = 40$ nm and $\varepsilon_d = 1.0$, varying with the length L. Dashed line stands for the simulated one with $L = 4$ μm employing FDTD method. (b) $L = 8$ μm and $\varepsilon_d = 1.0$, varying with the radius R. (c) $L = 8$ μm and $R = 20$ nm, varying with the dielectric constant ε_d
(For colored figure please scan the QR code on page 1)

indicates that the wider the NW is, the more modes it supports. Figure 5-69(c) indicates that the larger ε_d is, the more modes it supported. The intensities of the spectra decrease with the increasing ε_d. Besides, we also employ the FDTD method to simulate the transmitted spectrum of NW for comparison, as shown in Fig. 5-69(a) with dashed line. The simulated one agrees not very well with the model one, i.e., the gap of the resonance frequencies is imperfectly equal, and the former is a little smaller than the latter. The differences are mainly caused by the approximation in introducing parameter β. For example, for $R = 40$ nm, $k_0 R$ is from 0.628 to 3.14 (the wavelength is from 400 nm to 2000 nm), which crosses a large range in Fig. 5-67(b), resulting in the inaccuracy of the fixed parameter $\beta = -0.8$. In addition to this, the phenomenon that multiple modes are supported are consistent. These results provide help and deeper understanding in controlling the modes of a metal NW.

Furthermore, the results indicate that when the length of the NW is small enough (i.e., dozens of nm), resulting in a metal nanorod, the reducing number of the modes may make it a single mode in the transmitted spectrum. Figure 5-70 shows the transmitted spectra of Au nanorods with different conditions. The permittivity of Au is from Ref. [85]. The spectra show single mode at visible range. Figure 5-70(a) indicates that the peak of the mode red-shifts with an increasing intensity when the length of the nanorod increases. Figure 5-70(b) indicates that the peak blueshifts with a decreasing intensity when the radius of the nanorod increases. Figure 5-70(c) indicates that the peak red-shifts with an increasing intensity when the dielectric constant ε_d of the environment increases. These results agree well with other researchers' studies qualitatively, and quantitatively in certain degree[88]. Moreover, we also employ FDTD method to simulate the resonance spectra of Au nanorods in Fig. 5-70(a) with dashed lines. There are slight differences in the resonance peaks between the simulated and model ones, with several to dozen nm. The reason of the differences is almost the same as the one in Fig. 5-69(a), i.e., the approximation of β. For example, for $R = 10$ nm, $k_0 R$ is from 0.114 to 0.156 (the wavelength is from 550 to 750 nm). We notice that from the inset of Fig. 5-67(b), when $k_0 R < 0.2$, the results of our model will be away from the

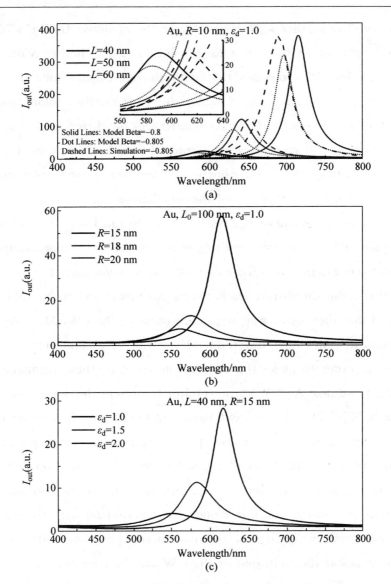

Fig. 5-70 The transmitted spectra of Au nanorods with different conditions. (a) $R = 10$ nm and $\varepsilon_d = 1.0$, varying with the length L. Solid and dot lines stand for the cases of $\beta = -0.8$ and $\beta = -0.805$, respectively. Dashed lines stand for the corresponding normalized simulated ones employing FDTD method. Inset is the zoom for the wavelength from 560 nm to 640 nm. (b) $L_0 = 100$ nm and $\varepsilon_d = 1.0$, varying with the radius R. (c) $L = 40$ nm and $R = 15$ nm, varying with the dielectric constant ε_d

(For colored figure please scan the QR code on page 1)

precise ones. The smaller $k_0 R$ is, the larger the error is. In Fig. 5-70(a), we change the value of β from -0.8 to -0.805 to reveal the behaviors of these resonance modes. It is obvious that with the increasing value of $|\beta|$, the resonance peak of the Au nanorod is blue shifted (solid lines to dot lines). The difference of $L = 60$ nm one becomes smaller, but the differences of $L = 40$ nm and $L = 50$ nm ones become larger. Besides, the shift is very sensitive with β. Hence, in addition to the parameter β, another reason why there are differences between our theory and the simulations may be the hypothesis of the output field at the end of the NW, as described in Eq. (5-177), which may be an approximate hypothesis, resulting in the inaccurate results compared with the simulated ones. In general, it may be the approximations of both β and Eq. (5-177) that cause the difference between our model and the simulations. In addition to this, they agree well with each other in the FWHM. This result is also of great help in controlling the modes of the metal nanorods.

Besides, in order to make the physical meaning of these resonance modes more clearly, we take Ag NW as an example to present the mode distributions, as shown in Fig. 5-71. The incident angle and the electric field intensity E_0 of the incident plane wave are 45° and 1 V/m, respectively, with p-polarization and illuminating from the left end of the Ag NW. The simulation step length is 2 nm for both x and y directions, and 100 nm for z direction. It indicates that, for a certain position z, multiple modes occur periodical with the wavelength λ, and for a certain wavelength λ, the mode occurs periodical with the position z. These modes at the right end of the NW are the strongest. It reveals that there exist standing waves that occur due to the interference between the PSPP and its reflective one, which agrees with the physical process that we assume.

Now we notice that the modes in metal NW and in metal nanorods are from the same equation, i.e., Eq. (5-178). Usually, we call the modes of metal NW as PSPP and the modes of metal nanorods as LSPR. Here, from the analyzing above, we find that LSPR and PSPP are the same in physics. The only difference is that the former is likely "localized" because of the small size, and

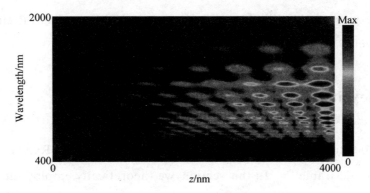

Fig. 5-71 Mode distributions inside the Ag NW along the z direction with $L = 4$ μm and $R = 40$ nm. The position of line along z direction is $r = 38$ nm and $\phi = 0$. The vertical coordinate stands for the wavelength in vacuum. The color bar stands for the intensity of the electric field $|E|$
(For colored figure please scan the QR code on page 1)

the latter is likely "propagated" because of the large size in length. However, the transmitted spectra in this study indicate that the mechanisms of LSPR and PSPP are the same, resulting from the propagation and the constructive interference of SPPs, no matter the size of the metal structure is. This illustrates that SPPs will also propagate (surface of the nanorod) in small size case and localize (tip of the NW) in large size case.

In summary, in this section, we employ Maxwell's Equations in both cylindrical and spherical coordinates to solve out the electric fields on the interface between the metal NW (or nanorod) and the surrounding medium. The propagation length is then analyzed and compared with other researchers' experimental data, resulting in good agreements. We obtain the transmitted spectra of both NW and nanorod, which show good agreements qualitatively with other studies and our previous study. The calculated spectra are based on the same formula so that we conclude the LSPR mode supported in metal nanorod and the PSPP mode supported in metal NW result from the same mechanism, which combines the propagating modes with their constructive interference effect. This work provides an innovative understanding on the SPP modes, i. e., unifies the LSPR and PSPP modes, which provides an alternative way to describe the phenomena of LSPR and PSPP simply and quantitatively.

5.6 Plexciton in TERS and in PSPs

New methods to improve the properties of surface-enhanced Raman scattering and PSP are desirable[78]. In this section, we theoretically employ the plasmon-exciton (plexciton) interaction to enhance the spectra in LSP and PSP. The tip-enhanced resonance Raman spectroscopy (TERRS) configuration is used to enhance both Stokes and anti-Stokes Raman spectra of rhodamine 6G simultaneously. We achieve a huge enhancement factor of $10^{13} \sim 10^{15}$. Furthermore, we obtain the plexciton by employing the Ag nanowire coated with rhodamine 6G, which improves the propagation properties.

The schematic of our calculated TERRS system is shown in Fig. 5-72. The three-dimensional finite-difference time-domain (3D-FDTD) method is employed to simulate the optical properties of the TERRS system with a cubical simulation region with size of 800 nm × 800 nm × 1000 nm. The boundaries utilize perfectly matched layer boundary conditions to avoid disturbance of the boundary reflection. TERRS consists of an Ag tip and an Ag substrate, and the distance between them is D. The R6G layer with the thickness of $d = 1$ nm is covered on Ag substrate. The apex radius and the full cone angle of the Ag tip is set as r and 20°, respectively. The nonuniform mesh step is 0.2 nm, and a smaller mesh step of 0.05 nm in z direction inside the gap between the tip and substrate is used. The permittivity of Ag and R6G are obtained from Ref. [85] and Ref. [89] (R6G monomer), respectively. A p-polarized plane wave with the incident angle of 60° and electric field amplitude of 1 V/m illuminates the system. The resonance Raman and normal Raman spectra of R6G are calculated by the density functional theory (DFT) and simulated by Gaussian 16 software.

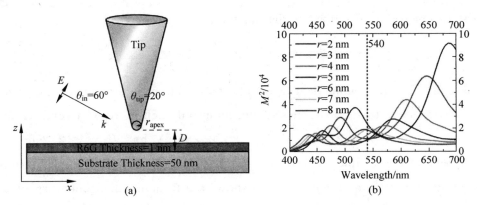

Fig. 5-72 (a) Schematic of TERRS configuration with Ag tip and Ag substrate. R6G layer with the thickness of 1 nm is covered on the substrate with the thickness of 50 nm. Tip-substrate distance and apex radius are D and r, respectively. The incident angle of the p-polarized plane wave and the full cone angle of the Ag tip are 60° and 20°, respectively. (b) The localized electric enhancement spectra varying with r. The vertical dashed line stands for the absorption peak of R6G

(For colored figure please scan the QR code on page 1)

In order to enhance the Raman signal from R6G effectively, the strong coupling between LSP and R6G, or plasmon and exciton, is necessary. Therefore, we tune the resonance the TERRS system close to the absorption of R6G by tuning the size of the structure. By defining the local electric enhancement $M^2 = |E_{\mathrm{TERRS}}/E_{\mathrm{vacuum}}|^2$, where $E_{\mathrm{vacuum}} = 1$ V/m, Fig. 5-72(b) shows that the electric field enhancement sharply decreases and LSP peaks blue shift with the increase of tip radius. Here, the distance is $D = 1$ nm. The LSP modes supported within the gap owe to the coupling between the tip and the substrate. The smaller the radius is, the stronger the "tip effect" is, resulting in larger electric field enhancement. Among these spectra, one of the modes of "$r = 8$ nm" matches the absorption of R6G (about 540 nm). Hence, in the rest of this work, all calculations are at the condition of $r = 8$ nm.

The LSP mode of the structure and the exciton mode of R6G couple strongly as shown in Fig. 5-73(a), resulting in the strong coupled plexciton. Two new modes of the plexciton are generated. One is the bright mode with lower energy E_- and the other is the dark mode with higher energy E_+. The energies of these two modes are described by the equation:

$$E_\pm = \frac{E_{plasmon} + E_{exciton}}{2} \pm \frac{\sqrt{(E_{plasmon} - E_{exciton})^2 + 4(\Delta E)^2}}{2}, \quad (5\text{-}179)$$

where ΔE is the couple degree of plasmon-exciton interaction, $E_{plasmon}$ and $E_{exciton}$ are the resonance energies of plasmon and exciton, respectively. Here, $E_+ = 2.436$ eV, $E_+ = 2.164$ eV, $E_{plasmon} = 2.289$ eV and $E_{exciton} = 2.296$ eV, resulting in the couple degree $\Delta E = 136$ meV. This indicates the strong coupling of plexciton. Figure 5-73(b) shows the Raman enhancement spectra. Here, the Raman enhancement factor is defined as M^4. When enhancing normal Raman spectra, the excitation wavelength is far away from the absorption band of R6G. In this case, generally, TERS system are employed corresponding to the red curve. The Raman enhancement factor reaches a maximum of about 7×10^8. However, TERS is only efficient for Stokes (anti-Stokes) Raman signals that are around the resonance energy, because anti-Stokes (Stokes) Raman signals are on the edge of the spectrum resulting in being far away from the resonance energy and little enhancement. While enhancing resonance Raman spectra, the excitation wavelength is near the absorption band of R6G, which is 540 nm here. The blue curve shows the resonant spectrum, with a dip around 540 nm and two peaks on both sides. The bright mode enhances Stokes Raman spectra with a maximum of about 6×10^9. Particularly, the dark mode enhances anti-Stokes Raman spectra with the enhancement factors around 4×10^8. This is the most important difference from the enhanced normal Raman. The anti-Stokes Raman signals are too weak to observe, whereas they can be highly enhanced by the dark mode of the plexciton. Besides, the anti-Stokes Raman signals are at the regime that the fluorescence barely exists, which results that the former is not disturbed by the latter. Therefore, it is significant for applications, especially for those in surface enhanced anti-Stokes Raman spectroscopy. Additionally, it has also been reported that the multipolar Raman scattering spectra exist in TERS, in which the surface plasmon gradient effect plays a significant role.

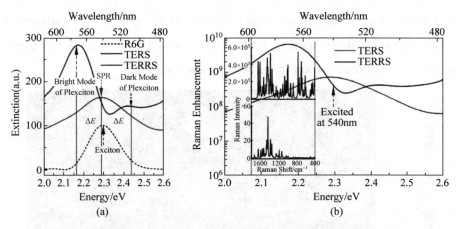

Fig. 5-73 (a) The enhancement spectra of TERS (red solid curve) and TERRS (blue solid curve), and the extinction spectrum of R6G (black dashed curve). (b) The Raman enhancement spectra of TERS (red solid curve) and TERRS (blue solid curve), respectively. The insets stand for the calculated resonance (up) and normal (down) Stokes Raman spectra, respectively. Here, for both (a) and (b), TERS and TERRS are calculated without and with R6G, respectively

(For colored figure please scan the QR code on page 1)

Furthermore, the resonance Raman spectra intensity is enhanced by a factor of about $10^4 \sim 10^6$ compared with the normal Raman spectra due to the chemical effect, as shown in the insets of Fig. 5-73(b). For example, the intensities of the Raman peaks at 1383 cm^{-1}, 1000 cm^{-1} and 616 cm^{-1} are enhanced by the factors of 2.4×10^4, 1.7×10^5 and 2.2×10^6, respectively. Combining the surface enhancement effects caused by TERRS with the chemical effects[90] caused by the resonance excitation, the total enhancement factor of Stokes Raman would be as high as about $10^{13} \sim 10^{15}$.

The behaviors of these modes are also investigated. Figure 5-74 shows the calculated intensity of the electric field distribution in the x-y plane of the bright, LSP and dark modes, respectively, varying with z, the distance from the apex of the tip. Here, $D = 1$ nm, indicating that the tip contacts with the R6G layer (if exists). Obviously, the bright mode displays the strongest intensity, while the dark mode displays the weakest one, and LSP mode displays the middle one. Besides, the multipoles of electric field occur in both bright and dark modes, which enhance the local electric field and do not radiate. The fields of them are more localized than those of LSP mode. As the distance from the apex increases, the intensities decrease rapidly.

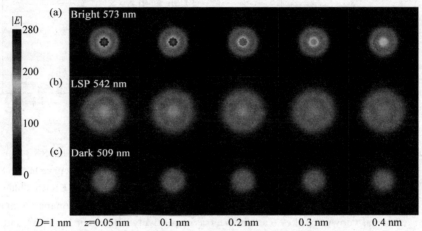

Fig. 5-74 Electric field distributions in xy-plane of bright (a), LSP (b) and dark (c) modes, respectively, varying with z, the distance from the apex. The peaks of the three modes are at 573 nm, 542 nm and 509 nm, respectively. The tip-substrate distance is $D = 1$ nm. The color bar stands for the electric field intensity which is in the unit of V/m. The size for each picture is 10 nm × 10 nm
(For colored figure please scan the QR code on page 1)

In order to get more information of the system, Fig. 5-75 shows charge density distributions in xz-plane ($y = 0$) for the same condition as Fig. 5-74. The charge density of LSP mode is the smallest, while the one of dark mode is the largest, which indicates that charge is most concentrated in dark mode. With the existence of R6G, the charge concentrates on both interfaces. Furthermore, accordingly, Fig. 5-76 shows the dipole moment of each mode varying with z. We notice that the dipole moment is obtained mostly on the interfaces ($z = -1$ and 0 nm). Particularly, the one of dark mode is not only nonzero but also the largest followed by the ones of bright and LSP modes. The total dipole moments are calculated for these modes through the integrals in Fig. 5-76, and give the results of 1.9 (bright), −3.64 (LSP) and −9.27 (dark). The negative sign corresponds to the phase difference of π. This indicates that "dark mode" is not dark, i.e., radiates as well as bright and LSP mode. This is the reason why dark mode can be employed to enhance the anti-Stokes Raman signals. Also notice that the dipole moment of LSP mode at $z = 0$ nm is negligible compared with the one at $z = -1$ nm.

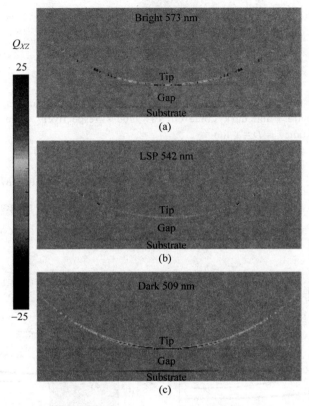

Fig. 5-75 Charge density distributions in xz-plane of bright (a), LSP (b) and dark (c) modes, respectively. The tip-substrate distance is $D = 1$ nm. The color bar stands for the normalized charge densities
(For colored figure please scan the QR code on page 1)

As shown more clearly in Fig. 5-77(a), the localized electric enhancement is stronger when it is closer to the apex. Besides, M^2 of the bright and dark modes decrease more rapidly than LSP mode does. M^2 in TERRS configuration is influenced by many factors. One of them is the tip-substrate distance D. Figure 5-77(b) shows M^2 of the bright mode with different D, and the thicknesses of R6G of them are all 1 nm. It is obvious that, the closer the tip is to the substrate, the larger M^2. Particularly, M^2 occurs a sudden change

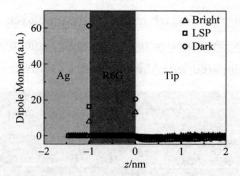

Fig. 5-76 Distance dependent dipole moments of bright (red triangles), LSP (black squares) and dark (blue circles) modes, respectively. The tip-substrate distance is $D = 1$ nm

(For colored figure please scan the QR code on page 1)

at the interface of R6G and vacuum, resulting in the smaller M^2 inside R6G, which prevents the applications of TERRS. Therefore, the largest M^2 occurs at the tip-substrate distance of $D = 1$ nm, corresponding to the contact between tip and R6G.

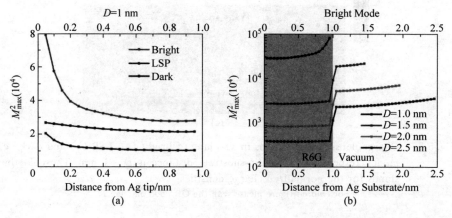

Fig. 5-77 (a) The localized electric enhancement factor of the bright (red), LSP (black) and dark (blue) modes, respectively, varying with z, the distance from the apex. The tip-substrate distance is $D = 1$ nm. (b) The localized electric enhancement factor of the bright mode of $D = 1$ (black), 1.5 (red), 2.0 (green) and 2.5 nm (blue), respectively, varying with the distance from the Ag substrate. Gray and white areas stand for R6G and vacuum, respectively

(For colored figure please scan the QR code on page 1)

Furthermore, as the gap distance is so small that the gradient of electric field (∇E) becomes large enough, which should not be neglected in enhancing of the

Raman signals. The coupling between the molecule and the electric field results in the perturbation Hamiltonian of the transitions in vibration levels: $H = -\mu \cdot E$, where μ is the dipole moment and it can be written as[91]

$$\mu_a = \mu_a^p + \alpha_{ab}E_b + \frac{1}{3}A_{abc}\frac{\partial E_b}{\partial c} + G_{ab}B_b + \cdots, \quad (5\text{-}180)$$

where $\{a, b, c\}$ are the permutation of the coordinates $\{x, y, z\}$, in which summing over repeated indices is implied, μ^p is the permanent dipole moment, B is the magnetic field, and α is the polarizability tensor. Among these, α, A and G are defined in Ref. [92].

Figure 5-78 shows the gradient of electric field of the plane $z = 0.05$ nm along z axis (E_z) and xy-plane ($E_{xy} = \sqrt{E_x^2 + E_y^2}$), respectively. The partial derivatives are along z $\left(\frac{\partial}{\partial z}\right)$ and r $\left(\frac{\partial}{\partial r}, \text{cylindrical coordinate}\right)$, respectively. It reveals that the gradient effect of bright mode is the largest and the one of the LSP mode is the smallest one (except for few hotspots). Particularly the partial derivatives of E (whether E_z or E_{xy}) along z are much larger than those along r. Besides, the intensities of the gradient are over 1200 V/m/nm at about $r = 1$ nm for the bright mode. Also, the ratio of electric field gradient over electric field $((\nabla E_{xy})/E)$ is calculated in Fig. 5-78. For dark modes, the most effective interval for the contribution of electric field gradient is around $r = 1$ nm. While for bright modes, the "hot ranges" are much wider. Particularly, some values of "hot ranges" are larger than 5 nm^{-1}. We notice that, although the electric field intensity enhancement of dark mode is smaller than that of LSP mode, the electric field gradient enhancement of dark mode is larger than that of LSP mode. This indicates that the considering of electric field gradient effect is necessary together with the intensity of E, especially for the bright and dark modes. Therefore, Stokes and anti-Stokes Raman intensities will further be strongly enhanced due to these three effects: resonance excitation effect, localized electric field intensity effect and electric field gradient effect.

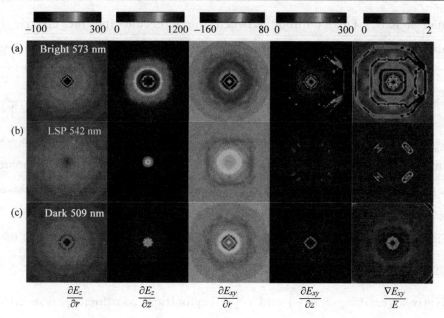

Fig. 5-78 Gradient of electric field distributions in xy-plane ($z = 0.05$ nm) for bright (a), LSP (b) and dark (c) modes, respectively. From left to right, the five columns stand for $\frac{\partial E_z}{\partial r}$, $\frac{\partial E_z}{\partial z}$, $\frac{\partial E_{xy}}{\partial r}$, $\frac{\partial E_{xy}}{\partial z}$, and $\frac{\nabla E_{xy}}{E}$, respectively. Here, the first four color bars are in the unit of V/(m/nm) and the last color bar is in unit of nm^{-1}. The distance is $D = 1$ nm

(For colored figure please scan the QR code on page 1)

The distance dependent ratios of electric field gradient over electric field of the three modes are shown in Fig. 5-79. Similarly, the ratio of bright mode is the largest one and they all decrease rapidly with the increase of the distance. The results of Fig. 5-77 and Fig. 5-79 indicate that both electric field enhancement and electric field gradient effects are focused near the upper surface. Consequently, we achieve high enhancement factors in both surface enhanced Stokes and anti-Stokes Raman spectra by employing the TERRS system.

Plexciton occurs not only in LSP but also in PSPP. Figure 5-80(a) shows the schematic of the Ag nanowire coated with R6G. When the incident plane wave couples into the Ag nanowire, the PSPP mode is allowed at the surface of the Ag cylinder, propagating along the z axis. Figure 5-80(b) shows the

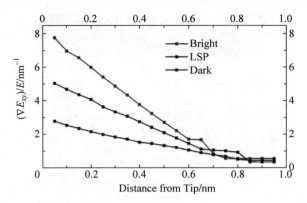

Fig. 5-79   The ratio of electric field gradient over electric field for bright (red), LSP (black) and dark (blue) modes, respectively, varying with the distance from tip. The tip-metal distance is $D = 1$ nm
(For colored figure please scan the QR code on page 1)

transmission spectra of this system, varying with the radii (R) of the Ag cylinders. When the Ag nanowire is in vacuum (uncoated with R6G), numerous modes occur at the wavelengths from 400 nm to 2000 nm. These modes are generated from the coherence of the electric field inside the nanowire. Particularly, the PSPP mode around 540 nm occurs when $R = 70$ nm, which marches the absorption band of R6G. Figure 5-80(c) shows the transmission spectra of the Ag nanowire ($R = 70$ nm) without (black) and with R6G (red), respectively. It is revealed that the PSPP mode at 540 nm splits into two modes (about 490 nm and 575 nm), which are known as the dark and bright modes of the plexciton, respectively. Hence, similar as the PSPP mode, plasmon-exciton couples and propagates at the surface of the Ag cylinder along the z axis. The coupled spectrum also reveals that the transmissions are enhanced for both dark and bright modes. Consequently, we obtain the plexciton in the Ag and R6G hybrid nanowire, which would improve the propagation properties of the PSPP modes. This is significant for the applications in the propagations of metal nanowire.

In summary, plexciton is obtained in both LSP and PSPP theoretically. Particularly, we employ TERRS configuration to enhance both Stokes and anti-Stokes Raman spectra simultaneously. Considering the resonance excitation

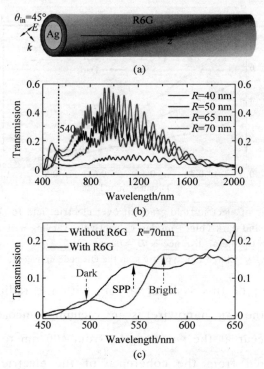

Fig. 5-80 (a) Schematic of the Ag nanowire coated with R6G. The length and radius of the Ag nanowire are 10 μm and R, respectively. The thickness of R6G is 4 nm. The p-polarized plane wave illuminates from one side of the nanowire with the incident angle of 45°. (b) Transmission spectra of Ag nanowire uncoated with R6G, varying with the radii R. The vertical dashed line stands for the absorption peak of R6G (540 nm). (c) Transmission spectra of Ag nanowire uncoated (black curve) and coated (red curve) with R6G. The radius of Ag nanowire is R = 70 nm

(For colored figure please scan the QR code on page 1)

effects together with SERS effect, we obtain a huge enhancement factor ($10^{13} \sim 10^{15}$). The bright and dark modes of the plexciton, as well as the LSP mode, are analyzed, and multipoles are observed in the formers. To obtain the highest enhancement factor of Raman spectra, the tip-substrate distance should be small enough, about 1 nm. Besides, the electric field gradient effect is also investigated, and it is large enough, especially the ones of bright and dark modes, that it cannot be neglected. Combining the resonance excitation effect, the electric field intensity effect and the electric field gradient effect, Stokes and anti-Stokes Raman spectra can be further strongly enhanced. Additionally, we employ the Ag and R6G hybrid nanowire to obtain the plexciton, and the propagation properties of the system would be improved.

References

[1] RITCHIE R H. Plasma losses by fast electrons in thin films[J]. Phys. Rev.,1957,106: 874-881.
[2] MUKHERJEE S, LIBISCH F, LARGE N, et al. Hot electrons do the impossible: plasmon-induced dissociation of H_2 on Au[J]. Nano Lett.,2013,13: 240-247.
[3] MARTIREZ J M P, CARTER E A. Prediction of a low-temperature N_2 dissociation catalyst exploiting near-IR-to-visible light nanoplasmonics[J]. Sci. Adv.,2017,3: 4710.
[4] LIU Z, HOU W, PAVASKAR P, et al. Plasmon resonant enhancement of photocatalytic water splitting under visible illumination[J]. Nano Lett.,2011,11: 1111-1116.
[5] MUBEEN S, LEE J, SINGH N, et al. An autonomous photosynthetic device in which all charge carriers derive from surface plasmons [J]. Nature Nanotechnol., 2013, 8: 247-251.
[6] HOU W, HUNG W H, PAVASKAR P, et al. Photocatalytic conversion of CO_2 to hydrocarbon fuels via plasmon-enhanced absorption and metallic interband transitions [J]. ACS Catal.,2011,1: 929-936.
[7] ZHANG Z, FANG Y, WANG W, et al. Propagating surface plasmon polaritons: towards applications for remote-excitation surface catalytic reactions[J]. Adv. Sci., 2016,3: 1500215.
[8] ZHAO W, WANG S, LIU B, et al. Exciton-plasmon coupling and electromagnetically induced transparency in monolayer semiconductors hybridized with Ag nanoparticles [J]. Adv. Mater.,2016,28: 2709-2715.
[9] MOSKOVITS M. Surface-enhanced spectroscopy[J]. Rev. Mod. Phys., 1985, 57: 783-826.
[10] BROCKMAN J M, NELSON B P, CORN R M. Surface plasmon resonance imaging measurements of ultrathin organic films[J]. Annu. Rev. Phys. Chem.,2000, 51: 41-63.
[11] DING Q, CHEN M, LI Y, et al. Effect of aqueous and ambient atmospheric environments on plasmon-driven selective reduction reactions[J]. Sci. Rep.,2015, 5: 10269.
[12] CUI L, WANG P, LI Y, et al. Selective plasmon-driven catalysis for para-nitroaniline in aqueous environments[J]. Sci. Rep.,2016,6: 20458.
[13] CAO E, LIN W, SUN M, et al. Exciton-plasmon coupling interactions: from principle

to applications[J]. Nanophotonics,2018,7: 145-167.

[14] CHENG Y, SUN M. Unified treatment for photoluminescence and scattering of coupled metallic nanostructures: I. Two-body system [J]. New J. Phys., 2022, 24: 033026.

[15] CHENG Y,ZHANG W, ZHAO J, et al. Understanding photoluminescence of metal nanostructures based on an oscillator model[J]. Nanotechnology,2018,29: 315201.

[16] GRIFFITHS D J, Introduction to electrodynamics [M]. 4th ed. New York: Pearson,2013.

[17] SU K H,WEI Q H, ZHANG X, et al. Interparticle coupling effects on plasmon resonances of nanogold particles[J]. Nano Lett.,2003,3: 1087-1090.

[18] SÖNNICHSEN C,REINHARD B M,LIPHARDT J,et al. A molecular ruler based on plasmon coupling of single gold and silver nanoparticles[J]. Nat. Biotechnol.,2005, 23: 741-745.

[19] HU H,DUAN H, YANG J K W, et al. Plasmon-modulated photoluminescence of individual gold nanostructures[J]. ACS Nano,2002,6: 10147-10155.

[20] ZHANG W,WEN T, YE L,et al. Influence of non-equilibrium electron dynamics on photoluminescence of metallic nanostructures[J]. Nanotechnology,2020,31: 495204.

[21] CHENG Y,SUN M. Unified treatment for scattering,absorption,and photoluminescence of coupled metallic nanoparticles with vertical polarized excitation[J]. New J. Phys.,2023,25: 033028.

[22] PASQUALE A J,REINHARD B M,DAL NEGRO L. Engineering photonic-plasmonic coupling in metal nanoparticle necklaces[J]. ACS Nano,2011,5: 6578-6585.

[23] VARNAVSKI O P,MOHAMED M B, EL-SAYED M A,et al. Relative enhancement of ultrafast emission in gold nanorods[J]. J. Phys. Chem. B,2003,107: 3101-3104.

[24] DULKEITH E, NIEDEREICHHOLZ T, KLAR T A, et al. Plasmon emission in photoexcited gold nanoparticles[J]. Phys. Rev. B,2004,70: 205424.

[25] VARNAVSKI O P, GOODSON T, MOHAMED M B, et al. Femtosecond excitation dynamics in gold nanospheres and nanorods[J]. Phys. Rev. B,2005,72: 235405.

[26] CHENG Y, SUN M. Physical mechanisms on absorption and photoluminescence properties of coupled plasmon-exciton (plexciton) systems[J]. The Journal of Physical Chemistry C,2023,127: 5457-5466.

[27] ESTEBAN R, BORISOV A G, NORDLANDER P, et al. Bridging quantum and classical plasmonics with a quantum-corrected model[J]. Nature Communications, 2012,3: 825.

[28] ASPELMEYER M,KIPPENBERG T J,MARQUARDT F. Cavity optomechanics[J]. Rev. Mod. Phys.,2014,86: 1391-1452.

[29] ZHANG Z,ZHANG S, GUSHCHINA I, et al. Excitation energy dependence of semiconductor nanocrystal emission quantum yields [J]. The Journal of Physical Chemistry Letters,2021,12: 4024-4031.

[30] SONG M,WU B,CHEN G,et al. Photoluminescence plasmonic enhancement of single

quantum dots coupled to gold microplates[J]. The Journal of Physical Chemistry C, 2014,118: 8514-8520.

[31] LIU X,LI Q,RUAN Q, et al. Borophene synthesis beyond the single-atomic-layer limit[J]. Nature Materials,2022,21: 35-40.

[32] LONGHI S. Parity-time symmetry meets photonics: A new twist in non-Hermitian optics[J]. Europhysics Letters,2017,120: 64001.

[33] LIMONOV M F,RYBIN M V,PODDUBNY A N,et al. Fano resonances in photonics [J]. Nature Photonics,2017,11: 543-554.

[34] ELKABBASH M,LETSOU T,JALIL S A,et al. Fano-resonant ultrathin film optical coatings[J]. Nature Nanotechnol. ,2021,16: 440-446.

[35] LIU N,LANGGUTH L,WEISS T, et al. Plasmonic analogue of electromagnetically induced transparency at the Drude damping limit [J]. Nature Materials, 2009, 8: 758-762.

[36] YU Y,XUE W,SEMENOVA E,et al. Demonstration of a self-pulsing photonic crystal Fano laser[J]. Nature Photonics,2017,11: 81-84.

[37] WANG M,WANG T,OJAMBATI O S, et al. Plasmonic phenomena in molecular junctions: principles and applications [J]. Nature Reviews Chemistry, 2022, 6: 681-704.

[38] FANO U. Effects of configuration interaction on intensities and phase shifts[J]. Phys. Rev. ,1961,124: 1866-1878.

[39] WU C,KHANIKAEV A B,ADATO R,et al. Fano-resonant asymmetric metamaterials for ultrasensitive spectroscopy and identification of molecular monolayers[J]. Nature Materials,2012,11: 69-75.

[40] MIROSHNICHENKO A E,FLACH S,KIVSHAR Y S. Fano resonances in nanoscale structures[J]. Rev. Mod. Phys. ,2010,82: 2257-2298.

[41] LUK'YANCHUK B,ZHELUDEV N I, MAIER S A, et al. The Fano resonance in plasmonic nanostructures and metamaterials[J]. Nature Materials,2010,9: 707-715.

[42] CHRISTOPHER P,MOSKOVITS M. Hot charge carrier transmission from plasmonic nanostructures[J]. Annu. Rev. Phys. Chem. ,2017,68: 379-398.

[43] SUN M,FANG Y,ZHANG Z,et al. Activated vibrational modes and Fermi resonance in tip-enhanced Raman spectroscopy[J]. Physical Review E,2013,87: 020401.

[44] MA J,CHENG Y, SUN M. Plexcitons, electric field gradient and electron-phonon coupling in tip-enhanced Raman spectroscopy (TERS)[J]. Nanoscale,2021,13: 10712-10725.

[45] JIANG N,FOLEY E T,KLINGSPORN J M,et al. Observation of multiple vibrational modes in ultrahigh vacuum tip-enhanced Raman spectroscopy combined with molecular-resolution scanning tunneling microscopy[J]. Nano Lett. ,2012,12: 5061-5067.

[46] SUN M,ZHANG Z,CHEN L, et al. Tip-enhanced resonance couplings revealed by high vacuum tip-enhanced Raman spectroscopy[J]. Advanced Optical Materials,2013, 1: 449.

[47] SUN M, ZHANG Z, CHEN L, et al. Plasmonic gradient effects on high vacuum tip-enhanced Raman spectroscopy[J]. Advanced Optical Materials, 2014, 2: 74-80.

[48] WANG C-F, EL-KHOURY P Z. Multimodal tip-enhanced nonlinear optical nanoimaging of plasmonic silver nanocubes[J]. The Journal of Physical Chemistry Letters, 2021, 12: 10761-10765.

[49] BOULESBAA A, BABICHEVA V E, WANG K, et al. Ultrafast dynamics of metal plasmons induced by 2D semiconductor excitons in hybrid nanostructure arrays[J]. ACS Photonics, 2016, 3: 2389-2395.

[50] CHOWDHURY R K, MUKHERJEE S, BHAKTHA S N B, et al. Ultrafast real-time observation of double Fano resonances in discrete excitons and single plasmon-continuum[J]. Phys. Rev. B, 2020, 101: 245442.

[51] CHOWDHURY R K, DATTA P K, BHAKTHA S N B, et al. Ultrafast investigation of individual bright exciton-plasmon polaritons in size-tunable metal-WS_2 hybrid nanostructures[J]. Advanced Optical Materials, 2020, 8: 1901645.

[52] LIN W, SHI Y, YANG X, et al. Physical mechanism on exciton-plasmon coupling revealed by femtosecond pump-probe transient absorption spectroscopy[J]. Materials Today Physics, 2017, 3: 33-40.

[53] CHEN C, LV H, ZHANG P, et al. Synthesis of bilayer borophene [J]. Nature Chemistry, 2022, 14: 25-31.

[54] YANG R, SUN M. Bilayer borophene synthesized on Ag (111) film: Physical mechanism and applications for optical sensor and thermoelectric devices[J]. Materials Today Physics, 2022, 23: 100652.

[55] YANG R, REN X, SUN M. Optical spectra of bilayer borophene synthesized on Ag (111) film [J]. Spectrochimica Acta Part A: Molecular and Biomolecular Spectroscopy, 2022, 282: 121711.

[56] JOHNSON P B, CHRISTY R W. Optical constants of the Noble metals[J]. Phys. Rev. B, 1972, 6: 4370-4379.

[57] WANG B, ZENG X-Z, LI Z-Y. Quantum versus optical interaction contribution to giant spectral splitting in a strongly coupled plasmon–molecules system[J]. Photon. Res., 2020, 8: 343-351.

[58] PURETZKY A A, LIANG L, LI X, et al. Twisted $MoSe_2$ bilayers with variable local stacking and interlayer coupling revealed by low-frequency Raman spectroscopy[J]. ACS Nano, 2016, 10: 2736-2744.

[59] SONG J, CAO Y, DONG J, et al. Superior thermoelectric properties of twist-angle superlattice borophene induced by interlayer electrons transport [J]. Small, 2023, 19: 2301348.

[60] MAK K F, LEE C, HONE J, et al. Atomically thin MoS_2: A new direct-gap semiconductor[J]. Phys. Rev. Lett., 2010, 105: 136805.

[61] NAJMAEI S, LIU Z, ZHOU W, et al. Vapour phase growth and grain boundary structure of molybdenum disulphide atomic layers[J]. Nature Materials, 2013, 12: 754-

759.

[62] YANG X, YU H, GUO X, et al. Plasmon-exciton coupling of monolayer MoS_2-Ag nanoparticles hybrids for surface catalytic reaction[J]. Materials Today Energy, 2017, 5: 72-78.

[63] DONG J, ZHANG Z, ZHENG H, et al. Recent progress on plasmon-enhanced fluorescence[J]. 2015, 4: 472-490.

[64] HAUBER A, FAHY S. Scattering of carriers by coupled plasmon-phonon modes in bulk polar semiconductors and polar semiconductor heterostructures[J]. Phys. Rev. B, 2017, 95: 045210.

[65] GUNST T, MARKUSSEN T, STOKBRO K, et al. First-principles method for electron-phonon coupling and electron mobility: Applications to two-dimensional materials[J]. Phys. Rev. B, 2016, 93: 035414.

[66] YANG X, ZHAI F, HU H, et al. Far-field spectroscopy and near-field optical imaging of coupled plasmon-phonon polaritons in 2D van der Waals heterostructures[J]. Adv. Mater., 2016, 28: 2931-2938.

[67] DING Q, SHI Y, CHEN M, et al. Ultrafast dynamics of plasmon-exciton interaction of Ag nanowire- graphene hybrids for surface catalytic reactions[J]. Sci. Rep., 2016, 6: 32724.

[68] WANG Y, CHEN H, SUN M, et al. Ultrafast carrier transfer evidencing graphene electromagnetically enhanced ultrasensitive SERS in graphene/Ag-nanoparticles hybrid [J]. Carbon, 2017, 122: 98-105.

[69] CHENG Y, WANG Y, SUN M. Unified treatment for photoluminescence and scattering of coupled metallic multi-nanostructures [J]. Results in Physics, 2022, 38: 105668.

[70] SALKUYEH D K. Positive integer powers of the tridiagonal Toeplitz matrices[J]. Int. Math. Forum, 2006, 1: 1061-1065.

[71] JIA J, SOGABE T, EL-MIKKAWY M. Inversion of k-tridiagonal matrices with Toeplitz structure[J]. Comput. Math. Appl., 2013, 65: 116-125.

[72] BARROW S J, FUNSTON A M, GÓMEZ D E, et al. Surface plasmon resonances in strongly coupled gold nanosphere chains from monomer to hexamer[J]. Nano Lett., 2011, 11: 4180-4187.

[73] CHANG D E, SØRENSEN A S, HEMMER P R, et al. Strong coupling of single emitters to surface plasmons[J]. Phys. Rev. B, 2007, 76: 035420.

[74] FRENKEL J. On the transformation of light into heat in solids. I[J]. Phys. Rev., 1931, 37: 17-44.

[75] ZHANG S, WEI H, BAO K, et al. Chiral surface plasmon polaritons on metallic nanowires[J]. Phys. Rev. Lett., 2011, 107: 096801.

[76] CHENG Y, ZHAI K, ZHU N, et al. Optical non-reciprocity with multiple modes in the visible range based on a hybrid metallic nanowaveguide[J]. Journal of Physics D: Applied Physics, 2022, 55: 195102.

[77] CHENG Y, SUN M. Unified treatments for localized surface plasmon resonance and propagating surface plasmon polariton based on resonance modes in metal nanowire [J]. Opt. Commun.,2021,499: 127277.

[78] CHENG Y, SUN M. Plexciton in tip-enhanced resonance Stokes and anti-Stokes Raman spectroscopy and in propagating surface plasmon polaritons [J]. Opt. Commun.,2021,493: 126990.

[79] FAN L, WANG J, VARGHESE L T, et al. An all-silicon passive optical diode[J]. Science,2012,335: 447-450.

[80] LIANG C, LIU B, XU A-N, et al. Collision-induced broadband optical nonreciprocity [J]. Phys. Rev. Lett.,2020,125: 123901.

[81] DEL BINO L, SILVER J M, WOODLEY M T M, et al. Microresonator isolators and circulators based on the intrinsic nonreciprocity of the Kerr effect[J]. Optica,2018,5: 279-282.

[82] CHENG Y, ZHANG Y, SUN M. Strongly enhanced propagation and non-reciprocal properties of CdSe nanowire based on hybrid nanostructures at communication wavelength of 1550 nm[J]. Opt. Commun.,2022,514: 128175.

[83] NINOMIYA S, ADACHI S. Optical properties of cubic and hexagonal CdSe[J]. J. Appl. Phys.,1995,78: 4681-4689.

[84] DODGE M J. Refractive properties of magnesium fluoride[J]. Appl. Opt.,1984,23: 1980-1985.

[85] PALIK E D. Handbook of optical constants of solids [M]. Boston: Academic Press,1985.

[86] SHEGAI T, HUANG Y, XU H, et al. Coloring fluorescence emission with silver nanowires[J]. Appl. Phys. Lett.,2010,96: 103114.

[87] SHEGAI T, MILJKOVIĆ V D, BAO K, et al. Unidirectional broadband light emission from supported plasmonic nanowires[J]. Nano Lett.,2011,11: 706-711.

[88] LEE K-S, EL-SAYED M A. Gold and silver nanoparticles in sensing and imaging: sensitivity of plasmon response to size, shape, and metal composition[J]. J. Phys. Chem. B,2006,110: 19220-19225.

[89] ZHAO J, JENSEN L, SUNG J, et al. Interaction of plasmon and molecular resonances for rhodamine 6G adsorbed on silver nanoparticles[J]. J. Am. Chem. Soc.,2007, 129: 7647-7656.

[90] MORTON S M, JENSEN L. Understanding the molecule-surface chemical coupling in SERS[J]. J. Am. Chem. Soc.,2009,131: 4090-4098.

[91] AYARS E J, HALLEN H D, JAHNCKE C L. Electric field gradient effects in Raman spectroscopy[J]. Phys. Rev. Lett.,2000,85: 4180-4183.

[92] SASS J K, NEFF H, MOSKOVITS M, et al. Electric field gradient effects on the spectroscopy of adsorbed molecules[J]. J. Phys. Chem.,1981,85: 621-623.

CHAPTER 6

Plasmon-Enhanced Fluorescence Spectroscopy

6.1 The principle of plasmon-enhanced fluorescence

Fluorescence, as well as photoluminescence (PL), is the property of some luminescent center, such as organic molecules and rare-earth (RE) ions, to absorb photon with high energy and to emit a photon with low energy subsequently. The processes occured between the absorption and the emission of photon could be demonstrated by a Jablonski diagram, which illustrates the electronic transition of fluorophore in excited states[1]. It is widely accepted that the excited electron transition from the upper energy level to ground level, often took place by two methods, one is emit photon, and the other is relaxation of energy into phonon[1]. With the condition of weak excitation, that is, far from saturation of the excited state, the fluorescence spontaneous emission rate γ_{em}^0 can be regarded as the excitation from ground state to excited state and the subsequent relaxation back to the ground state via fluorescence emission[1], i.e.

$$\gamma_{em}^0 = \gamma_{exc}^0 Q_i^0, \qquad (6\text{-}1)$$

where Q_i^0 and γ_{exc}^0 is the quantum yield and the excitation rate, respectively, and the superscripts "0" specify that the molecule is in free space and does not couple to the local environment. The subscript "i" indicates that the quantum yield is defined by the intrinsic properties of the molecule.

As indicated before, Q_i^0 is the probability of relaxing from excited to ground states by emitted a fluorescence photon. Here, the radiative and non-radiative decay rates can be defined as γ_r^0 and γ_{nr}^0, respectively, so the intrinsic quantum yield for Q_i^0 could be defined more generally as

$$Q_i^0 = \gamma_r^0 / (\gamma_r^0 + \gamma_{nr}^0). \tag{6-2}$$

If the local environment of the fluorophore has been changed, the excitation and decay rates will be changed correspondingly. Then the Eqs. (1) and (2) can modified as

$$\gamma_{em} = \gamma_{exc} Q, \tag{6-3}$$

and

$$Q = \frac{\gamma_r}{\gamma_r + \gamma_{nr}} = \frac{\gamma_r}{\gamma_r + \gamma_{nr}^0 + \gamma_{abs} + \gamma_m}, \tag{6-4}$$

where, γ_{abs} accounts for dissipation to heat in the environment, and γ_m accounts for coupling to non-radiative electromagnetic modes, such as the emitted energy transferred into heat through the interaction between the electron and lattice. The whole decay rate $\gamma = \gamma_r + \gamma_{nr}$ defines the lifetime $\tau = 1/\gamma$ of the excited state. In general, the fluorescence emission is not only dependent on the molecular properties but also the external parameters accounting for the local environment of the fluorophore.

For a conventional fluorescence measurement, pursuing the brighter and more stable signals of the fluorophore performed at the minimization of the internal and environmentally conditioned non-radiative processes, increasing higher spontaneous-emission rate mainly depending on its nature properties[1]. Though the fluorescence is known as one of the best choice optical method for the detection of biological and chemical species, typical disadvantage for the conventional fluorescence technique is the relatively low signal-to-noise (background) ratio which restrict its applications to important areas of medical diagnostics, food control, and security, particularly the detection of a target fluorescent molecules on single molecule level. So the exploring a proper solution to overcome the disadvantage above has attracted great attention.

Since Purcell's pioneering work[2], many researches have been demonstrated that the excited atomic state lifetime is critically dependent on the inner properties of the atom and its environment both from experimental and theoretical aspect[1]. Due to the interaction of the fluorophore with its environment nearby, the fluorescence processes, both the excitation and the emission, can be modulated through modifying its local EM field. As a result, obtaining the limit sensitivity detection of fluorescence signal, can be performed by control of the local EM field around the fluorophores. With the help of the coupled effect of light with localized surface plasmons (LSPs-supported by nanoparticles(NPs) or nanoperodical typed metallic nanostructure) and SPPs-traveling along periodical typed metal nanostructure), which can provide strong confinement of electromagnetic field intensity. These confined EM fields can interact with fluorophores at their wavelengths absorption λ_{ab} and emission λ_{em} which alter respective transitions between the ground and higher-excited states (Fig. 6-1). It is reported that the internal conversion (IC) process relies critically on the electronic configuration of the fluorophore through the overlap of its wave-functions. Thus, it is important to investigate the energy transfer process between the fluorophore and the metallic nanostructure, which will help us to study the contributed effect of the modified relaxation rate of fluorophore towards the radiative emission rate.

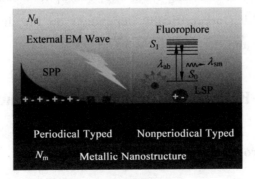

Fig. 6-1　Schematic of confined field of SPP and LSP modes coupled with a fluorophore excited with external EM field

(For colored figure please scan the QR code on page 1)

Based on the EM mechanism, the cross section of PEF from the metal nanostructure is represented as $\sigma_{PEF}(\lambda_L, \lambda, d_{av})$, where λ is the emission wavelengths, λ_L is the excitation wavelengths, and d_{av} is the average distance of fluorescent molecule from metal substrate. With the help of the coupled effect between the plasmon resonance with the incidence wave and fluorescence emission, the total enhancement factor(EF) can be depicted as[3-4]

$$\sigma_{PEF}(\lambda_L, \lambda, d_{av}) = |M_{EM}(\lambda_L, \lambda, d_{av})|^2 \times \frac{\sigma_{FL}(\lambda_L, \lambda)}{|M_d(\lambda, d_{av})|^2}$$

$$= \left|\frac{E_{loc}(\lambda_L, d_{av})}{E_{in}(\lambda_L)}\right|^2 \times \left|\frac{E_{loc}(\lambda, d_{av})}{E_{in}(\lambda)}\right|^2 \times \frac{\sigma_{FL}(\lambda_L, \lambda)}{|M_d(\lambda, d_{av})|^2}$$

$$= |M_1(\lambda_L, d_{av})|^2 |M_2(\lambda, d_{av})|^2 \times \frac{\sigma_{FL}(\lambda_L, \lambda)}{|M_d(\lambda, d_{av})|^2}, \quad (6\text{-}5)$$

where $|M_{EM}|^2$ is total EF; while $|M_1|^2$ and $|M_2|^2$ are EFs related with the coupled effect between SPR to the incident wave and fluorescence emission, respectively; $\sigma_{FL}(\lambda_L, \lambda)$ and $\sigma_{PEF}(\lambda_L, \lambda)$ are the scattering cross section of fluorohpre in the free space and local field, respectively, and $|M_{d_{av}}|^2$ is factor that depict the energy transfer from fluorophore to metal substrate. From Eq. (6-5), the fluorescence-enhanced effect shows a balance between several processes including the excitation rate increasing by the local EM field, an radiative decay rate enhancement by SPCE, and the quenched effect due to the nonradiative energy transferring from the fluorophore center to the metallic substrate, all of which are critically dependent on the distance between the fluorophore center and the metallic substrate[5].

6.2 Plasmon-Enhanced Upconversion Luminescence

6.2.1 Brief introduction

Upconversion(UC) is the optical process transferring energy from the low energy to the higher energy, which is so-called anti-Stokes process. Usually, it is more efficient compared to nonlinear harmonic generation, therefore, it has attracted special attention from researchers in related fields, and it is promising for the applications of solar harvesting, because photons lacking of sufficient energy can also be translated into higher energy photons through that process[6-8]. The pumping process about UC materials has higher efficiency than that of two or more photons. The former is driven by direct irradiation with near-infrared(NIR)lasers which are cheap and continuous, however the latter is using ultrashort-pulse lasers which are largely expensive and in which the mismatching phase[9-10]. may exists The research of RE/ lanthanide-doped UC materials have become the trend since 1960s[11], Auzel confirmed that UC luminescence was real, and they prepared UC luminescent materials with Er^{3+} and Tm^{3+} as luminescent centers, Yb^{3+} as sensitizer, $NaYWO_3$ as matrix, and excited them under infrared(IR)light source[12]. RE UC luminescent materials usually consist of activators, sensitizers and matrix materials. Activators and sensitizers are made up of RE elements. Activators are the center of luminescence, and sensitizers transfer their absorbed energy to activators to improve the luminescent efficiency of activators[13], while matrix materials themselves are not luminescent. But it can provide suitable lattice sites for activators[13]. Matrix materials include fluorides, oxides, halides and sulfur compounds[14]. Among them, fluorides have the advantages of wide transmittance range, high thermal stability and low phonon energy[13], therefore, is the most commonly used. At present, $NaYF_4$ is considered as the most efficient matrix material[13]. In 2004, Auzel summarized the mechanism of UC luminescence after extensive experiments which can be classified into three kinds: excited-state absorption(ESA)[10], energy-transfer UC[15], and photo avalanche[16]. However, the trend of UC phosphors has become more evident in recent years because they have potential to significantly improve photovoltaic(PV) device efficiency due to their higher thermal and photo stability, larger anti-Stokes shifts, and longer fluorescence lifetime[17-18]. But there are many limitations when energy levels of lanthanide ions are embedded into the host crystal lattice, which include low quantum efficiency, limited NIR light harvesting, and the uncontrollable nonradiative processes[19], which hinder seriously the future development of lanthanide-doped NP nanocrystals.

So far, many approaches have been provided to solve the above issues of the applications of NP, including the structure, size and surface control about UC materials[20-22], and the design of core-shell structure[23], the regulation of doping ions and UC matrices[24], and the modulation of localized light field accompanied with SP effect[25] etc. By coupling with incident light, plasmonic effect can touch off strong localized electromagnetic field, which results in more efficient nonlinearity of metal itself or its surrounding dielectric materials[26]. The principle above has been used to promote photochemical reactions[27] and enhance luminescence from optical launchers nearby[28]. Maybe there are two mechanisms for plasmon-enhanced UC luminescence (PE-UCL). Firstly, the excitation rate of lanthanide sensitizer ions can be increased by plasma-enhanced local electric field near metal NP. Secondly, during UC emission, the decay rate of lanthanide-activated ions can also be increased by the effect of plasmon[29].

6.2.2 Physical principle and mechanism

The Hamiltonian of quantum electrodynamics can describe the interaction between photon and electron, and it can be written as[30]

$$H_{int} = -\frac{e}{m}\boldsymbol{p} \cdot \boldsymbol{A} + \frac{e^2}{2m}|\boldsymbol{A}|^2 + e\phi, \qquad (6\text{-}6)$$

where, e is the electronic charge, m is the electronic mass, $\boldsymbol{p}$ is electronic gauge momentum, $\boldsymbol{A}$ is vector potential and ϕ is the scalar of the electromagnetic field. Note that only the first term needs to be considered for the dipole transitions[31]. The rate of electronic transition between two states can be calculated as follows by Fermi's golden rule[32],

$$W = \frac{2\pi}{h} | \langle \psi_f | H_{int} | \psi_i \rangle | \rho(\omega_f - \omega_i), \tag{6-7}$$

where, the subscript i and f represent the initial and final state respectively, and ρ represents the density of states. It is necessary to define the multipolar Hamiltonian when the optical field's spatial variation is far more than electronic wave functions' spatial scope[33]:

$$H_{int} = -\boldsymbol{d} \cdot \boldsymbol{E}_0 - \boldsymbol{m} \cdot \boldsymbol{B}_0 - [(\ddot{\boldsymbol{Q}} \nabla) \cdot \boldsymbol{E}]_0 + \cdots, \tag{6-8}$$

where, $\boldsymbol{d}$ and $\boldsymbol{m}$ are electric dipole and magnetic dipole, respectively, and $\boldsymbol{Q}$ is the electric quadrupole moments. What's more, the subscript 0 stands for the values calculated at atomic nucleus.

The process of RE ions' UCL is realized by the transition between their $4f$ electron levels and found that the process in the bulk of crystals is also dipole transitions[32]. However, this case can significantly change the rate of transition despite it has no effect on the energy levels. The line strength has derived by Judd and Ofelt through the theory of second-order perturbation:

$$S_{ED}(J_i, J'_f) = e^2 \sum_{t=2,4,6} \Omega_t | \langle f^n[\gamma SL]J || U(t) || f^n[\gamma'S'L']J' \rangle |^2, \tag{6-9}$$

where, Ω_t is the intensity parameters usually determined by experiment[32], and the symbols in the brackets are the tensors operators. Considering the Coulomb interaction and using the theory of time-dependent perturbation when active ions (A and B) are very close to each other, the rate of energy transfer would be written as[34]

$$W_{ET} = \frac{2\pi}{h} | \langle H_{AB} \rangle |^2 \int g_A(E) g_B(E) dE, \tag{6-10}$$

where $g_A(E)$ and $g_B(E)$ are respectively the two ions' emission and absorption spectra, respectively, which are normalized.

As we mentioned in the introduction, the process of UC can be classified into three categories. ESA is a continuous multi-photon absorption process of single ion, which is the most basic process to achieve UC. The principle of luminescence is shown in Fig. 6-2[13]. The same RE ion transit energy from ground-state level to excited-state level through continuous two-or multi-photon absorption, and then the energy is radiated in the form of light and release the process of returning to the ground-state level.

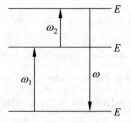

Fig. 6-2 The energy level diagram of ESA[13]

The energy-transfer upconversion (ETU) is a phenomenon of UC luminescence resulting from the interaction between the same or different ions. The principle of UC luminescence is that the energy is transferred by cross relaxation in the form of non-radiative coupling of ions. According to the different modes of transmission, ETU is divided into three main types: serial energy transfer (SET), cross relaxation (CR) and cooperative upconversion (CU), and its UC luminescence principle is shown in Fig. 6-3[13].

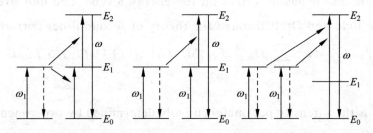

Fig. 6-3 The energy level diagrams are SET, CR and CU, respectively[13]

The photo avalanche (PA) is the result of the interaction of ESA and ETU (CR). The principle of UC is shown in Fig. 6-4[8]. The energy difference between excited light ω_1, E_2 and E_3 are the same. Ions located at E_2 level absorb the energy and are excited to the excited state level E_3 by cross-relaxation process, and the luminous ions at E_1 and E_3 levels are excited. All of them converge on the E_2 level, so the number of photons on the E_2 level increases sharply, just like an avalanche, hence called the "photon avalanche" process[8].

ETU is acknowledged as the most efficient among the above UC mechanisms[35]. Therefore, it is necessary for us to build the equations about rate including entirely states and all transitions involved. As shown in Fig. 6-5, Yb^{3+} possesses 11 electrons in shell $4f$, and forms the double level systems[32]. The resonant occurs between the excited state($^2F_{5/2}$) of Yb^{3+} and the level($^4I_{11/2}$) of Er^{3+}, so there is a highly efficient energy transfer, which is so-called first step of ETU[32].

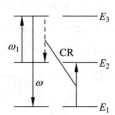

Fig. 6-4 The energy level diagram of photon avalanche[13]

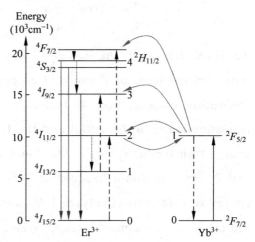

Fig. 6-5 Energy levels of Yb^{3+} and Er^{3+} ions relevant to the energy transfer UC process[32]

(For colored figure please scan the QR code on page 1)

Note that the three-photon process and the $^4F_{7/2}$ level's population are omitted in the following rate equations, because the former has litter effect compared to that of two-photon and the latter's relaxation is extremely fast[36-39]. Moreover, the level of $^2H_{11/2}$ and $^4S_{3/2}$ are seen as a single level due to their distance close enough. Finally, the rate equations are as follows[32]

$$\frac{dN_{D1}}{dt} = \sigma\Phi N_{D0} - W_{D10}N_{D1} + c_{Bd2}N_{A2}N_{D0} - c_{Fd2}N_{D1}N_{A0} -$$
$$c_{d3}N_{D1}N_{A1} - c_{d4}N_{D1}N_{A1} - c_{d4}N_{D1}N_{A2}, \tag{6-11}$$

$$\frac{dN_{A1}}{dt} = W_{A21}N_{A2} - W_{A10}N_{A1} - c_{d3}N_{D1}N_{A1}, \tag{6-12}$$

$$\frac{dN_{A2}}{dt} = c_{Fd2}N_{D1}N_{A0} - c_{Bd2}N_{A2}N_{D0} - c_{d4}N_{D1}N_{A2} - W_{A20}N_{A2} - W_{A21}N_{A2},$$

$$\tag{6-13}$$

$$\frac{dN_{A3}}{dt} = W_{A43}N_{A4} + c_{d3}N_{A1}N_{D1} - W_{A30}N_{A3}, \quad (6\text{-}14)$$

$$\frac{dN_{A3}}{dt} = c_{d4}N_{D1}N_{A2} - W_{A40}N_{A4} - W_{A43}N_{A4}, \quad (6\text{-}15)$$

$$N_D = N_{D0} + N_{D1}, \quad (6\text{-}16)$$

$$N_A = N_{A0} + N_{A1} + N_{A2} + N_{A3} + N_{A4}, \quad (6\text{-}17)$$

where, N_i represents the ions'density of energy level i and subscripts D and A represent the level of donor and acceptor, respectively. σ is the donor ion's absorption cross section, Φ is incident light flux, W is the decay rate of transition between the initial and the final states, c_d means the coefficient of energy transfer between the the donor and the acceptor and the subscripts F and B represent donor to acceptor(forward) and acceptor to donor(backward) energy transfer, espectively, and N_D and N_A represent the doping densities of the donor and acceptor[27], respectively.

Note that decay processes dominates the weak excitation, however, the processes of UC dominates the strong excitation[32]. The emission rate of green and red photons in the weak(Eqs. (6-17),(6-18))/strong(Eqs. (6-19),(6-20)) excitation can be calculated as[32]

$$\Phi_G = \frac{W_{A40} c_{d4} c_{Fd2} N_A N_D}{W_{A4} c_{Bd2} W_{D10}^2} \cdot (\sigma\Phi)^2, \quad (6\text{-}18)$$

$$\Phi_R = \left(\frac{W_{A43} c_{d4}}{W_{A4}} + \frac{W_{A21} c_{d3}}{W_{A10}}\right) \frac{c_{Fd2} N_A N_D}{c_{Bd2} W_{D10}^2} \cdot (\sigma\Phi)^2, \quad (6\text{-}19)$$

$$\Phi_G = \frac{W_{A40}}{W_{A4}} \frac{N_{D0}}{2} \sigma\Phi, $$

$$\Phi_R = \frac{W_{A43}}{W_{A4}} \frac{N_{D0}}{2} \sigma\Phi, \quad (6\text{-}20)$$

where W_{A4} is the total decay rate about the A4 energy level. From Eqs. (6-18) and (6-19), the intensities of upconverted luminescence linearly increase with the increasing of the donor's and acceptor's densities, and also have the linear relations with c_{d4} and c_{Fd2}, but inversely with c_{Bd2} and W_{D10}^2 [27]. We can find easily that the intensity of UC luminescence no longer depend on the energy transfer coefficients because the rate of energy transfer is so quickly that the process of UC is restricted by donor's absorption in strong excitation limit[32].

Plasmonics is a new promising field, using the metal nanostructures' unique optical properties which derive from the ability to provide collective electron excitations(called SPs)to realize the nanoscale manipulation to light[40]. In the case of the diameter of metal particles $d \ll \lambda$, the polarization charges are collected on the particle surface due to the electrons inside particle all move toward the same phase with the excitation of plane-wave radiation. These charges allow a resonance occur at the dipole plasmon frequency. That is, a homogeneous resonantly enhanced field is allowed to occur inside the particle, and a dipolar field is built up outside the particle. This contributes to the strong enhancement of near field besides the particle surface[41]. The SP can affect the cross section σ, the rate W and the coefficients c of Eq. (6-18)—Eq. (6-20). To the σ, it can be calculated using the Poynting's theorem which indicates the process will cause power dissipation. For the monochromatic light, the power dissipation is written as

$$P_{abs} = \frac{1}{2}\int_V \omega(\varepsilon'' |E|^2 + \mu'' |H|^2)dV, \qquad (6\text{-}21)$$

where, ε'' is the permittivity's imaginary part and μ'' is the permeability's imaginary part. For the optically active ions, the power dissipation is written as[32]

$$P_{abs} = \frac{1}{2}\int_V \text{Re}\{j^* \cdot E\}dV, \qquad (6\text{-}22)$$

where j is the current density and it can be expressed by d which is induced dipole moment:

$$j = -i\omega d\delta(r - r_0), \qquad (6\text{-}23)$$

So

$$P_{abs} = \frac{\omega}{2}\text{Im}\{d^* \cdot E(r_0)\}, \qquad (6\text{-}24)$$

$$d = \alpha E, \tag{6-25}$$

$$P_{abs} = \frac{\omega}{2}\text{Im}\{\alpha\}\,|\hat{n}_d \cdot E(r_0)|^2, \tag{6-26}$$

where $\hat{n}_d$ is unit vector along the d. Thus, the σ can be written as[26,36]

$$\sigma = \frac{P_{abs}}{I_{inc}} = c\mu_0\omega\text{Im}\{\alpha\}\frac{|\hat{n}_d \cdot E(r_0)|^2}{|E_0|^2}, \tag{6-27}$$

where E_0 is the amplitude of incident[32]. The Eq. (6-10) can be changed as follows in the approximation of dipole[30]

$$W = \frac{\pi\omega}{3\hbar\varepsilon}|d|^2\rho(r_0,\omega), \tag{6-28}$$

where r_0 is the emitter's position. According to Poynting's theorem[32] the radiation rate is given by

$$P_{rad} = -\frac{1}{2}\int_V \text{Re}\{j^* \cdot E\}dV, \tag{6-29}$$

$$P_{rad} = \frac{\omega}{2}\text{Im}\{d^* \cdot E(r_0)\}. \tag{6-30}$$

In fact, the Eq. (6-30) is different from Eq. (6-24) in the way of physical meaning although they have the same forms. The former represents incident field while the latter represents dipole field and also can be expressed by dyadic Green function[32]

$$P_{rad} = \frac{\omega^3}{2c^2\varepsilon}[d \cdot \text{Im}\{\vec{\vec{G}}(r_0,r_0;\omega)\} \cdot d]. \tag{6-31}$$

From Eq. (6-28) and Eq. (6-31), we can also use dyadic Green function to express the density of states[32] as fellows

$$\rho(r_0,\omega) = \frac{6\omega}{\pi c^2}[\hat{n}_d \cdot \text{Im}\{\vec{\vec{G}}(r_0,r_0;\omega)\} \cdot \hat{n}_d], \tag{6-32}$$

Now, for a plasmonic nanostructure, it's promising to forecast radiative decay rate enhancement using dyadic Green function according to Eq. (6-28) and Eq. (6-32). Although the nonradiative decay is not effected by plasmonic field, the additional decay channels can be induced and calculated as[32]

$$\frac{W_{nr}}{W_0} = \frac{1}{P_0}\frac{1}{2}\int_V \text{Re}\{j^* \cdot E\}dV, \tag{6-33}$$

where, W_0 is decay rate about free space, P_0 is radiation rate about classical dipole, j and E are current density and the electvic field emitted by emitter's dipole, respectively, and can be expressed as[32]

$$P_0 = \omega^4 |d|^2 / 12\pi\varepsilon_0 c^3, \tag{6-34}$$

$$j = \omega\varepsilon'' E, \tag{6-35}$$

$$E(r) = \frac{\omega^3}{2c^2\varepsilon} \overleftrightarrow{G}(r_0, r_0; \omega) \cdot d. \tag{6-36}$$

Above that the energy transfer rate in the absorbing and dispersive media can be given by[39]

$$W_{ET} = \int d\omega g_A(\omega) g_D(\omega) \tilde{w}(\omega), \tag{6-37}$$

where, $g_A(\omega)$ and $g_D(\omega)$ are the acceptor and donor ions' normalized absorption and emission spectra in free-space, respectively; $\tilde{w}(\omega)$ is the transition rate and can be calculated as[32]

$$\tilde{w}(\omega) = \frac{2\pi\omega^2}{\hbar^2 \varepsilon_0 c^2} |d_{A^a} \cdot \overleftrightarrow{G}(r_A, r_D, \omega) \cdot d_D|^2, \tag{6-38}$$

where, d_A and d_D represent acceptor and donor ions' dipole moments, and r_A and r_D are acceptor and donor positions, respectively. For the plasmonic nanostructure, the radiative efficiency can be expressed as[32]

$$\left(\frac{W_{A40}}{W_{A4}}\right)_{sp} = \frac{F_P W_{A40}}{F_P W_{A40} + W_{A43} + W_{nr}}, \tag{6-39}$$

where F_P represents the green emission's Purcell factor. Such factor is usually written as[38]

$$F_P = \frac{3Q}{4\pi^2 V}\left(\frac{\lambda}{n}\right)^3 \text{ with } V = \frac{\int \varepsilon |E|^2 dr}{\max(\varepsilon |E|^2)}. \tag{6-40}$$

And the radiative efficiency can be expressed as[32]

$$F_{rad} = \frac{\left(\dfrac{W_{A40}}{W_{A4}}\right)_{sp}}{\dfrac{W_{A40}}{W_{A4}}} = \frac{W_{A40} + W_{A43}}{W_{A40} + (W_{A43} + W_{nr})/F_P}. \tag{6-41}$$

Apparently, only in the case of $W_{A43} + W_{nr} \gg W_{A40}$, the radiative efficiency can be enhanced significantly. For the UC enhancement, the absorption enhancement F_a makes the major contribution. For the limit of strong excitation[32],

$$F_{\text{strong}} \approx F_a. \qquad (6\text{-}42)$$

While for the limit of weak excitation, energy transfer coefficients must be considered which can be seen from Eq. (6-18). Therefore[32],

$$F_{\text{weak}} \approx \frac{F_{d4} F_{Fd2} F_a^2}{F_{Bd2} F_{D10}^2}, \qquad (6\text{-}43)$$

where F_{dx} represents the enhancement factor according to the coefficients c, and their subscripts correspond to each other. The F_{D10} corresponds to the donor ion's IR decay rate W_{D10}. Note that $F_{Bd2} = F_{Fd2}$ because the donor's and acceptor's energy levels are resonant nearly, so the Eq. (6-35) can be shortened as[32]

$$F_{\text{weak}} \approx \frac{F_{d4} F_a^2}{F_{D10}^2}. \qquad (6\text{-}44)$$

Moreover, in the region of weak excitation, we use another simplified system to describe the two-photon process UCL (Fig. 6-6)[41]. Assuming the emitter polarizability istnot affected by the environment, so the excitation enhancement (f_{ex}) can be written as[41]

$$f_{ex} = \frac{|\mu \cdot E|^4}{|\mu \cdot E_0|^4} \times \frac{W_1}{W_1^M}, \qquad (6\text{-}45)$$

where W_1 and W_1^M represent total decay rate from state $|1\rangle$ in the case of absenting and presenting metal NPs, respectively. Similarly, the enhancement of emission for the transition from $|2\rangle$ to $|0\rangle$ may be written as[41]

$$f_{em} = \frac{\eta_2^M(\omega)}{\eta_2(\omega)}, \qquad (6\text{-}46)$$

where, $\eta_2 = \dfrac{W_{2R}}{W_{2R} + W_{2NR}}, \eta_2^M = \dfrac{W_{2R}^M}{W_{2R}^M + W_{2NR}^M}$; W_{2R} and W_{2R}^M are the radiative and nonradiative decay rates from $|2\rangle$ in the case of presenting metal. Therefore, the total UC enhancement can be written as[41]

$$f_{\text{total}} = f_{\text{ex}} \times f_{\text{em}} = \frac{|\mu \cdot E|^4 \times W_1 \times \eta_2^M(\omega)}{|\mu \cdot E_0|^4 \times W_1^M \times \eta_2(\omega)}. \quad (6\text{-}47)$$

From Eq. (6-47), we can conclude that the effect of PE-UCL is mainly ascribed to the enhancement of excitation and emission. Moreover, it's better to occur plasmon resonance at the wavelength of excitation than emission[32]. The intensity of UC luminescence has a closer relationship with the absorption enhancement in the case of both weak and strong excitation[32].

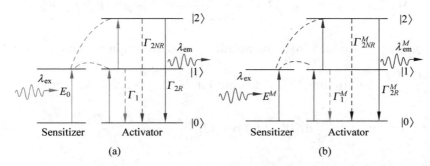

Fig. 6-6 The simplified system for describing the energy transfer for the process of two-photon UCL, where $|0\rangle$, $|1\rangle$ and $|2\rangle$ are the level of ground, intermediate and emission, respectively[41]

(For colored figure please scan the QR code on page 1)

6.3 Principle of Plasmon-Enhanced FRET

The principle of plasmon-enhanced FRET(PE-FRET) is a promising field and many studies have shown the plasmon effects on FRET[42].

Considering Coulomb criterion and minimum coupling[43], and studying plasmon-coupled resonance energy transfer (PC-RET) in the quantum electrodynamics, the rate of RET can be described as[44]

$$W_{\text{ET}} = \frac{2\pi}{\hbar} \sum_{d,d'} \sum_{a,a'} \int dE P_{d'} P_a |\langle \phi_{a'} | \phi_a \rangle|^2 |\langle \phi_d | \phi_{d'} \rangle|^2 \cdot$$
$$|M(E, r_D, r_A)|^2 \delta(E_{d'} - E_d - E) \delta(E_{a'} - E_a - E), \quad (6\text{-}48)$$

where, a' and d' represen the states of vibration related with the excited state of the acceptor and donor, respectively; d and a represen the same meanings as the d′ and a′ butare related with ground state; $P_{d'}$ and P_a denote the probability distribution at thermal equilibrium with respect to donor and acceptor; $|\langle \phi_{a'} | \phi_a \rangle|^2$ and $|\langle \phi_d | \phi_{d'} \rangle|^2$ are the Franck Condon factors of acceptor and donor, respectively; E is photon energy; and $M(E, r_D, r_A)$ is transition amplitude, depending on the positions of donor and acceptor and it can be written as[45]

$$M(E, r_D, r_A) = -\mu^D \mu^A \frac{e_A \cdot E^D(r_A, E)}{p_{ex}(E)} = -\mu^A \cdot \widetilde{E^D}(r_A, E), \quad (6\text{-}49)$$

where, μ^A can be divided into magnitude μ^A and direction e_A, and $\widetilde{E^D}(r_A, E) = \frac{\mu^D E^D(r_A, E)}{p_{ex}(E)}$. Substituting Eq. (6-49) into Eq. (6-48), W_{ET} can be written as[44]

$$W_{ET} = \frac{\hbar}{2\pi} \int dE \, W_{em}(E) W_{abs}(E) \left| \frac{e_A \cdot E^D(r_A, E)}{p_{ex}(E)} \right|^2, \quad (6\text{-}50)$$

where,

$$W_{em}(E) = (2\pi/\hbar) \sum_{a,a'} P_a |\mu^A|^2 |\langle \phi_{a'} | \phi_a \rangle|^2 \cdot$$

$$\delta(E_{a'} - E_a - E), \quad (6\text{-}51)$$

$$W_{abs}(E) = (2\pi/\hbar) \sum_{d,d'} P_{d'} |\mu^D|^2 |\langle \phi_d | \phi_{d'} \rangle|^2 \cdot$$

$$\delta(E_{d'} - E_d - E). \quad (6\text{-}52)$$

Also $E = \hbar \omega$ and $\delta(\hbar \omega) = \delta(\omega)/\hbar$, so Eq. (6-50) can be written as[44]

$$W_{ET} = \frac{9 c^4 W_r^D}{8\pi} \int d\omega \frac{\sigma(\omega) I(\omega)}{\omega^4} \left| \frac{e_A \cdot E^D(r_A, E)}{p_{ex}(E)} \right|^2, \quad (6\text{-}53)$$

where, c is the speed of light in vacuum, $\sigma(\omega)$ is the cross-section of absorption, $I(\omega)$ is the emission spectra of normalization in Gaussian units, and the latter two can be written as[43-44]

$$\sigma(\omega) = \frac{4\pi^2 \omega}{3 \hbar \alpha_a c} \sum_{a,a'} P_a |\mu^A|^2 |\langle \phi_{a'} | \phi_a \rangle|^2 \delta(\omega_{a'} - \omega_a - \omega), \quad (6\text{-}54)$$

$$I(\omega) = \frac{\eta(\omega)}{W_r^D} = \frac{4 \alpha_e \omega^3}{3 \hbar c^3 W_r^D} \sum_{d,d'} P_{d'} |\mu^D|^2 |\langle \phi_d | \phi_{d'} \rangle|^2 \delta(\omega_{d'} - \omega_d - \omega),$$

$$(6\text{-}55)$$

where W_r^D is the total emission rate of the donor without plasmonic media, and can be calculated using[44]:

$$W_r^D = \int \eta(\omega) d\omega, \qquad (6\text{-}56)$$

where $\eta(\omega)$ is the emission spectra of donor which is absence of plasmonic media and can be calculated on[44]

$$\eta(\omega) = (4\alpha_e \omega^3 / 3\hbar c^3 \sum_{d,d'} P_{d'} |\mu^D|^2 |\langle \phi_d | \phi_{d'} \rangle|^2 \cdot$$
$$\delta(\omega_{d'} - \omega_d - \omega), \qquad (6\text{-}57)$$

where α_a and α_e are the real refractive indices to check the cross sections of absorption and the emission spectra of normalization, respectively[43,44]. The coupling factor of Eq. (6-53) is associated with the second-order Green's function $\bar{\bar{g}}(r,r',\omega)$[45]:

$$\frac{e_A \cdot E^D(r_A, E)}{p_{ex}(E)} = -\frac{\omega^2}{c^2 \varepsilon_0} e_A \cdot \bar{\bar{g}}(r_A, r_D, \omega) \cdot e_D \quad \text{(SI Units)},$$

$$= -\frac{4\pi\omega^2}{c^2} e_A \cdot \bar{\bar{g}}(r_A, r_D, \omega) \cdot e_D \quad \text{(CGS Units)},$$

$$(6\text{-}58)$$

where, $\bar{\bar{g}}(r_A, r_D, \omega)$ satisfies

$$\left(\frac{\varepsilon_r(r,\omega)\omega^2}{c^2} - \nabla \times \nabla \times \right) \bar{\bar{g}}(r,r',\omega) = \delta(r - r'). \qquad (6\text{-}59)$$

Above Eq. (6-58) demonstrates that computing $E^D(r_A, E)$ and $p_{ex}(E)$ are equal to solving the second-order Green's function $\bar{\bar{g}}(r,r',\omega)$[44].

In fact, the following forms are usually used in the process of experiments[44]:

$$I(\omega) = \frac{I(\bar{v})}{2\pi c},$$

$$W_r^D = \frac{\phi_D}{\tau_D},$$

$$\varepsilon(\bar{v}) = \frac{N_A}{\ln 10} \sigma(\bar{v}) \quad (\text{mol}^{-1} \cdot \text{cm}^2),$$

$$= \frac{N_A}{10^3 \ln 10} \sigma(\bar{v}) \quad (\text{M}^{-1} \cdot \text{cm}^{-1}).$$

So,

$$W_{ET} = \frac{\phi_D 9000\ln 10}{\tau_D 128\pi^5 N_A} \int d\bar{v} \frac{\varepsilon(\bar{v})I(\bar{v})}{\bar{v}^4} \left| \frac{e_A \cdot E^D(r_A, \bar{v})}{p_{ex}(\bar{v})} \right|^2$$

$$= \frac{\phi_D}{\tau_D}(8.785 \times 10^{-25} \text{ mol} \times \tilde{J}), \qquad (6-60)$$

$$\tilde{J} \equiv \int d\bar{v} \frac{\varepsilon(\bar{v})I(\bar{v})}{\bar{v}^4} \left| \frac{e_A \cdot E^D(r_A, \bar{v})}{p_{ex}(\bar{v})} \right|^2 \ (\text{mol}^{-1}). \qquad (6-61)$$

where, $\bar{v} = (2\pi c)^{-1}\omega$. Equation (6-60) clearly demonstrated that the PC-RET rate is controlled by factors including coupling factor, coefficient of molecular absorption, and emission spectrum of normalization[44]. Also

$$p_{ex}(\bar{v}) = (2\pi c)^{-1} p_0 \delta(\bar{v}_{d'd} - \bar{v}), \qquad (6-62)$$

$$E^D(r_A, \bar{v}) = (2\pi c)^{-1} \varepsilon_r^{-1} R^{-3}(3e_R(e_R \cdot e_D) - e_D) p_0 \delta(\bar{v}_{d'd} - \bar{v}). \qquad (6-63)$$

So,

$$\left| \frac{e_A \cdot E^D(r_A, \bar{v})}{p_{ex}(\bar{v})} \right|^2 = \frac{(e_A \cdot e_D - 3(e_A \cdot e_R)(e_R \cdot e_D))^2}{\varepsilon_r^2 R^6} \equiv \frac{\kappa^2}{\varepsilon_r^2 R^6}. \qquad (6-64)$$

Evidently, the coupling factor of Förster theory is not dependent on frequency, therefore, it can be removed from the integral in Eq. (6-64)[44]. Finally, for better understanding, the comparison between PC-RET and FRET is drawn as Table 6-1[44].

Table 6-1 Comparison between theories of PC-RET and FRET[44]

Formula Form	PC-RET	FRET
Line Shape Expression	$W_{ET} = \frac{\hbar}{2\pi} \int dE W_{em}(E) W_{abs}(E) \times \left\| \frac{e_A \cdot E^D(r_A, E)}{p_{ex}(E)} \right\|^2$	$W_{ET} = \frac{\hbar}{2\pi} \frac{k^2}{\varepsilon_r^2 R^6} \times \int dE W_{em}(E) W_{abs}(E)$
Absorption Cross Section Expression (CGS)	$W_{ET} = \frac{9c^4 W_r^D}{8\pi} \int d\omega \frac{\sigma(\omega)I(\omega)}{\omega^4} \times \left\| \frac{e_A \cdot E^D(r_A, \omega)}{p_{ex}(\omega)} \right\|^2$	$W_{ET} = \frac{9c^4 k^2 W_r^D}{8\pi \varepsilon_r^2 R^6} \int d\omega \frac{\sigma(\omega)I(\omega)}{\omega^4}$
Föster's Expression (CGS)	$W_{ET} = \frac{\phi_D 9000\ln 10}{\tau_D 128\pi^5 N_A} \int d\bar{v} \frac{\varepsilon(\bar{v})I(\bar{v})}{\bar{v}^4} \times \left\| \frac{e_A \cdot E^D(r_A, \bar{v})}{p_{ex}(\bar{v})} \right\|^2$	$W_{ET} = \frac{\phi_D 9000\ln 10 k^2}{\tau_D 128\pi^5 N_A n^4 R^6} \times \int d\bar{v} \frac{\varepsilon(\bar{v})I(\bar{v})}{\bar{v}^4}$

References

[1] LAKOWICZ J R. Principles of fluorescence spectroscopy[M]. 3rd Edition. New York: Springer-Verlag,2006.
[2] PURCELL E M. Spontaneous emission probabilities at radio frequencies[J]. Phys Rev, 1946,69: 681.
[3] GALLOWAY C M,ETCHEGOIN P G,LE RU E C. Ultrafast nonradiative decay rates on metallic surfaces by comparing surface-enhanced Raman and fluorescence signals of single molecules[J]. Phys Rev Lett,2009,103: 063003.
[4] JOHANSSON P, XU H X. Surface-enhanced Raman scattering and fluorescence near metal nanoparticles[J]. Phys Rev B,2005,72: 035427.
[5] ITOH T,IGA M,TAMARU H, et al. Quantitative evaluation of blinking in surface enhanced resonance scattering and fluorescence by electromagnetic mechanism[J]. J Chem Phys,2012,136: 024703.
[6] HUANG X,HAN S,HUANG W,et al. Enhancing solar cell efficiency: the search for luminescent materials as spectral converters[J]. Chem. Soc. Rev. ,2013,42: 173-201.
[7] HAGSTROM A L, DENG F, KIM J H. Enhanced triplet-triplet annihilation upconversion in dual-sensitizer systems: translating broadband light absorption to practical solid-state materials[J]. ACS Photonics,2017,4: 127-137.
[8] FRAZER L,GALLAHER J K,SCHMIDT T. Optimizing the efficiency of solar photon upconversion[J]. ACS Energy Letters,2017,2: 1346-1354.
[9] LIU Q,GUO B, RAO Z, et al. Strong two-photon-induced fluorescence from photostable,biocompatible nitrogen-doped graphene quantum dots for cellular and deep-tissue imaging[J]. Nano Lett. ,2013,13: 2436-2441.
[10] FARRER R A,BUTTERFIELD F L,CHEN V W,et al. Highly efficient multiphoton-absorption-induced luminescence from gold nanoparticles[J]. Nano Lett. , 2005, 5: 1139-1142.
[11] KAISER W, GARRETT C. Advanced fluorescence microscopy techniques—FRAP, FLIP,FLAP,FRET and FLIM[J]. Phys. Rev. Lett. ,1961,7: 229-231.
[12] MATERIALS A F. Devices using double-pumped phosphors with energy transfer[J]. Proceedings of the IEEE. ,1973,61: 758-786.
[13] HUANG D. Theoretical and application of rare earth upconversion luminescent materials[J]. Journal of Chifeng University,2017,33: 7-8.
[14] LUO B C,WANG H B,ZHUO N Z. Overview of rare earth upconversion luminescent materials [J]. China Lighting Electrical Appliances,2015,6: 10-14.
[15] ZOU X, IZUMITANI T. Spectroscopic properties and mechanisms of excited state absorption and energy transfer upconversion for Er^{3+}-doped glasses[J]. Non-Cryst. Solids,1993,162: 68-80.

[16] SHPAISMAN H, NIITSOO O, LUBOMIRSKY I, et al. Can up-and down-conversion and multi-exciton generation improve photovoltaics[J]. Sol. Energy Mater. Sol. Cells,2008,92: 1541-1546.

[17] NIE S, EMORY S R. Probing single molecules and single nanoparticles by surface-enhanced Raman scattering[J]. Science,1997,275: 1102-1106.

[18] ZOU W Q, VISSER C, MADURO J A, et al. Broadband dye-sensitized upconversion of near-infrared light[J]. Nat. Photonics,2012,6: 560.

[19] WANG F, WANG J, LIU X. Direct evidence of a surface quenching effect on size-dependent luminescence of upconversion nanoparticles[J]. Angew. Chem.,2010,122: 7618-7622.

[20] BAI X, SONG H, PAN G, et al. Size-dependent upconversion luminescence in Er^{3+}/Yb^{3+}-codoped nanocrystalline yttria: saturation and thermal effects[J]. J. Phys. Chem. C,2007,111: 13611-13617.

[21] WANG J, SONG H, XU W, et al. Phase transition, size control and color tuning of $NaREF_4$: Yb^{3+}, Er^{3+} (RE = Y, Lu) nanocrystals [J]. Nanoscale,2013,5: 3412-3420.

[22] CHEN G, SHEN J, OHULCHANSKYY T Y, et al. (α-$NaYbF_4$: Tm^{3+})/CaF_2 core/shell nanoparticles with efficient near-infrared to near-infrared upconversion for high-contrast deep tissue bioimaging[J]. ACS Nano,2012,6: 8280-8287.

[23] WANG J, DENG R, MACDONALD M A, et al. Enhancing multiphoton upconversion through energy clustering at sublattice level[J]. Nat. Mater.,2014,13: 157-162.

[24] SCHIETINGER S, AICHELE T, WANG H Q, et al. Plasmon-enhanced upconversion in single $NaYF_4$: Yb^{3+}/Er^{3+} co-doped nanocrystals [J]. Nano Lett.,2010,10: 134-138.

[25] CHEN G, DING C, WU E, et al. Tip-enhanced upconversion luminescence in Yb^{3+}-Er^{3+} co-doped $NaYF_4$ nanocrystals[J]. J. Phys. Chem. C,2015,119: 22604-22610.

[26] WU B, UENO K, YOKOTA Y, et al. Enhancement of a two-photon-induced reaction in solution using light-harvesting gold nanodimer structures[J]. J. Phys. Chem. Lett.,2012,3: 1443-1447.

[27] AYALA-OROZCO C, LIU J G, KNIGHT M W, et al. Fluorescence enhancement of molecules inside a gold nanomatryoshka [J]. Nano Lett.,2014,14: 2926-2933.

[28] HE J, ZHENG W, LIGMAJER F, et al. Plasmonic enhancement and polarization dependence of nonlinear upconversion emissions from single gold nanorod@ SiO_2 @ CaF_2: Yb^{3+}, Er^{3+} hybrid core-shell satellite nanostructures[J]. Light: Science & Applications,2017,6: e16217.

[29] MING T, CHEN H J, JIANG R B, et al. Plasmon-controlled fluorescence: beyond the intensity enhancement. [J]. J Phys Chem. Lett,2012,3: 191-202.

[30] NOVOTNY L, HECHT B. Principles of nano-optics [M]. Cambridge: Cambridge University Press,2012.

[31] KLIMOV V V, DUCLOY V, LETOKHOV V S, et al. Spontaneous emission of an atom in the presence of nanobodies [J]. Quantum Electron,2001,31: 569.

[32] PARK W, LUA D, AHNA S. Plasmon enhancement of luminescence upconversion [J]. Chem. Soc. Rev. ,2015,44: 2940.

[33] BARRON L D, GRAY C G. Baxter states in the XY model [J]. J. Phys. A: Math. , Nucl. Gen. ,1973,6: 59.

[34] JUDD B R. Optical absorption intensities of rare-earth ions [J]. Phys. Rev. ,1962, 127: 750.

[35] AUZEL F. Upconversion and anti-Stokes processes with f and d ions in solids[J]. Chem. Rev. ,2004,104: 139-174.

[36] CANTELAR E, MUNOZ J A, SANZ-GARCI' A J A, et al. Yb^{3+} to Er^{3+} energy transfer in $LiNbO_3$ [J]. J. Phys. : Condens. Matter,1998,10: 8893-8903.

[37] KINGSLEY J D, FENNER G E, GALGINAITIS S V. Kinetics and efficiency of infrared-to-visible conversion in LaF_3: Yb, Er [J]. Appl. Phys. Lett. , 1969, 15: 115-117.

[38] SIMONDI-TEISSEIRE B. Yb^{3+} to Er^{3+} energy transfer and rate-equations formalism in the eye safe laser material Yb: Er: $Ca_2Al_2SiO_7$ [J]. Opt. Mater. ,1996,6: 267-274.

[39] YEH D C, SIBLEY W A, SUSCAVAGE M, et al. Multiphonon relaxation and infrared-to-visible conversion of Er^{3+} and Yb^{3+} ions in barium-thorium fluoride glass [J]. J. Appl. Phys. ,1987,62: 266-275.

[40] MAIER S A. Plasmonics: fundamentals and applications[M]. New York: Springer Science &Business Media,2007.

[41] XU W, CHEN X, SONG H. Upconversion manipulation by local electromagnetic field [J]. Nano Today,2017,17: 54-78.

[42] BLUM C, ZIJLSTRA N, LAGENDIJK A, et al. Nanophotonic control of the Förster resonance energy transfer efficiency[J]. Phys. Rev. Lett. ,2012,109: 203601.

[43] DUNG H T, KNÖLL L, WELSCH D G. Intermolecular energy transfer in the presence of dispersing and absorbing media[J]. Phys. Rev. A: At. ,Mol. ,Opt. Phys. , 2002,65: 043813.

[44] HSU L Y, DING W D, SCHATZ G C. Plasmon-coupled resonance energy transfer[J]. J. Phys. Chem. Lett. ,2017,8: 2357-2367.

[45] DING W, HSU L Y, SCHATZ G C. Plasmon-coupled resonance energy transfer: a real-time electrodynamics approach[J]. J. Chem. Phys. ,2017,146: 064109.

CHAPTER 7

Plasmon-Enhanced Raman Scattering Spectra

7.1 Surface-Enhanced Raman Scattering Spectroscopy

7.1.1 Brief history of SERS spectroscopy

In 1923, Adolf Smekal first applied theoretical methods to predict the presence about light inelastic scattering. In German, it was also called Smekal-Raman effect. As early as 1922, the Indian Physicist C. V. Raman (Fig. 7-1 (a)) published his book *Molecular Diffraction of Light*, which was the primary step in his series of investigations with co-workers. Eventually, on February 28, 1928, the radiation effect was discovered and named after his name, called Raman scattering. In 1928, Raman and his student K. S. Krishnan (Fig. 7-1 (b)) discovered this effect in the liquid, and then Grigory Landsberg and Leonid Mandelstam discovered this effect in the crystal. In addition, Raman won the Physics Nobel Prize for his outstanding contribution in the field of scattered light in 1930. Until more than thirty years later, C. V. Raman and K. S. Krishnan were able to record spectra using mercury lamps and photographic plates.

In 1974, Fleischmann (Fig. 7-2 (a)) observed Surface-enhanced Raman Scattering(SERS) for the first time by observing the Raman spectrum about pyridine adsorbed on the silver electrode[1]. In 1977, two independent teams noted that the concentration of scattering materials could not explain the

CHAPTER 7 Plasmon-Enhanced Raman Scattering Spectra

Fig. 7-1 (a) Sir Chandrasekhara Venkata Raman(1888—1970); (b) Kariamanickam Srinivasa Krishnan(1898—1961), are Indian physicists who discovered the Raman scattering effect, and in 1930, jointly won the Nobel Prize in Physics

Fig. 7-2 (a) Martin Fleischmann and(b) Richard P. van Duyne

phenomenon of enhanced signals. Therefore, each team proposed a mechanism to enhance the signal, and Jeanmaire and Van Duyne (Fig. 7-2(b)) proposed electromagnetic effects[2], while Albrecht and Creighton proposed a charge transfer effect[3]. Subsequently, Rufus Ritchie predicted the presence of SP[4].

Xu firstly observed the single molecular SERS spectra, and interpreted huge electromagnetic enhancement with the couple of two nanoparticles at a distance about 2 nm, and SERS enhancement can be reach up to about 10^{10}, see Fig. 7-3[5-6].

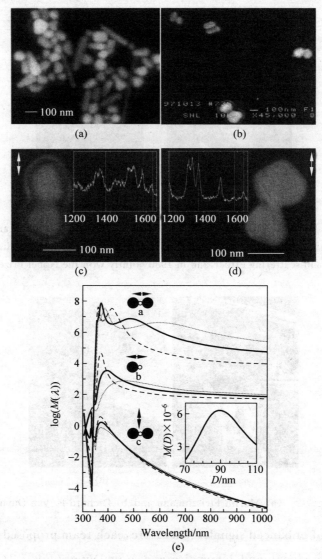

Fig. 7-3 SEM images of immobilized Ag particles. The pictures show (a) Overview of Ag particle shapes and sizes, (b) Ag particle dimers observed after incubation with 1×10^{-11} M Hb for 3 h, and (c), (d) Hot dimers and corresponding single Hb molecule spectra. The double arrows in (c) and (d) indicate the polarization of the incident laser field. Calculated electromagnetic enhancement factor for the midpoint between two Ag spheres separated by $d = 5.5$ nm and for a point $d/2$ outside a single sphere. The solid and open circles indicate the position of the Ag spheres and the Hb molecule, respectively, in relation to the incident polarization vector (double arrows). The calculations have been performed for spheres of diameters $D = 60$ mm (dashed curves), 90 mm (solid curves) and 120 nm (dotted curves). Inset shows the enhancement versus D for $\lambda_l = 514.5$ nm and a Stokes shift of 1500 cm^{-1} for configuration a

Tian et al. studied a plasmon magnetic silica nanotube, a core-shell structure with ultrathin silica or alumina-coated Au NPs[2], and its simple structural diagram was shown in Fig. 7-4.

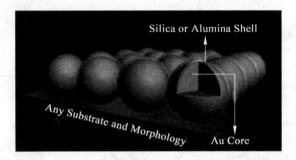

Fig. 7-4   Simple schematic covers Au NPs with ultrathin alumina and silica layers[2]
(For colored figure please scan the QR code on page 1)

This structure had many advantages. First, in order to prevent aggregation of metal NP, the structure was covered with an ultrathin layer. Second, the material contacted with the substrate directly. Third, the NP were able to meet the contours of different substrates. In addition, using this technique to measure yeast cells, high-quality molecular Raman spectroscopy could be collected, a simple schematic of which was shown in Fig. 7-5. In short, this technology was not only used to detect yeast cells, but also greatly promoted research in materials and life sciences, including food safety testing, detection of environmental pollutants, observation of cancer cells, etc[7].

7.1.2   Physical mechanism of SERS spectroscopy

SERS spectroscopy was mainly developed in the metal plasmon field. In general, the SERS effect was divided into physical enhancement and chemical enhancement. Its mechanisms included SP excitation, chemical enhancement related to charge transfer, or charge transfer resonance Raman. The physical mechanism of laser irradiation onto the metal surface was shown in Fig. 7-6. When the laser was illuminated onto the surface about Au or Ag, the nanostructure of the metal compressed the light into a relatively small volume between the two

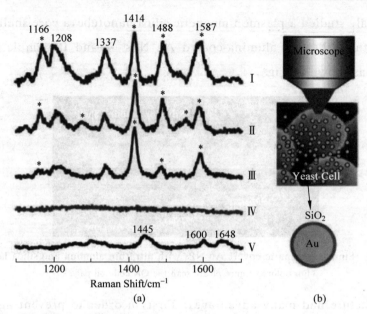

Fig. 7-5 Schematic of the biological structure of yeast cells detected *in situ* by shell-isolated NP-enhanced Raman spectroscopy(SHINERS)

(a) Under this mechanism, the molecular Raman spectra(curves I-III) obtained with different hotspots. In curve IV, there is no molecular adsorption substrate. Curve V shows the Raman spectra of yeast cells themselves. (b) A SHINERS experiment on living yeast cells[2]

(For colored figure please scan the QR code on page 1)

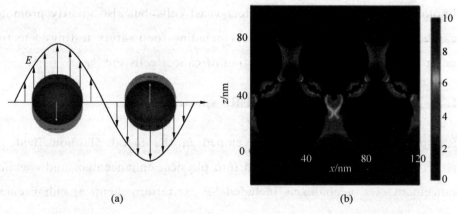

Fig. 7-6 Sketch of the mechanism of laser irradiation to Au or Ag nanostructured surfaces

(a) image shows LSP caused by collective oscillations of charge on the surface of metal NP; (b) image is the distribution of the EM field intensity, which is simulated by the finite element method[7]

(For colored figure please scan the QR code on page 1)

NPs, thereby greatly enhanced strength of the EM field, and this led to the emergence of hot spots.

Local surface plasmon resonance (LSPR) effect was caused by the collective oscillation of valence electrons resonating with the incident light frequency, where the valence electron was in the coinage metal NP. And the illustration about the LSPR was shown in Fig. 7-7(a). Here, a spherical NP having a radius a was used, and was irradiated with z-polarized light, and the wavelength was λ. Assuming that the electric field around the NPs was uniform, the external electromagnetic field of the NP could be described as

$$E_{out}(x,y,z) = E_0 \hat{z} - \alpha E_0 \left[\frac{\hat{z}}{r^3} - \frac{3z}{r^5}(x\hat{x} + y\hat{y} + z\hat{z}) \right], \quad (7\text{-}1)$$

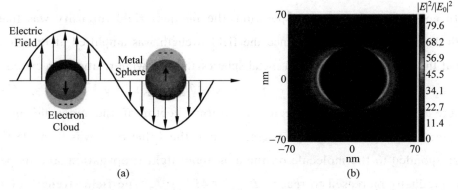

Fig. 7-7 (a) Illustration of the localized SP resonance effect; (b) Extinction efficiency (ratio of cross section to effective area) of a spherical Ag NP of 35 nm radius in vacuum $|E|^2$ contours for a wavelength corresponding to the plasmon extinction maximum. Peak $|E|^2 = 85$[8]

(For colored figure please scan the QR code on page 1)

where, x, y, z were Cartesian coordinates; and $\hat{x}, \hat{y}, \hat{z}$ were unit vectors; r and α were the radial distance and metal polarizability, respectively. And α could be expressed as

$$\alpha = ga^3, \quad (7\text{-}2)$$

where, a was the radius of spherical NP, and g could described as

$$g = \frac{\varepsilon_{in} - \varepsilon_{out}}{\varepsilon_{in} + 2\varepsilon_{out}}, \quad (7\text{-}3)$$

where ε_{in} and ε_{out} were dielectric constant about the metal NPs and external environment, respectively. Due to the strong wavelength dependence on the real part about the dielectric constant for the metal NPs, the enhancement amplitude was determined by the wavelength. When $\varepsilon_{in} \approx -2\varepsilon_{out}$, the maximum enhancement was produced.

According to Mie theory, the extinction spectrum about randomly shaped NP could be described as

$$E(\lambda) = \frac{24\pi^2 Na^3 \varepsilon_{out}^{3/2}}{\lambda \ln(10)} \left[\frac{\varepsilon_i(\lambda)}{(\varepsilon_r(\lambda) + \chi\varepsilon_{out})^2 + \varepsilon_i(\lambda)^2} \right], \qquad (7\text{-}4)$$

where, $\varepsilon_r(\lambda)$ and $\varepsilon_i(\lambda)$ were imaginary and real parts about ε_{in}, the function of metal dielectric, respectively, and χ was the shape factor, for a general sphere, the value of χ was usarally 2, but its value could be as high as 20.

In general, in the Raman spectrum, the incident field intensity was linear with the scattering intensity. Since the field strength was amplified at the surface about the NPs, $r = a$. For small metal spheres, the following relationship exists:

$$|E_{out}|^2 = E_0^2 [|1-g|^2 + 3\cos^2\theta(2\text{Re}(g) + |g|^2)], \qquad (7\text{-}5)$$

where θ was the angle between the position vector of the molecule on the surface and the incident field vector. When the value of θ was 0° or 180°, it corresponded to the molecule on the axis about light propagation, and its peak was gradually increased to reach $|E_{out}|^2 = 4E_0^2|g|^2$. The field strength of the spherical particle position function was shown in Fig. 7-7, where the ratio about maximum intensity to minimum intensity was 4 and the average intensity in the radial direction was $|\bar{E}_{out}|^2 = 2E_0^2|g|^2$.

In the Raman scattering effect, the applied field induced an oscillating dipole in the molecules at the surface. The following relationship exists:

$$EF = \frac{|E_{out}|^2 |E'_{out}|^2}{|E_0|^4} = 4|g|^2|g'|^2, \qquad (7\text{-}6)$$

where EF was the enhancement factor of SERS EM. When the Stokes shift was small, g and g' were at the same wavelength, then EF was approximately g^4. In previous studies, this was called the fourth power about field enhancement at the NPs surface, also known as E^4 enhancement. For small

CHAPTER 7 Plasmon-Enhanced Raman Scattering Spectra

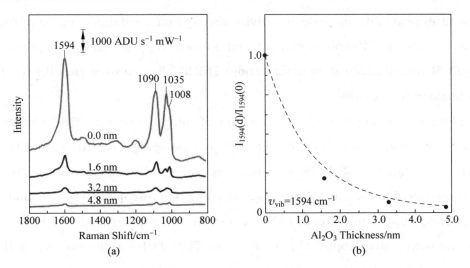

Fig. 7-8 (a) Surface-enhanced Raman spectra of pyridine adsorbed to silver film over nanosphere samples treated with various thicknesses of alumina (0.0 nm, 1.6 nm, 3.2 nm, and 4.8 nm). ex = 532 nm, P = 1.0 mW, and acquisition time = 300 s. (b) Plot of surface-enhanced Raman spectroscopy intensity as a function of alumina thickness for the 1594 cm^{-1} band(circles)[8]

(For colored figure please scan the QR code on page 1)

spheres, the value of g was about 10, EF could reach 10^4 or even 10^5, and for high-order Ag NPs, the enhancement factor could reach 10^8.

However, in practical applications, the EF of the SERS system could be expressed as,

$$EF = \frac{[I_{SERS}/N_{surf}]}{[I_{NRS}/N_{vol}]}, \qquad (7-7)$$

where, I_{SERS} was the SERS intensity, I_{NRS} was the normal Raman intensity, N_{surf} was the molecular number, and N_{vol} was the molecular number in the excitation volume[8].

7.2 Tip-Enhanced Raman Scattering Spectroscopy

7.2.1 Brief introduction of TERS spectroscopy

In 1985, Wessel proved TERS firstly. This technology combined a scanning probe microscope (SPM) with surface Raman spectroscopy, which used metal tips to dramatically increase sensitivity and spatial resolution. In 2000, four groups Anderson, Zenobi, Pettinger and Kawata reported independently the result of tip-enhanced Raman scattering(TERS). This proved that the TERS technology was feasible[9].

TERS spectroscopy was a combination of apertureless near-field scanning optical microscopy and Raman spectroscopy having high sensitivity and high spatial resolution. TERS was a new technology associated with multiple microscopes, for example, scanning tunneling microscopy (STM), scanning near-field optical microscopy (SNOM), shear force microscopy (SFM), and atomic force microscopy (AFM). For the TEM-TERS technique, when the laser was illuminated at the apex about a metal-coated or metal tip, it was capable of exciting the tip apex around the LSP. When the tip was close to the substrate, a hot spot could be created in the gap between the substrate and the tip. The strong Raman signal generated by this hot spot could be scattered to the far field. Figure 7-9 showed a simplified schematic of the TERS technique. Studies had shown that when light enhancement was limited, the hot spot belonged to the near-field light at the tip and could not spread to the far field in large quantity. When the tip scanned the sample surface, the signal of the tip was collected at each point to improve spatial resolution rate, thus beyond the limits of light diffraction. In general, light diffraction depended on the size and shape of the tip apex.

7.2.2 Physical mechanism of TERS spectroscopy

In general, the spatial resolution of a typical optical microscope was limited by optical diffraction:

$$\Delta x = 0.61\lambda/\mathrm{NA}, \qquad (7\text{-}8)$$

where NA was the abbreviation for numerical aperture. Also, the resolution was determined by the NA and the wavelength of the incident light. Studies had shown that in the visible range, the resolution could be as high as 200 nm.

CHAPTER 7 Plasmon-Enhanced Raman Scattering Spectra

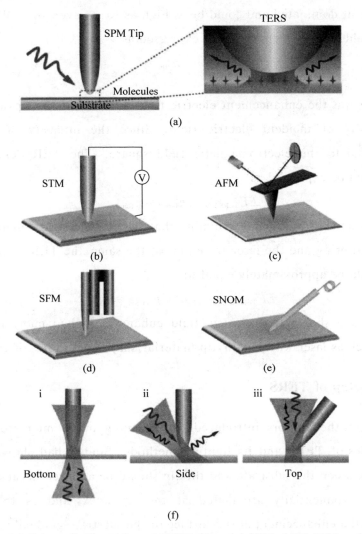

Fig. 7-9 (a) Schematic illustration of TERS. (b)～(e) STM,AFM,SFM,and SNOM. (f) Three different modes of excitation and collection: (i) Side excitation and side collection, (ii) Bottom excitation and bottom collection, and (iii) Top excitation and top collection[9]

(For colored figure please scan the QR code on page 1)

However, for current nanooptics, the resolution was still too low.

Enhancement mechanisms for TERS technology included chemical enhancement and electric field enhancement. Precious metals Au, Ag, etc. were preferred as the tip or substrate of the STM. In general, electric field enhancement dominated and could be as high as 10^9 or even 10^{11}. Wherein the electric field enhancement could be expressed as

$$g = \frac{E_{tip}}{E_0}, \qquad (7\text{-}9)$$

where, E_{tip} was the enhancement electric field intensity at the tip, and E_0 was the intensity of incident electric field. Since the intensity of light was proportional to the electromagnetic field square. The TERS enhancement factor could be expressed as

$$EF_{\text{TERS-EM}} = g_{laser}^2 \times g_{Raman}^2. \qquad (7\text{-}10)$$

If the Raman shift was small, assuming that the electric field enhancement of Raman scattering and the laser were almost the same, the TERS enhancement factor could be approximately equal to

$$EF_{\text{TERS-EM}} = g_{laser}^2 \times g_{Raman}^2 \approx g^4. \qquad (7\text{-}11)$$

Studies had shown that electric field enhancement had many influencing factors, such as laser wavelength, tip material, shape, and radius of curvature.

7.2.3 Setup of TERS

In this part, the authors introduced the scanning probe microscopy (SPM) technologies of TERS and its lighting methods. Studies had shown that the distance between the substrate and the tip should be adjusted because the near field was exponentially attenuated at the tip apex, and as the distance increased, the enhancement at the surface of the substrate gradually decreased. For many SPM technologies, SFM, SNOM, AFM and STM were mainly used to construct TERS devices with different characteristics. STM belonged to an atomic resolution surface imaging instrument as shown in Fig. 7-9(b). The technology had atomic resolution and high control accuracy and was mainly used to study single molecules. However, in low-temperature ultrahigh vacuum UHV environments, TERS-STM devices were not ideal for the detection of biological samples. Figure 7-9(c) showed a simplified schematic of the AFM. In AFM, no especial treatment or sample was essential, and the technique was not limited to conductors, it could operate on any surface. In addition, the technology could be used in liquid environments, and was widely used in the research of organic molecules and organisms. However, the spatial resolution of this technique was not high, only a few nanometers. This technology was also popular for semiconductor and 2D materials. For SFM systems(Fig. 7-9(d)), it could be used for the detection of any sample, as well as in liquid environments, which was consistent with AFM. However, compared to AFM, the metal tip used by SFM was more stable. And its operation was more complicated, and the intensity of its lateral resolution was determined by the amplitude of the oscillation of the tuning fork. In addition, Fig. 7-9 (e) shown another significant microscope, a simplified schematic of the SNOM.

Figure 7-9(f) showed the optical design, which was a key part of TERS technology. Note that this technique required consideration of both optical efficiency and SPM performance, enabling the laser to be concentrated to the tip apex relatively easily, and collecting Raman signals efficiently. As could be seen from the figure, the device was usually illuminated in three view, top, side and bottom. Studies had shown that different structures of SPM systems chose different illumination methods. For bottom view, mainly for AFM-TERS systems, the substrate was transparent, and the objective lens had a high numerical aperture, so that more signals could be collected.

Side view designs were typically used in SFM and STM systems because of the preferred use of transparent samples in bottom view, which greatly limited the variety of samples. Side view enabled p-polarized light at the tip axis to excite a relatively strong electric field at the tip, which compensated for the reduced numerical aperture of the objective while creating a greater enhancement factor. For top view, both opaque and transparent samples could be used. This type of illumination was typically used in SNOM and AFM systems. The top view combined the advantages of both of the above view methods, enabling the use of high numerical aperture objectives, allowing the laser to gather and collect more signals. However, the device was more demanding on the tip, requires an especial tip, or requires a tip holder to prevent the tip from blocking the laser and scattered light, and its operation was complicated. Therefore, in low-temperature and UHV environments, top lighting was rarely used. In addition, there was a special top-view device that was configured by a parabolic mirror and mainly used in STM systems.

In previous studies, TERS technology was mainly carried out in air, and the sample was easily contaminated by impurities in the air. Later, the HV and even the UHV environment were effectively combined with the TERS technology, which greatly improved the cleanliness of the operating environment. The HV environment effectively prevented sample contamination, provided a clean environment for TERS measurements, and prevented the Ag tip from oxidizing in the air.

In 2007, Pettinger first proposed a UHV TERS device, see Fig. 7-10, (a) the schematic of this device. (b) A schematic diagram of the same kind of device

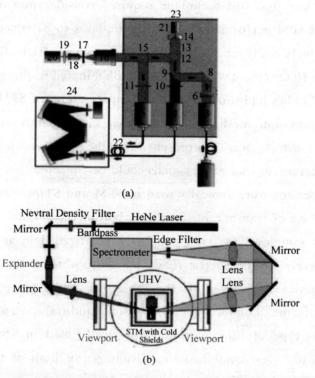

Fig. 7-10 Schematic of four typical UHV-TERS setups
(a) Pettinger's group setup[10]; (b) Van Duyne's group setup[11]; (c) HV-TERS setup[12];
(d) The setup in Dong's group[13]
(For colored figure please scan the QR code on page 1)

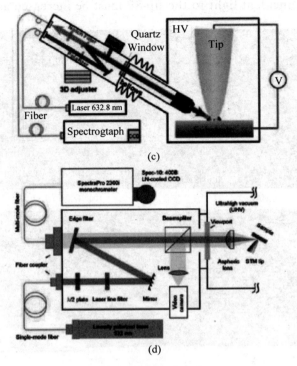

Fig. 7-10(Continued)

proposed by Van Duyne. (c) A HX TERS device. (d) A low-temperature ultrahigh vacuum (LT-UHV) TERS device designed by Dong et al., which included four sections, a laser source, a LT-UHV-STM chamber, a lens, and a spectrometer equipped a CCD detector with high sensitivity.

Next, the author briefly introduced the preparation of the tip. In the TERS measurement, SPM played a key role. The diameter, material and shape of the tip not only determined the enhancement factor of TERS, but also the imaging quality and rate of spatial resolution of SPM. So, producing high-quality tips was the key, resulting in greater enhancement factors. In Fig. 7-11, (a) showed a typical silver plated AFM tip. In the UV-Vis region, the Ag tip produced a higher enhancement than the Au tip. (b) and (c) showed SEM images about typical Ag and Au tips etched, respectively. And in TERS, in order to collect a relatively strong signal, or to obtain an ideal signal-to-noise ratio, the coupling effect about the incident light to the tip SP must be increased as shown in (d).

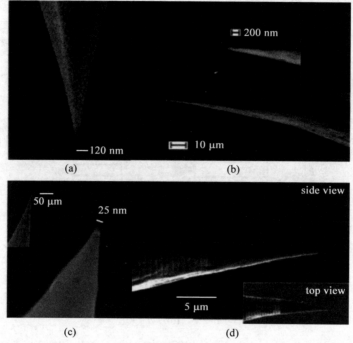

Fig. 7-11 SEM image
(a) A silver-coated metalized cantilever tip[14]. (b) A typical etched Ag tip[15]. (c) A typical etched Au tip[16]. (d) A conical metallic tip with a grating coupler on the shaft[17]

7.3 Remote-Excitation SERS

The schematic and physical mechanism of remote-excitation SERS can be seen from Fig. 1-12[18-19] (a) and (b). The laser focused on terminus B and emission at terminus A. The polarization of incident laser in (a) is parallel to the NW (the arrows mean polarization of the incident laser). It is very clear that terminus A emitted light intensively. However, there is little light while the polarization of the incident laser is perpendicular to the NW in Fig. 7-12(b). In Fig. 7-12(c) and (d), the laser is focused on terminus A and emission is at terminus B. The situation is almost the same. Figure 7-12(e) shows the incident polarization dependence on emission in detail. Data A(black squares)are collected when the emission is at terminus A. And data B(red spots) mean the emission is at the other terminus. The θ(described in Fig. 7-12(d), θ = 60° represents polarization of the incident laser parallel to the NW) is used and normalized the intensity of emission to plot polar diagram. We find the experimental data A and B coincide with the optical image in Fig. 7-12. The SEM picture in Fig. 7-12(f) shows that the target NW is 5.5 μm long and has a diameter of about 200 nm. The spherical termini are indicated by a magnified SEM image with high resolution. The near-field distributions in the $x - z$ plane (laser along the z axis) are also plotted in logarithmic scale in Fig. 7-12(g) and (h) for parallel and for perpendicular incident polarization, respectively. It is very clear that there are standing waves on the NW. It is obvious that at the emission terminus the intensity of the electric field with parallel incident polarization is much greater than that with perpendicular incident polarization. From the electromagnetic mechanism of SERS($I_{SERS} \propto |E|^4$) and our results, we know that the best Raman signal can be acquired with parallel incident polarization. In quasi-spherical terminal Au NW, the fundamental m = 0 SPP mode is excited with parallel incident polarization and the m = 1 SPP mode is excited with the perpendicular. The large difference of electric field intensity is shown in Fig. 7-12(g) and (h) shows that the propagating efficiency of the m = 0 SPP mode is much larger than that of m = 1 SPP mode. The emission of this kind of Au NW mainly depends on propagation of the m = 0 SPP mode, which achieves harvest with parallel incident polarization.

The remote excitation SERS of MGITC are measured on single Au NW with quasi-spherical termini. The polarization of the incident laser is parallel to NW to receive a maximum emission at the remote terminus according to our result in Fig. 7-13. The sketch of the remote excitation SERS technique is in Fig. 7-13 (a). The red arrows indicate the incident laser. The blue arrow representes propagating SPP on the Au NW. Orange arrows mean the Raman signal of MGITC on the terminus of the Au NW. The SEM image of this Au NW with

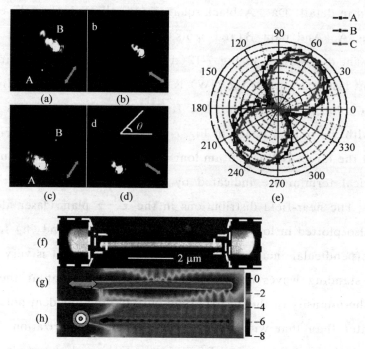

Fig. 7-12　(a) Excitation with parallel incident polarization at terminus B. (b) Excitation with perpendicular incident polarization at terminus B. (c) Excitation with parallel incident polarization at terminus A. (d) Excitation with perpendicular incident polarization at terminus A. (e) Incident polarization dependence of emission. Data A(black squares) Emission at terminus A. Data B(red spots) Emission at terminus B. Data C(green triangles)Simulation results by FDTD. (f) SEM image of the Au NW. The enlarged area indicated the shape of termini, whose scale bar was 100 nm. Distribution of electric field intensity was simulated and plotted in logarithmic scale with parallel incident polarization (g) and with perpendicular incident polarization(h)

(For colored figure please scan the QR code on page 1)

quasi-spherical termini in Fig. 7-13(b), its length is 6.5 μm and diameter is 210 nm. The optical image of propagation of SPP on an Au NW is presented in Fig. 7-13(c). The top spectrum(red line)in (d) is remote excitation SERS. The detection point is at terminus A, but the excitation point is at terminus B. The bottom spectrum(black line)in Fig. 7-13(d) is local SERS, whose excitation and detection points are at terminus A. We can see that the intensity of local SERS is about 150 times greater than that of remote-excitation SERS. We also have

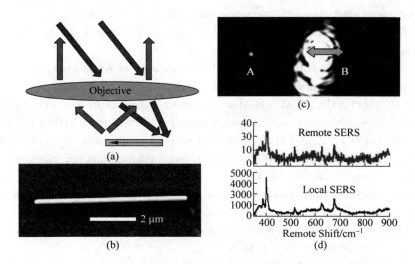

Fig. 7-13 The polarization of the incident laser, is parallel to NW
(a) The sketch of the remote excitation SERS technique; (b) The SEM image of this Au NW with quasi-spherical termini; (c) The optical image of propagation of SPP on an Au NW; (d) Remote-excitation SERS

(For colored figure please scan the QR code on page 1)

tried to measure remote-exitation SERS with perpendicular incident polarization. It is very hard to acquire remote-exitation SERS spectra due to the very low electric intensity. This result indicates that we have successfully applied this novel technique on Au NW. The antioxide property of Au material under ambient conditions ensured the Au NW could work for a much longer time compared to an Ag nanostructure. The simple structure of a single Au NW also makes it more convenient for application.

References

[1] MICHAELS A, JIANG J, BRUS L. Ag nanocrystal junctions as the site for surface-enhanced Raman scattering of single rhodamine 6G molecules[J]. J Phys Chem B, 2000, 104: 11965-11971.

[2] LI J F, HUANG Y F, DING Y, et al. Shell-isolated nanoparticleenhanced Raman spectroscopy[J]. Nature, 2010, 464: 392-395.

[3] CAO E, LIN W H, SUN M T, et al. Exciton-plasmon coupling interactions: from

principle to applications[J]. Nanophotonics,2018,7(1): 145-167.

[4] RITCHIE R H. Plasma losses by Fast Electrons in thin films[J]. Phys. Rev.,1957, 106: 874-881.

[5] XU H,BJERNELD E J,KÄLL M,et al. Spectroscopy of single hemoglobin molecules by surface enhanced Raman scattering[J]. Phys. Rev. Lett.,1999,83: 4357.

[6] XU H,AIZPURUA J,KÄLL M,et al. Electromagnetic contributions to single-molecule sensitivity in surface-enhanced Raman scattering[J]. Phys. Rev. E,1992,62: 4318.

[7] KANG L,CHU J,ZHAO H,et al. Recent progress in the applications of graphene in surface-enhanced Raman scattering and plasmon-induced catalytic reactions[J]. J. Mater. Chem. C,2015,3: 9024-9037.

[8] PAUL L S,JON A D,NILAM C S,et al. Surface-enhanced Raman spectroscopy[J]. Annual Rev. Anal. Chem,2008,1: 601-626.

[9] ZHANG Z L,SHENG S X,WANG R M,et al. Tip-enhanced Raman spectroscopy[J]. Anal. Chem.,2016,88: 9328-9346.

[10] STEIDTNER J,PETTINGER B. High-resolution microscope for tip-enhanced optical processes in ultrahigh vacuum[J]. Rev. Sci. Instrum,2007,78: 103104-103108.

[11] JIANG N,FOLEY E T,KLINGSPORN J M,et al. Observation of multiple vibrational modes in ultrahigh vacuum tip-enhanced Raman spectroscopy combined with molecular-resolution scanning tunneling microscopy [J]. Nano Lett., 2012, 12: 5061-5067.

[12] SUN M T, ZHANG Z L, ZHENG H R, et al. In-situ plasmon-driven chemical reactions revealed by high-vacuum tip-enhanced Raman spectroscopy[J]. Sci. Rep., 2012,2: 647.

[13] ZHANG R,ZHANG Y,DONG Z C,et al. Chemical mapping of a single molecule by plasmon-enhanced Raman scattering[J]. Nature,2013,498: 82-86.

[14] HAYAZAWA N, INOUYE Y, SEKKAT Z, et al. Near-field Raman scattering enhanced by a metallized tip[J]. Chem. Phys. Lett.,2001,335: 369-374.

[15] ZHANG W H, YEO B S, SCHMID T, et al. Single molecule tip-enhanced Raman spectroscopy with silver tips[J] J. Phys. Chem. C,2007,111: 1733-1738.

[16] XU G Z,LIU Z H,XU K,et al. Constant current etching of gold tips suitable for tip-enhanced Raman spectroscopy[J]. Rev. Sci. Instrum.,2012,83: 103708.

[17] ROPERS C, NEACSU C C, ELSAESSER T, et al. Grating-coupling of surface plasmons onto metallic tips: a nanoconfined light source[J]. Nano Lett.,2007,7: 2784-2788.

[18] HUANG Y Z,FANG Y R,SUN M T. Remoteexcitation of surface-enhanced Raman scattering on single Au Nanowire with quasi-spherical Termini[J]. J. Phys. Chem. C, 2011,115: 3558-3561.

[19] SUN M T,ZHANG Z L,WANG P J. et al. Remotely excited Raman optical activity using chiral plasmon propagation in Ag Nanowire[J]. Light: Science & Applications, 2013,2: e112.

CHAPTER 8

High-Vacuum Tip-Enhanced Raman Scattering Spectroscopy

8.1 Brief Introduction

It was understood from the above that TERS combined scanning probe microscopy and Raman spectroscopy. TERS technology was widely used in nanoscience and technology and was an important ultra-sensitive spectral analysis technology owing to its high spatial resolution and sensitivity, where its spatial resolution exceeded the light diffraction limit[1-46]. However, for spectral measurements about molecules that were very sensitive to the environment, the use of TERS alone will not achieve the goal of the experiment and requires a high vacuum to complete[21-22]. High-vacuum tip-enhanced Raman Scattering (the HV-TERS) spectrum was based on scanning tunneling microscopy (STM), which provided structural analysis about adsorbed samples on the surface of a clean, single crystal at low temperatures[21,33], or local spectral data about surface catalytic reactions and atomic resolution STM imaging[47].

In 2012, Sun et al. fabricated the HV-STM-TERS device as shown in Fig. 8-1. Where, the object was in the high-pressure chamber at a position approximately 10.5 mm near the tip/substrate, had a magnification of 50 × and a numerical aperture (NA) of 0.5[37]. This will greatly increase the probability of collecting Raman signals, as well as the HV-TERS spectra of Stokes and anti-Stokes.

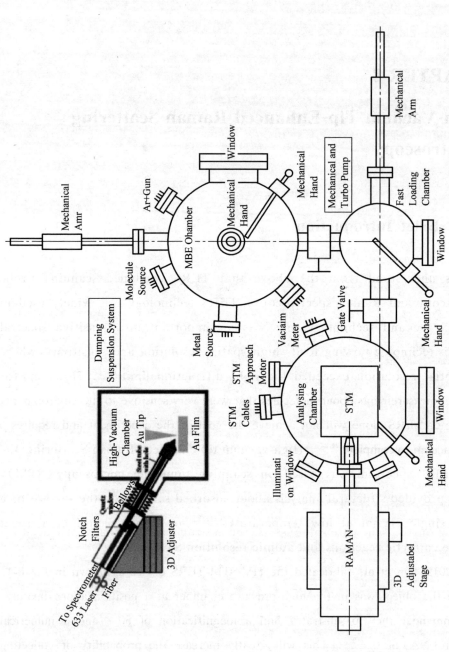

Fig. 8-1　Scheme of HV-TERS spectroscopy system with three chambers for MBE preparation, fast loading and TERS analysis[147]
(For colored figure please scan the QR code on page 1)

8.1.1 Brief description of setup of HV-TERS

A schematic about the HV-TERS spectroscopy scheme designed by Sun et al. was shown in Fig. 8-1 and consisted of a fast loading chamber, a molecular beam epitaxy(MBE)chamber, and an analysis chamber. The analysis chamber consisted of a Raman spectrometer and a scanning tunneling microscope. The MBE chamber, contained an argon(Ar^+) gun and two replaceable metal and molecular sources, respectively. Among them, the argon gun was mainly used to clean the surface, two replaceable molecular sources were capable of depositing target molecules on the substrate, and two replaceable metal sources were mainly used to make different films. Therefore, the MBE chamber was capable of fabricating a single-crystal metal film and depositing a single layer of molecules. For fast loading chambers, the samples made in the MBE chamber were moved to the analysis chamber, or the externally produced samples were transferred to the analysis chamber. The STM in the analysis chamber primarily images the sample at atomic resolution. In the STM, the objective lens concentrated the laser at the tip of the STM and collected signals from the gap between the substrate and the tip, where the 3D adjustable stage was primarily used for focusing. As could be seen from Fig. 8-1, the three chambers were each separated by a gate valve, which effectively prevented the use of one of the chambers alone, opening the other chambers to lose the vacuum environment.

Based on the schematic of Fig. 8-1, the instrument was finally produced by several companies, and the actual product of the instrument was shown in Fig. 8-2. Subfigure(a) showed the exterior and overall view of the instrument, with three chambers, each with an ion pump and an additional titanium(Ti) sublimation pump in the analysis chamber[19]. So this chamber was more space and had more components. Moreover, the pressure in the chamber was 10^{-7} Pa, and the necessary HV environment could reduce the contamination of the

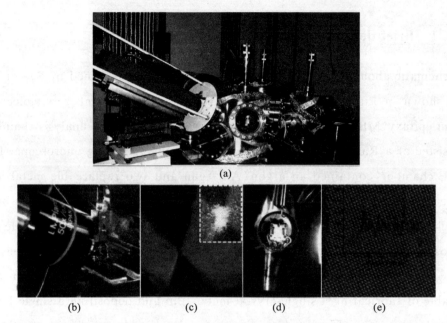

Fig. 8-2 (a) Picture of the actual finished HV-TERS spectroscopy system. (b) The core part of TERS objective and STP tip. (c) Image of the tip and the reflected tip by the substrate. The inset is an image of the tip illuminated with a 633 nm laser. (d) The Ar^+ gun cleaning and MBE part in the MBE chamber. (e) The atomic resolved image of graphite(7 nm × 7 nm). The inset is a Raman spectrum from a sample with p-aminothiophenol molecules on a gold surface[47]

(For colored figure please scan the QR code on page 1)

sample, so that the experimental environment was purer. What was more important was that this keeps the gap between the STM tip and the substrate in a purely tunneled state. When the laser was focused on the tip, the objective lens and the STM tip were displayed in (b), with a magnified image about the tip of the objective lens shown in (c). As could be seen from the figure, the objective lens was in the chamber, and the adjustable 3D platform and Raman optics were outside the chamber. There was a hole in the target tube and the objective lens for releasing the gas into the vacuum chamber. The tube connected the objective lens to the external component, and in an optical path, the chamber was sealed with a quartz window that enables the tube to move in 3D space. Here the He-Ne laser was with 632.8 nm, the tip axis was at a 60°

angle to the side view, and the 3D piezoelectric platform was used for sample and tip operation. In the MBE chamber, the annealing system and the Ar^+ gun were shown in (d), and the final adjustment results of the HV-TERS spectroscopy device were shown in (e), and the parameters could be set in Table 8-1.

Table 8-1 Parameters of the home-built V-TERS setup[47]

Configuration	Vacuum	Tip	Laser	STM Spatial Resolution	Raman Spatial Resolution	Specialization	Speciallzation Samples
Side Illumination, 60 Degrees off Tip Axis	10^{-7} Pa	Au Tip with 25 nm Tip Radius	632.8nm	0.1nm	30 nm	Objective Inside the Vacuum Chamber	In situ measurement

STM, scanning tunneling microscopy.

8.1.2 Detailed description of setup of HV-TERS

As seen from Fig. 8-1 of the HV-TERS spectroscopy system, the instrument comprised three parts: an analysis chamber, an MBE chamber and a fast loading chamber, wherein the most central part was the analysis chamber. It was obvious from the structure that, the MBE chamber and the analysis chamber were connected by a fast loading chamber. There were two magnetic transfer arms between them, one was between the fast load chamber and the analysis chamber, and onother between the fast load chamber and the MBE chamber, primarily used to transfer samples to different chambers. The prepump of the fast load chamber connection system, including the turbo pump and the mechanical dry pump, was connected to the vent channel since the fast load chamber needed to be opened frequently. In addition, each chamber was equipped with a separate ion pump to maintain a continuous vacuum environment. There was also an additional Ti sublimation pump in the analysis chamber, plus various optical components, transmission electron microscopy (TEM) and objective lens, which made the analysis chamber very large. Studies had shown that the three chambers were closed by quartz windows and could be used alone without affecting each other. If the vacuum environment was lost, the other chambers could remain to work normally, and the pressure in the chamber was 10^{-7} Pa. The Raman detector and chamber were supported by a very rigid steel frame. The entire device was suspended by a spring that isolates vibration noise above 1 Hz and the entire suspension frame was placed on the damping block, thereby achieving the purpose of further reducing noise. In addition, the three chambers were made of non-magnetic stainless steel, which greatly reduced the gas release rate.

 Next, the author introduced the details of the chamber. Figure 8-3(a)~(c) showed various details of the analysis chamber, which mainly included Raman spectroscopy and STM. The entire STM was in a HV environment with four damping pads on its feet. And the STM tip was fixed on the *xyz* piezoelectric scanner, and the sample holder could only be close to the *z* direction. There were two magnetic transfer arms between the chambers for transferring the sample. The objective lens fixed to the tube holder was connected to a sealed quartz window. The holes in the tube and the objective lens served two purposes, one was mainly for releasing gas to maintain the vacuum environment, and the other was to balance the air pressure and prevent the objective lens from being deformed due to excessive gas pressure in the room. For other structures about the Raman detector, such as notch filters, and fibers. , were outside the chamber and connected to a closed quartz window, as shown in Fig. 8-3(b). The bellows tube connected the sealing window to the chamber and the entire Raman detector was placed on a 3D adjustable platform for focusing. A He-Ne laser with 632.8 nm was used for the experiment. The angle between the laser and the tip axis was 60°. Professor Sun suggested that analyzers and polarizers could be added to the pre-detector box in order to perform more sensitive experiments. To reduce system vibration, the pre-detector was coupled to the spectrometer and laser through a fiber. And the device used a CCD camera to optical bright-field image at the tip region. Figure 8-3(c) showed the appearance of the analysis chamber.

CHAPTER 8 High-Vacuum Tip-Enhanced Raman Scattering Spectroscopy

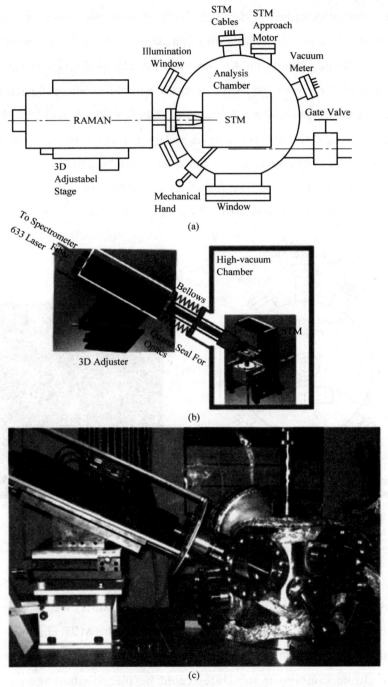

Fig. 8-3 Schematic of the analysis chamber and photograph of the instruments
(a) The analysis chamber; (b) The core section of the system; (c) The analysis chamber[47]
(For colored figure please scan the QR code on page 1)

Next, the author described in detail the MBE chamber as shown in Fig. 8-4(a), which contained an Ar^+ gun, and two replaceable molecular and metal sources, respectively. Among them, the Ar^+ gun was mainly used to clean the surface, and the emitted Ar^+ beam was shot onto the sample at an optimized angle of 25° for cleaning purpose. Two replaceable metal sources could be heated to 3000℃ to evaporate the metal. Here, a mica substrate was used, and a single-crystal metal film could be ready in the MBE chamber, for example, a single-crystal Au film could be used for the study of TERS. In addition, Fig. 8-4(b) showed the appearance of the MBE chamber.

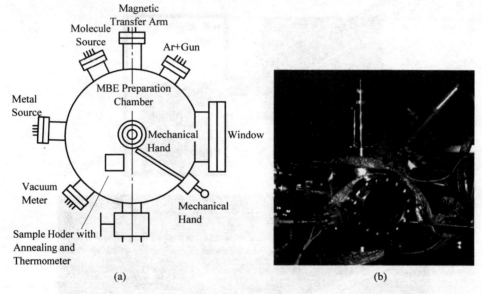

Fig. 8-4 (a) Schematic of the MBE chamber; (b) Its photograph[47]
(For colored figure please scan the QR code on page 1)

In Fig. 8-5, (a) showed the details of the fast loading chamber, which was mainly used to transfer the externally prepared sample, i.e. *ex situ*, to the analysis chamber, or to transfer the sample ready in the MBE chamber, i.e., *in situ* to the analysis chamber. To reduce the number of times the chamber was opened, multiple samples or substrates could be placed simultaneously on the sample holder. In addition, (b) showed the appearance of the fast loading chamber.

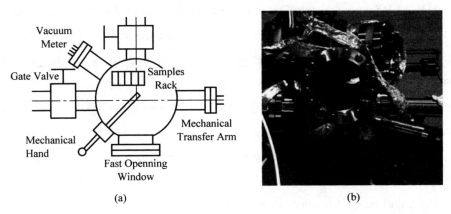

Fig. 8-5 (a) Schematic of the fast loading chamber; (b) Its photograph[47]
(For colored figure please scan the QR code on page 1)

The HV-TERS spectroscopy device produced by Professor Sun had many advantages. For example, only the objective lens was in the HV chamber, and other optical components were outside the cavity, which made it convenient to contact the optical components during the experiment. And, the numerical aperture of the objective lens was large, so that a non-deformed beam could be obtained. Another important advantage was that the device provided a HV preparation chamber, and the preparation of samples in a sufficiently clean environment could greatly reduce contamination.

In the future, further improvements to the HV-STM-TERS spectroscopy instrument will be required, for example, a femtosecond laser using time-resolved ultra-HV-STM-TERS spectroscopy, which will be the most important device for studying surface catalytic reactions.

8.2 The Application of HV-TERS Spectroscopy in *in situ* Plasmon-Driven Chemical Reactions

In previous studies, Sun et al. rationally combined TERS spectroscopy having high spatial resolution and high spectral sensitivity with a HV environment, which was a great success in chemical analysis and spectroscopy. The authors used HV-TERS spectroscopy to study plasmon-driven chemical reactions. to controll the chemical reaction by changing the tunneling current, the bias voltage, and the intensity of the incident laser for adjusting the plasmon intensity. The reaction temperature could be calculated by observing the peaks of the Stokes and anti-Stokes HV-TERS spectra. And the system had broadly expanded the research field of chemical reactions, at the same time, the research in the field of catalytic reaction had risen to a new platform[37].

A simplified design of the HV-TERS spectrum was shown in Fig. 8-6. Its advantages had been described above. In Fig. 8-7, (a) and (b) showed the HV-TERS spectrum about 4-nitrobenzenethiol(4NBT)adsorbed on the Ag film when the bias voltage was set to -0.5 V and 0.5 V, respectively. For reference, (c) showed the normal Raman spectrum about the 4NBT powder. It could be seen from the figure that a strong Raman peak appears in the wavenumber range of 1000 cm^{-1} to 1700 cm^{-1}, and the normal Raman spectrum about 4NBT was significantly different from its HV-TERS spectrum. By comparing (a) with (b), it could be found that the profiles of the two were almost identical, indicating that the influence of the polarity of the voltage on the HV-TERS spectrum was negligible. For Fig. 8-7 (c), the outline was significantly different. At the wavenumber of 1332 cm^{-1}, the Raman peak of the $-NO_2$ stretch mode of 4NBT appears in Fig. 8-7(c). It was hardly observed in Fig. 8-7(a) and (b). However, at the wave numbers of 1142 cm^{-1}, 1387 cm^{-1} and 1432 cm^{-1}, strong Raman peaks were clearly observed in (a) and (b), but were hardly observed in (c). It was known that DMAB had two production methods in the experiment, which were from PATP and 4NBT. In addition, (d) and (e) showed the HV-TERS spectrum in the wavenumber range from 700 cm^{-1} to 1050 cm^{-1}, while (f) showed the normal Raman spectrum of the 4NBT in the same wavenumber segment. As could be seen from the figure, at the wavenumber of 852 cm^{-1}, a strong Raman peak appeared in (f), and was hardly observed in (d) and (e). This proved that the DMAB here was generated by 4NBT.

Next, the authors measured the HV-TERS spectrum of laser power

CHAPTER 8   High-Vacuum Tip-Enhanced Raman Scattering Spectroscopy

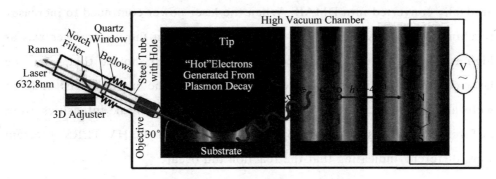

Fig. 8-6 The schematic of HV-TERS. Plasmon-driven chemical reaction measured in HV-TERS[37]

(For colored figure please scan the QR code on page 1)

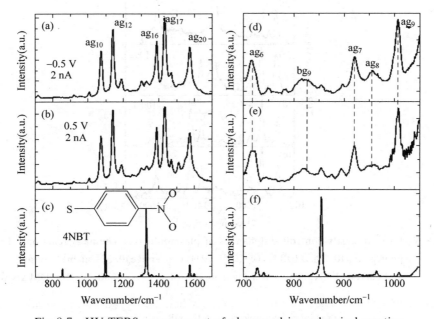

Fig. 8-7 HV-TERS measurement of plasmon-driven chemical reaction
(a) and (b) The measured HV-TERS of DMAB in the region of 700 cm^{-1} to 1700 cm^{-1}, and (c) The normal Raman spectrum of 4NBT powder. (d)~(f) The zoom-in spectra of (a)~(c) from 700 cm^{-1} to 1050 cm^{-1}[37]

(For colored figure please scan the QR code on page 1)

dependent 4NBT. In Fig. 8-8(a)~(d) showed the HV-TERS spectrum about 4NBT at the laser power of 0.5%, 3%, 10%, and 100%, respectively. As could be seen from the figure, when the laser power was greater than 3%, 4NBT was gradually converted into DMAB. When the laser power continued to increase, reaching 10%, DMAB and 4NBT could coexist. When the laser power was as high as 100%, the reaction proceeded completely. (e) showed that when the laser power was again at 0.5%, compared with (a), the curves of (d) and (e) were almost identical, it is expected that the Raman peak intensities were different. So the laser power drops back to 0.5%, but the HV-TERS spectram were different, indicating that the reaction did occur.

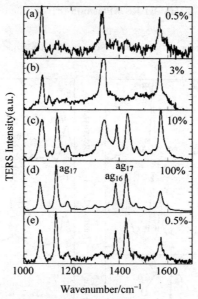

Fig. 8-8 Laser-intensity-controlled dynamics of plasmon-driven chemical reactions. Laser power at(a)0.5%,(b)3%,(c)10%,(d)100% and(e)0.5% at bias voltage 1 V, and current 1 nA. The vibrational modes of $a_{g_{12}}$, $a_{g_{16}}$ and $a_{g_{17}}$ for DMAB were assigned in(d)[37]

(For colored figure please scan the QR code on page 1)

In the HV-TERS spectrum, the author adjusted the local electromagnetic field enhancement by adjusting the bias voltage and tunneling current to adjust the distance between the metal tip and substrate. In Fig. 8-9, (a) showed that when the bias voltage was constant, the distance between the metal tip and the substrate decreased as the tunnel current increased, and the Raman peak at the wavenumber of 1336 cm^{-1} gradually decreased. So studies had shown that the

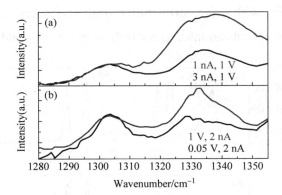

Fig. 8-9 The current and voltage-dependent plasmon-driven chemical reaction (a) the current and (b) The voltage dependence of TERS of DMAB. The Raman intensities were normalized at 1304 cm^{-1} for comparison[37]

(For colored figure please scan the QR code on page 1)

reduction in the distance between the metal tip and the substrate could increase the local electromagnetic field effectively, and could excite stronger SP resonance at the metal tip, thus converting more 4NBT into DMAB. As could be seen from (b), while the bias voltage was lowered, the intensity of the Raman peak with wave number 1336 cm^{-1} gradually decreased. The bias voltage was reduced, resulting in a reduced distance between the substrate and the metal tip, hence obtaining a stronger plasmon resonance.

Studies had shown that the plasmon-driven catalytic reaction in the HV-TERS spectroscopy system was significantly faster than the SERS system. In SERS, the Raman peak of 4NBT at a wavenumber of 1336 cm^{-1} was not observed until 140 minutes after the reaction, while for the HV-TERS spectrum, it only took a few minutes. The reduction in the distance between the substrate and the tip could manufacture a tremendous plasmon resonance, thereby greatly facilitating the plasmon-driven catalytic reaction. As was demonstrated in Fig. 8-10, the Stokes and anti-Stokes HV-TERS spectra further demonstrated that the 4NBT reaction produced DMAB. And the four vibration modes of DMAB could be obviously observed in the spectrum, which were $a_{g_{10}}$, $a_{g_{12}}$, $a_{g_{16}}$ and $a_{g_{17}}$, respectively. Studies had shown that this plasmon-driven chemical reactions with high probability and efficiency. In addition, the temperature about the plasmon-driven catalytic reaction could be found by the

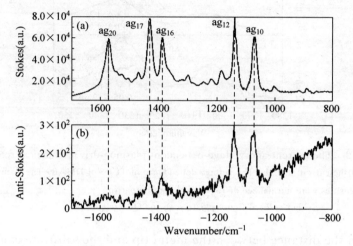

Fig. 8-10 The Stokes and anti-Stokes HV-TERS peaks
(a) The Stokes TERS, and (b) The anti-Stokes TERS at the bias voltage 1 V and the tunneling current 2.5 nA[37]
(For colored figure please scan the QR code on page 1)

following relationship

$$I_s/I_{as} = a \times e(\hbar \omega / k_B T) \quad (8\text{-}1)$$

where, I_s was the intensity of Stokes, I_{as} was the intensity of anti-Stokes, $\hbar$ was the Planck constant, k_B was the Boltzmann constant, T was the experimental temperature, and a was a constant. Finally, after calculation, the experimental temperature T was 327(± 4)K, and the constant a was 2.06(± 0.19).

8.3 Plasmonic Gradient Effect

In previous studies, electric field gradient effects and multiple vibrational modes were observed in the TERS spectra in HV and ultra-HV (UHV) environments due to the high-order terms of the Hamiltonian of the Raman spectrum. Among them, such effect and modes reflected two features of the high-order terms of the Hamiltonian. The Hamiltonian of a Raman spectrum of a molecule in a non-uniform electric field could be described as,

$$H = H_0 + H_1 + H_2 = \left(\alpha_{\alpha\beta}E_\beta + \frac{1}{3}A_{\alpha,\beta\gamma}\frac{\partial E_{\beta\gamma}}{\partial r}\right)E_\alpha +$$

$$\frac{1}{3}\left(A_{\gamma,\alpha\beta}E_\gamma + C_{\alpha\beta,\gamma\delta}\frac{\partial E_{\gamma\delta}}{\partial r}\right)\frac{\partial E_{\alpha\beta}}{\partial r} + \cdots, \qquad (8-2)$$

where, $\alpha_{\alpha\beta}$ was the electric dipole-dipole polarizability, $A_{\alpha,\beta\gamma}$ was the electric dipole-quadrupole polarizability, and $C_{\alpha\beta,\gamma\delta}$ was the electric quadrupole-quadrupole polarizability. In addition, E_α was the external electric field, and $\partial E_{\alpha,\beta\gamma}/\partial r$ was the electric field gradient. And the first item was the coupling between the molecular polarizability, the scattering field and the incident field. And second item was the coupling between the electric dipole-quadrupole polarizability, scattered field gradient and incident field. Then, the third item was the coupling between electric dipole-quadrupole polarizability, scattered field and incident field gradient. The coupling between the electric quadrupole-quadrupole polarizability was the last item, the scattering field gradient and the incident field gradient[48].

Studies had shown that in the HV-TERS spectroscopy system, the arrival of the Darling-Dennison resonance and the Fermi resonance had been observed owing to the near-field gradient effect. As was demonstrated in Fig. 8-11, the HV-TERS spectroscopy about the pyrazine adsorbed on Ag film was revealed.

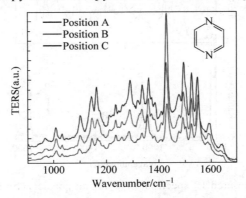

Fig. 8-11 The near-field gradient effect, the number of IR-active mode peaks in TERS
(For colored figure please scan the QR code on page 1)

Studies had also shown that the Raman spectrum about pyrazine adsorbed on the Ag film belonged to the SERS spectrum, which was significantly different from the TERS spectrum. Because of the near-field gradient effect, the number of IR-active mode peaks in TERS was greater than in the SERS spectrum.

For comparison, the authors observed the IR spectrum about the pyrazine powder, and the HV-TERS spectrum and the Raman spectrum shown in Fig. 8-12(a)~(c), respectively. Almost all Raman modes about pyrazine powder could be detected in the HV-TERS spectrum, and there were various additional strong enhancement modes, which might be due to chemical enhancement in the HV-TERS spectroscopy system, maybe or other causes. For example, the IR activity modes in HV-TERS, it was necessary to measure the IR vibration spectrum about the pyrazine powder(the black curve in (a)),

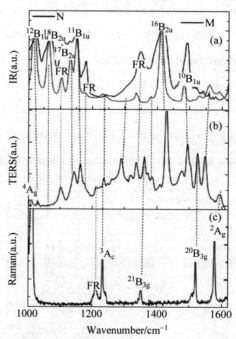

Fig. 8-12 (a) The measured IR spectrum of pyrazine powder, (b) HV-TERS of pyrazine adsorbed on Ag film, and (c) Raman spectrum of pyrazine powder[4]

(For colored figure please scan the QR code on page 1)

and its IR spectrum was depicted by a red line. Comparing (a) with (b), it could be concluded that the strong enhanced vibrational peaks presented might be caused by the IR activity modes of the pyrazine powder.

Figure 8-13 showed the SERS and TERS spectra of pyrazine without the use of an Ag film. It could be clearly observed that the number of Raman peaks appearing in the TERS spectrum was obviously larger than that in the SERS spectrum. Thus, the advantages of the HV-TERS system were obtained, and since the prepared Ag substrate was rough, it might cause an IR-activity pattern to be observed in the HV-TERS and SERS spectra. Studies had shown that the concentration of molecules in SERS was 10^{-2} M, while in HV-TERS was 10^{-4} M, but in HV, the TERS signal was significantly higher than the SERS signal. This obviously revealed the significant advantages of HV for the analysis of ultrasensitive spectra.

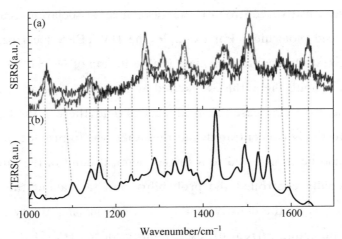

Fig. 8-13 (a) SERS and (b) TER spectra of pyrazine[4]
(For colored figure please scan the QR code on page 1)

In summary, in this section, the authors studied the Raman activity, IR-activity and overtone mode of pyrazine adsorption on the Ag film in the HV-TERS spectroscopy system. This was due to near-field gradient effects and nonlinear vibrations. The advantages of the HV-TERS spectrum were evident. HV environments could observe nonlinear vibration and near-field gradient effects at very low molecular concentrations. Therefore, the HV-TERS spectroscopy system had raised the research of ultrasensitive molecular spectroscopy to a new platform.

8.4 Plasmonic Nanoscissors

In previous studies, photochemical methods had been widely used in many fields, such as photosynthesis, photocatalysis, and degradation of plastics. In chemical reactions, the authors preferred to be able to break a particular chemical bond or control a particular mode to achieve a reasonable control of the chemical reaction. However, due to the absorption of weak light in the molecule and the redistribution of vibrational energy, it was difficult to destroy a single chemical bond for molecules adsorbed on the surface. Zhang et al. proposed a new method to solve the above problem, that is, plasmon scissor could achieve sensitive control of the resonance dissociation reaction about surface-adsorbed molecules. For example, the HV-TERS spectroscopy system had been successfully used to dissociate the excitation of NC_2H_6 from malachite green(MG). Among them, the SP excited at the metal tip had two important functions, one was to increase the local electric field, thereby the light from the laser was effectively concentrated, and the other was to provide hot electrons, its energy could be transferred to a single chemical bond in the molecule. The method rationally controlled the probability and rate about the dissociation reaction by adjusting the density of the hot electrons, mainly by adjusting the bias voltage, tunneling current or laser intensity in the HV-TERS spectroscopy system[2].

Studies had shown that hot electrons produced by plasmon decay played a vital role in surface catalytic reactions. Plasmon scissors had been successfully applied to molecular design in SERS and TERS experiments. The hot electrons generated by plasmon decay could break the weakest chemical bond —N=N— of DMAB. And the hot electrons were able to transfer energy to a molecule that adsorbed on the substrate, dissociating the electronically driven bonds.

The authors had successfully used the HV-TERS spectroscopy device manufactured by Sun et al. to dissociate MG adsorbed on the surfaces of Au and Ag by hot electrons produced by plasmon decay. The necessary conditions for this experiment were the laser signal and the hot electron. The hot electron was mainly used to provide the energy required for the reaction, overcame the reaction barrier, and initiate the chemical reaction. The laser was mainly used to promote the electronic transition of the molecular bond in the MG. But the crux of the problem was that the life of the hot electrons was too short, just a few femtoseconds, so it was very difficult to complete the experiment. In addition, it was worth noting that the resonance dissociation here was not an electrical means, but an optical one.

In Fig. 8-14, (a) showed the optical absorption spectrum of MG. It could be seen that a strong absorption peak appeared at a wavelength of 632.8 nm, and no absorption peak was observed at a wavelength of 785 nm. There was a weaker absorption peak at a wavelength of 514.5 nm. Figure 8-14(b) showed the SERS spectrum of MG in an Ag sol at frequencies of 514.5 nm, 632.8 nm, and 785 nm. It was known that normal Raman was excited at a wavelength of 785 nm, while resonance excitation was at a wavelength of 632.8 nm, and excitatien at 514.5 nm was likely to excitate at 785 nm. As could be seen from (b), at a wavelength of 632.8 nm, the Raman peaks A—D were selectively excited by the resonance electron transition. (c) showed that the Raman peaks A—D were associated with the vibration modes in the MG and with the —NC_2H_6 fragments. Studies had shown that MG exhibits C_2 symmetry and had two modes of vibration, a and b, respectively. The five selective enhanced vibration modes from low frequency to high frequency that had been marked in the figure are a_{39}, a_{45}, b_{49}, b_{61} and a_{58} respectively. The vibration mode of the Raman E appearing in (b) was related to the C—C expansion mode of benzenyl. Studies had shown that the resonance excitation energy at a wavelength of 632.8 nm was concentrated in the A—D modes, so that the Raman peak corresponding to these modes could be selectively excited.

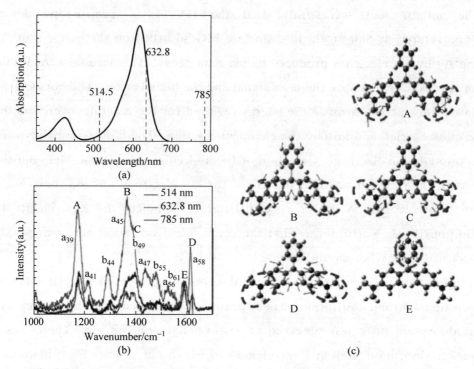

Fig. 8-14 Absorption and Raman spectra and calculated normal modes of MG
(a) Absorption spectrum of MG in water, (b) SERS and surface-enhanced resonance Raman scattering(SERRS)spectra of MG in Ag sol, and (c) Vibrational modes A—E of MG[2]
(For colored figure please scan the QR code on page 1)

Next, the authors demonstrated that the resonance dissociation process was measured using a laser with a wavelength of 632.8 nm under HV conditions using the time dependence of the tip-enhanced resonance Raman spectroscopy (TERRS). In Fig. 8-15, subfigures (a)~(c) were a time-sequential TERRS spectrum of MG adsorbed on the surface of Ag using a laser having a wavelength of 632.8 nm under high vacuum. (a) showed the TERRS spectrum measured at the beginning of the reaction. It could be seen that the intensity of the Raman peaks A—D in the TERRS spectrum was significantly enhanced compared to the non-resonant SERS spectrum. Studies had shown that the resonant excitation energy selectively excited the four vibration modes of A—D during the electronic transition. The TERRS spectrum measured by the

CHAPTER 8   High-Vacuum Tip-Enhanced Raman Scattering Spectroscopy   349

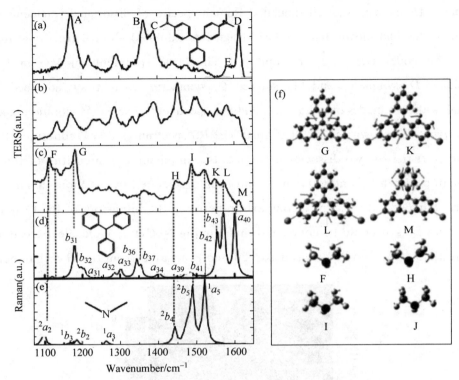

Fig. 8-15   Time-sequential TERRS and the vibrational modes of MG
(a) The Initial, (b) Intermediate(20 minutes after continuous radiation using a laser), and (c) The final spectra(40 minutes after continuous radiation using a laser). The tunneling current and the bias voltage are 1 nA and 1 V, respectively. (d) and (e) Simulated Raman spectra of fragments(see the insets). (f) The vibrational modes of dissociated fragments of MG (corresponding to the experimental peaks in (d) and (e) as calculated by density functional theory)[2]
(For colored figure please scan the QR code on page 1)

authors directly proved the time-dependent reaction kinetics. As the reaction time prolonged, the intensity of the A—D Raman peaks was suppressed, and a new Raman peak was generated, keeping the laser intensity constant, resulting in an unstable and complex vibrational spectrum as displayed in (b). The spectrum shown in (c) was a relatively stable spectrum at 40 minutes of reaction. A 2D image of its corresponding TERS spectrum was shown in Fig. 8-16, further demonstrating that (c) here was the final spectrum of the reaction. The A—D Raman peak appearing in (a) was now disappearing, and the new Raman peak F—M had appeared, as shown in (c), and the spectrum at this time was the most stable. (d) and (e) showed simulated Raman spectra of dissociated fragments of MG. Studies had shown that HN(CH$_3$) fragments exhibited C_{2v} symmetry, and four vibration modes are a_1, a_2, b_1, and b_2. The large fragments of MG in (d) exhibited D_3 symmetry, and (f) showed C_2 symmetry when it was adsorbed on metal, with two modes of vibration, a and b, respectively. The F—M vibration mode in (c) was also shown in (f). The final TERRS spectrum (see (c)) contained the characteristics of two dissociated fragments, comparing the intensities of the C Raman peaks in (a) and (c), which had been dissociated by plasmon scissors by more than 60%. In (c), by a comparison between the intensities of the G and I Raman peaks, it could be concluded that for the N(CH$_3$)$_2$ molecule, exceeding more than 40%, and 20% was desorbed and evacuated from the HV chamber.

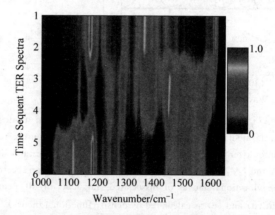

Fig. 8-16 Time-sequential TERS spectra, and the color bar in the right of the figure. The time interval is 20 minutes for each TERS spectrum[2]
(For colored figure please scan the QR code on page 1)

Figure 8-16 is the time-sequential TERS spectra. To further understand the role of plasmon scissors, the authors conducted a similar experimental study, which measured the TERRS spectra about MG adsorbed on the Au film as displayed in Fig. 8-17. Several vibration modes of A—E appear in Fig. 8-17(a) at the beginning of the reaction. As the reaction progressed for 40 minutes (Fig. 8-17(b)), the previously occurring Raman peak did not completely disappear, and a new Raman peak F—M appeared. Therefore, it was concluded that on the Au substrate the chemical reaction was slower compared to the Ag substrate as shown in Fig. 8-15(c). Studies had shown that if hot electrons were attached to a molecule, the electron transfer process took place in femtoseconds, and whether the electrons successfully attaching to the molecule was determined by the density of the electrons.

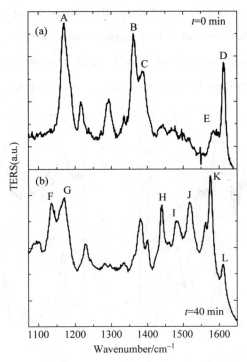

Fig. 8-17 Time-sequential TERRS and the vibrational modes of MG adsorbed on the Au film. (a) The initial and (b) The spectra at 40 minutes after continuous radiation using a laser. The tunneling current and the bias voltage are 1 nA and 1 V, respectively[2]

As mentioned above, in the HV-TERS spectroscopy system, the probability and rate of the reaction were adjusted by adjusting the bias voltage, the tunneling current and the laser intensity. The simplest method was to change the intensity of the plasmon by adjusting the intensity of the laser for purpose of controlling the probability and rate of the dissociation reaction. In Fig. 8-18, (a) and (b) showed that when the laser power was reduced by 10% of the total, the dissociation rate was slower after 40 minutes of reaction, further demonstrating the results of Fig. 8-14. Figure 8-18(b) displayed that the A—C Raman peak which appeared at the beginning of the reaction still existed after 40 minutes of the reaction, and a new Raman peak appeared. Figure 8-18(c) and (d) showed that the collected TERRS spectra were consistent when the tunneling current was constant and laser illumination was not used. Therefore, it was concluded that the tunneling current and the energy provided by the laser do not dissociate the molecules, and the laser must be contacted to provide a denser thermal electron for dissociating the molecules.

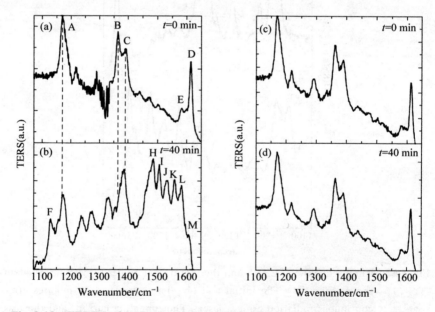

Fig. 8-18 TERRS of MG adsorbed on the Ag film at $t = 0$ and $t = 40$ minutes. (a),(b) Sample irradiated with 10% lower laser power compared to the sample analyzed in Fig. 8-14(a) and (c), showing evidence for a slower dissociation rate and, (c) and (d) Without continuous laser-irradiation of the sample, showing that the tunneling current alone is not sufficient to induce dissociation[2]

(For colored figure please scan the QR code on page 1)

Next, the author introduced the physical mechanism of the resonance dissociation reaction. The dissociation reaction mainly depended on the thermal electrons produced by the decay of the plasmon. In the HV-TERS spectroscopy system, by irradiating the gap between the substrate and the tip with a laser, thereby producing a dense plasmon, the hot electrons generated by the decay of the plasmon had higher kinetic energy, and the laser and kinetic energy together initiated the dissociation reaction. The schematic of the principle was shown in Fig. 8-19. Process A demonstrated that the resonant excitation of the laser concentrates the energy on the vibrational mode related with the $—NC_2H_6$ fragments of the MG, causing its energy to be enhanced, thereby exciting the MG molecule in the excited-state potential energy surface (ESPES), which was produced by plasmon decay. The thermal electrons changed the ESPES, from neutral to negative ion-excited states, and then, the kinetic energy was moved from the hot electron to the vibrational dissociation energy in the molecule, as shown in process C. Therefore, studies had shown that lasers were required to continuously illuminate the sample during the experiment to ensure that a large density of plasmon was generated, thereby generating a large amount of hot electrons.

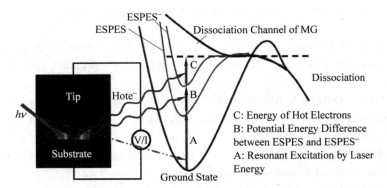

Fig. 8-19 The mechanism of resonant dissociation by plasmon scissors. Three main energetic components driving the chemical reaction: (A) Resonant absorption of the laser light, enhancing the vibrational modes associated with $—NC_2H_6$ and exciting the MG molecules to an ESPES; (B) Hot electrons, temporarily changing the molecules' ESPES from a neutral to a negative ion-excited states; (C) The kinetic energy of the hot electrons, converted into intramolecular vibrational thermal energy[2]

(For colored figure please scan the QR code on page 1)

In short, the role of plasmon scissors and its working principle were introduced. Plasmon scissors were the important tool for controlling the dissociation of adsorbed molecules on the surface. It was also proved that the HV-TERS spectroscopy system was an effective technique for measuring nanoscale *in situ* surface chemical analysis.

References

[1] MAUSER N, HARTSCHUH A. Tip-enhanced near-field optical microscopy[J]. Chem. Soc. Rev. ,2014,43: 1248-1262.
[2] ZHANG Z, SHENG S, ZHENG H, et al. Molecular resonant dissociation of surface-adsorbed molecules by plasmonic nanoscissors[J]. Nanoscale,2014,6: 4903-4908.
[3] SUN M T, ZHANG Z L, CHEN L, et al. Plasmon-driven selective reductions revealed by tip-enhanced Raman spectroscopy[J]. Adv. Mater. Interf. ,2014,1: 2196-7350.
[4] SUN M T, ZHANG Z L, CHEN L, et al. Plasmonic gradient effects on high-vacuum tip-enhanced Raman spectroscopy[J]. Advanced Optical Materials,2014,2: 74-80.
[5] SONNTAG M D, CHULHAI D, SEIDEMAN T, et al. The origin of relative intensity fluctuations in single-molecule tip-enhanced Raman spectroscopy[J]. J. Am. Chem. Soc. ,2013,135: 17187-17192.
[6] PAULITE M, BLUM C, SCHMID T, et al. Full spectroscopic tip-enhanced Raman imaging of single nanotapes formed from β-amyloid(1-40)peptide fragments[J]. ACS Nano,2013,7: 911-920.
[7] SCHMID T, OPILIK L, BLUM C, et al. Nanoscale chemical imaging using tip-enhanced Raman spectroscopy: a critical review [J]. Angew. Chem. Int. Ed. , 2013, 52: 5940-5954.
[8] ZHANG R, ZHANG Y, DONG Z C, et al. Chemical mapping of a single molecule by plasmon-enhanced Raman scattering[J]. Nature,2013,498: 82-86.
[9] SUN M, ZHANG Z, KIM Z, et al. Plasmonic scissors for molecular design[J]. Chem. Eur. J. ,2013,19: 14958-14962.
[10] ZHANG Z L, SUN M T, RUAN P P, et al. Electric field gradient quadrupole Raman modes observed in plasmon-driven catalytic reactions revealed by HV-TERS [J]. Nanoscale,2013,5: 4151-4155.
[11] ZHANG Z L, CHEN L, SUN M T, et al. Insights into the nature of plasmon-driven catalytic reactions revealed by HV-TERS[J]. Nanoscale,2013,5: 3249-3252.
[12] STOCKLE R M, SUH Y D, DECKERT V, et al. Nanoscale chemical analysis by tip-enhanced Raman spectroscopy[J]. Chem. Phys. Lett. ,2000,318: 131-136.
[13] ANDERSON M S. Locally enhanced Raman spectroscopy with an atomic force

microscope[J]. Appl. Phys. Lett. ,2000,76: 3130-3132.

[14] HAYAZAWA N,INOUYE Y,SEKKAT Z,et al. Metallized tip amplification of near-field Raman scattering[J]. Opt. Commun. ,2000,183: 333-336.

[15] PETTINGER B,PICARDI G,SCHUSTER R,et al. Surface-enhanced and STM tip-enhanced Raman spectroscopy of CN^- ions at gold surfaces[J]. J. Electroanal. Chem. ,2003,554: 293-299.

[16] PETTINGER B,REN B,PICARDI G,et al. Nanoscale probing of adsorbed species by tip-enhanced Raman spectroscopy[J]. Phys. Rev. Lett. ,2004,92,96101.

[17] REN B,PICARDI G,PETTINGER B,et al. Preparation of gold tips suitable for tip-enhanced Raman spectroscopy and light emission by electrochemical etching[J]. Rev. Sci. Instrum. ,2004,75: 837-841.

[18] ANDERSON N,HARTSCHUH A,CRONIN S,et al. Nanoscale vibrational analysis of single-walled carbon nanotubes[J]. J. Am. Chem. Soc. ,2005,127: 2533-2537.

[19] DOMKE K F,ZHANG D,PETTINGER B, et al. Toward Raman fingerprints of single-dye molecules at atomically smooth Au(111)[J]. J. Am. Chem. Soc. ,2006, 128: 14721-14727.

[20] ZHANG W H,YEO B S,SCHMID T,et al. Nanoscale roughness on metal surfaces can increase tip-enhanced Raman scattering by an order of magnitude[J]. Nano Lett. , 2007,7: 1401-1405.

[21] STEIDTNER J,PETTINGER B. High-resolution microscope for tip-enhanced optical processes in ultrahigh vacuum[J]. Rev. Sci. Instrum,2007,78: 103104.

[22] STEIDTNER J,PETTINGER B. Tip-enhanced Raman spectroscopy and microscopy on single-dye molecules with 15 nm resolution [J]. Phys. Rev. Lett. , 2008, 100: 236101.

[23] BAILO E,DECKERT V. Tip-enhanced Raman scattering[J]. Chem. Soc. Rev. , 2008,37: 921-930.

[24] HARTSCHUH A, ANDERSON N, NOVOTNY L, et al. Near-field Raman spectroscopy using a sharp metal tip[J]. J. Microsc. ,2003,210: 234-240.

[25] CHEN J N,YANG W S, DICK K, et al. Tip-enhanced Raman scattering of p-thiocresol molecules on individual gold nanoparticles[J]. Appl. Phys. Lett. , 2008, 92: 093110.

[26] SUN M T,FANG Y R,YANG Z L,et al. Chemical and electromagnetic mechanisms of tip-enhanced Raman scattering [J]. Phys. Chem. Chem. Phys. , 2009, 11: 9412-9419.

[27] YANO T,VERMA P,SAITO Y,et al. Pressure-assisted tip-enhanced Raman imaging at a resolution of a few nanometers[J]. Nat. Photonics,2009,3: 473-477.

[28] ZHANG D,HEINEMEYER U,STANCIU C,et al. Nanoscale spectroscopic imaging of organic semiconductor films by plasmon-polariton coupling[J]. Phys. Rev. Lett. , 2010,104: 056601.

[29] STADLER J,SCHMID T, ZENOBI R,et al. Nanoscale chemical imaging using top-

illumination tip-enhanced Raman spectroscopy[J]. Nano Lett. ,2010,10: 4514-4520.

[30] FANG Y R, ZHANG Z L, SUN M T, et al. High vacuum tip-enhanced Raman spectroscope based on a scanning tunneling microscope[J]. Rev. Sci. Instrum. ,2016, 87: 033104.

[31] DOMKE K F, PETTINGER B. Studying surface chemistry beyond the diffraction limit: 10 years of TERS[J]. Chem Phys Chem,2010,11: 1365-1373.

[32] KIM H,KOSUDA K M,VAN DUYNE R P,et al. Resonance Raman and surface-and tip-enhanced Raman spectroscopy methods to study solid catalysts and heterogeneous catalytic reactions[J]. Chem. Soc. Rev. ,2010,39: 4820-4844.

[33] UETSUKI K, VERMA P, YANO T, et al. Experimental identification of chemical effects in surface-enhanced Raman scattering of 4-aminothiophenol [J]. J. Phys. Chem. C,2010,114: 7515-7520.

[34] DECKERT-GAUDIG T, RAULS E, DECKERT V, et al. Aromatic amino acid monolayers sandwiched between gold and silver: a combined tip-enhanced Raman and theoretical approach[J]. J. Phys. Chem. C,2010,114: 7412-7420.

[35] WOOD B R,BAILO E,KHIAVI M A,et al. Tip-enhanced Raman scattering(TERS) from hemozoin crystals within a sectioned erythrocyte[J]. Nano Lett. ,2011,11: 1868-1873.

[36] YANG Z L,LI Q H, FANG Y R, et al. Deep ultraviolet tip-enhanced Raman scattering[J]. Chem. Commun. ,2011,47: 9131-9133.

[37] SUN M T,ZHANG Z L, ZHENG H R, et al. In-situ plasmon-driven chemical reactions revealed by high-vacuum tip-enhanced Raman spectroscopy[J]. Sci. Rep. , 2012,2: 647.

[38] JIANG N,FOLEY E T,KLINGSPORN J M,et al. Observation of multiple vibrational modes in ultrahigh-vacuum tip-enhanced Raman spectroscopy combined with molecular-resolution scanning tunneling microscopy [J]. Nano Lett. , 2012, 12: 5061-5067.

[39] VAN SCHROJENSTEIN L E M, DECKERT-GAUDIG T, MANK A J G, et al. Catalytic processes monitored at the nanoscale with tip-enhanced Raman spectroscopy [J]. Nat. Nanotechnol. ,2012,7: 583-586.

[40] KAZEMI-ZANJANI N,CHEN H H,GOLDBERG H A,et al. Label-free mapping of osteopontin adsorption to calcium oxalate monohydrate crystals by tip-enhanced Raman spectroscopy[J]. J. Am. Chem. Soc. ,2012,134: 17076-17082.

[41] KUROUSKI D, DECKERT-GAUDIG T, DECKERT V, et al. Structure and composition of insulin fibril surfaces probed by TERS[J]. J. Am. Chem. Soc. ,2012, 134: 13323-13329.

[42] PETTINGER B,SCHAMBACH P, VILLAGOMEZ C J, et al. Tip-enhanced Raman spectroscopy: near-fields acting on a few molecules[J]. Annu. Rev. Phys. Chem. , 2012,63: 379-399.

[43] STADLER J,SCHMID T,ZENOBI R,et al. Developments in and practical guidelines

for tip-enhanced Raman spectroscopy[J]. Nanoscale,2012,4: 1856-1870.
[44] SONNTAG M D, KLINGSPORN J M, GARIBAY L K, et al. Single-molecule tip-enhanced Raman spectroscopy[J]. J. Phys. Chem. C,2012,116: 478-483.
[45] SUN M T, FANG Y R, ZHANG Z Y, et al. Activated vibrational modes and Fermi resonance in tip-enhanced Raman spectroscopy[J]. Phys. Rev. E,2013,87: 020401.
[46] SUN M T, ZHANG Z L, CHEN L, et al. Tip-enhanced resonance couplings revealed by high vacuum tip-enhanced Raman spectroscopy[J]. Adv. Opt. Mater. ,2013,1: 449-455.
[47] SUN M T. Chapter 6-high-vacuum tip-enhanced Raman spectroscopy[M]. Elsevier: Micro and Nano Technologies,2017,129-140.
[48] FANG Y, ZHANG Z, CHEN L, et al. Near-field plasmonic gradient effects on high-vacuum tip-enhanced Raman spectroscopy[J]. Phys. Chem. Chem. Phys. ,2015,17: 783-794.

CHAPTER 9

Physical Mechanism of Plasmon-Exciton Coupling Interaction

9.1 Brief Introduction of Plexcitons

The strong coupling of plasmon and excitons produced plexcitons, also known as plasmon-exciton polaritons, which were quasiparticles that combined pronounced optical nonlinearity with nanoscale energy confinement. The plasmon was a collective electron oscillation, the excitons were excited electrons, and combined with the holes generated by the excitation. Strong coupling produced plexcitons from metal films (plasmon) and organic molecular layers (excitons). Enhanced control of the dispersion relationship for the propagation of plexcitons could facilitate the study of coherent and collective coupling about distant emitters. In previous studies, plexcitons contributed to the direct energy flow in exciton energy transfer, which traveled 20 microns and was the length equivalent to the width of a human hair. The authors demonstrated the strong coupling between the excitons of carbon nanotubes through experiments, and proved that the spatially extended plasmon mode was formed by the diffraction coupling about periodically arranged gold nanoparticles (Au NPs), nanorods or nanodisks. The plexcitons had a fairly long life span of more than 100 fs, yet only a highly directional propagation of only 20 microns. Studies had shown that the compatibility of electrical excitation with this type of the plexitons system had opened up new avenues for many studies, such as cavity-assisted energy transfer, ultra-fast active plasmon devices, and low-power optoelectronic components[1].

In addition, plexitons could be implemented in a variety of materials, such as quantum dots[2-3], quantum wells[4], organic semiconductors[5-12], transition metal disulfides[4,13-14], and carbon nanotubes[15].

Plexitons could be used as coupled Lorentz oscillators. The linear extinction spectrum of gold nanorods coated with J-aggregating dye molecules could be accurately reproduced.

In previous reference, the interaction of plasmon-exciton coupling with exciton-plasmon-photon conversion had been extensively studied theoretically and experimentally[16-37]. Such studies had shown that there were two types of plasmon-exciton coupling, the strong and the weak coupling, which exhibited different optical properties. For strong plasmon coupling, the two new states of matter and light were separated by energy in Rabi splitting, thereby exhibiting the characteristic anti-crossing behavior of exciton-LSP energy coordination. This produced plexitons, a quasiparticle with unique properties in neither original particle. For weak plasmon-exciton coupling interactions, SERS, SP-enhanced absorption, SP-enhanced fluorescence, and fluorescence quenching were also included[38].

Surface plasmon resonance (SPR) was generated by the collective oscillation caused by the photoelectromagnetic field and was mainly used to study the interaction of matter and light over the diffraction limit[38].

These studies revealed the nature of light-material interactions and drove the applications of plasmon-exciton hybridization such as energy, environment, catalysis, and optical communication[38].

9.2 Plasmon-Exciton Coupling Interaction

9.2.1 Strong plasmon-exciton coupling interaction

When the energy transferred between excitons and light was greater than their average dissipation, a strong coupling of plasmon-excitons was formed, resulting in a new type of polariton, also a quasi-particle, plexcitons. In the strong coupling interaction, the new model contained the characteristics of two resonators, excitons and photons, and the author assumed that it was in a semiconductor resonator, and the model was widely used in different fields. The formation of plexcitons could lead to Bose-Einstein condensation[39]. Strong coupling interaction played an important role in quantum information processing, which was due to the strong coupling to achieve quantum coherent oscillation. Another important application was that it could change the electromagnetic field environment of the fluorophore, thereby controlling the reaction rate by adjusting the threshold of the chemical reaction. In addition, the strong coupling of microwave and superconducting qubits was also widely used, such as photon detectors[40], routers[41], and single-photon switches[42].

Studies had shown that strong coupling was mainly determined by three important parameters[43] κ, g and γ, corresponding to the rate at which light escapes from the cavity, the energy transfer rate between matter and light, and the rate at which the material lost its polarization, respectively. When the value of g was greater than the other two rates, there was a strong coupling interaction, and the energy exchange between matter and light was periodic. When the microcavity and the matter resonated, two new resonant frequencies could be observed from the transmission and reflection spectra, as shown in Fig. 9-1, where the resonant frequency could be described by,

$$\omega_{\text{exciton}} = \omega_{\text{cavity}} = \omega_0 \pm g \text{(new frequency)}. \tag{9-1}$$

In strongly coupled interactions, anti-cross-dispersion could also be referred

CHAPTER 9 Physical Mechanism of Plasmon-Exciton Coupling Interaction   361

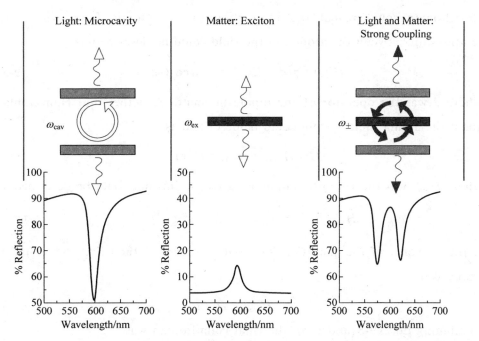

Fig. 9-1 The schematic of the reflection spectra simply depicting the result of a strong coupling interaction between light and matter[44]

(For colored figure please scan the QR code on page 1)

to as Rabi splitting. Here is a brief introduction to the Rabi division[45].

Rabi oscillations

It was a common phenomenon that when a laser illuminated a material, there was a population fluctuation between the levels, wherein the oscillation frequency of the fluctuation was the Rabi frequency. Note that the pulse time must be less than the relaxation time about the medium, taking into account the damping for understanding the lifetime pulse time of the excited state particles. The Rabi oscillation had the periodicity occurred between energy transition and the stationary state about the two-state quantum system near the oscillating drive field, wherein the field could be described as

$$E(t) = E_0 \cos(\omega t). \tag{9-2}$$

And the frequency was close to the resonant frequency $|\omega - \omega_0| \ll \omega_0$, The

Hamiltonian could reveal the interaction of the atomic field, where the relationship between the atom and the field could be described as

$$\hat{H}^I(t) = \hat{d} \cdot E(t) = -\hat{d}\cos(\omega t), \tag{9-3}$$

where $\hat{d}$ was the operator of the dipole moment. And the total Hamiltonian about the atomic field system could be described as

$$\hat{H} = \hat{H}_{atom} + \hat{H}_{field} + \hat{H}^I(t), \tag{9-4}$$

where $\hat{H}_{atom}$ was the Hamiltonian about a free atom, and could be described as

$$\hat{H}_{atom} = \frac{1}{2}\hbar\omega_{eg}(|e\rangle\langle e| - |g\rangle\langle g|). \tag{9-5}$$

If the influence of the quantum field could be ignored, then Eq. (9-4) could be changed to

$$\hat{H} = \hat{H}_{atom} + \hat{H}^I(t). \tag{9-6}$$

Combining Eqs. (9-3) and (9-5), the final Hamiltonian was

$$\hat{H} = \hbar\omega_{eg}|e\rangle\langle e| - \hat{d}\cos(\omega t). \tag{9-7}$$

The vector of system state was

$$|\Psi(t)\rangle = \sum_k c_k(t) e^{\frac{-iE_k t}{\hbar}} |k\rangle = c_g(t)|g\rangle + c_e(t) e^{-i\omega_{eg} t}|e\rangle. \tag{9-8}$$

Substituting Eq. (9-8) into the Schrödinger equation, which was time-dependent, the following relationship could be obtained:

$$i\hbar\frac{\partial|\Psi(t)\rangle}{\partial t} = \hat{H}|\psi(t)\rangle. \tag{9-9}$$

Combined Eq. (9-8) with Eq. (9-9), the coupled first-order differential equation could be deduced as

$$\dot{C}_e = -\frac{i}{\hbar}E_0\cos(\omega t) d_{eg}^* e^{-i\omega_{eg} t} C_g, \tag{9-10}$$

$$\dot{C}_g = -\frac{i}{\hbar}E_0\cos(\omega t) d_{eg} e^{i\omega_{eg} t} C_e. \tag{9-11}$$

Extending the $\cos(\omega t)$ term, since the time caused by the applied field evolves slowly compared to ω_0, the quickly rotating term was ignored and detuning was introduced, $\Delta = \omega_{eg} = -\omega$, so, one obtained

$$\dot{C}_e = -\frac{i}{2\hbar}E_0 d_{eg}^* e^{-i\Delta t}C_g,\qquad(9\text{-}12)$$

$$\dot{C}_g = 0.\qquad(9\text{-}13)$$

Eq. (9-3) could be integrated into

$$C_e = -\frac{i d_{eg}^* E_0}{\hbar}e^{\frac{i\Delta t}{2}}\sin\left(\frac{\Omega_R t}{2}\right),\qquad(9\text{-}14)$$

$$\Omega_R = \sqrt{\Delta^2 + \frac{(i d_{eg}^* E_0)^2}{\hbar^2}},\qquad(9\text{-}15)$$

where Ω_R and Δ were the Rabi frequency and splitting energy, respectively.

Rabi splitting

The strong coupling mechanism had attracted more and more attention. Its research field mainly focused on two aspects, one was solid physics, which mainly studied semiconductors, the modified photon gap in 3D structure[46] and spontaneous emission[47-48], and the other was atomic physics, it mainly studied the interaction about atomic cavities[49].

In previous studies, the authors proposed the solid-state quantum electrodynamics(QED)effects, the placement of atoms in optical microcavities could produce atomic cavity systems, and the splitting phenomena was observed from the transmission resonance about the empty cavity[50] and atomic fluorescence spectra[51]. One of them, called the vacuum Rabi splitting(VRS), described the quantum properties about the EM field[52].

QED indicated that the spontaneous emission about atoms was caused by the interaction between atoms in the vacuum field. In free space, there were various EM couplings between atoms, and the emissivity of spontaneous was irreversible[53]. The coherent system that the author once studied consists of the cavity QED and the external environment. There was an interaction between the external environment and system. If the external environment was neglected, the consistency of the system itself will be reduced. This was also called the decorrelation mechanism, it weakens the energy exchange between the EM field and the atoms until it disappeared. Therefore, reducing the influence of cavity EM was a major challenge. At present, many methods existed to reduce the coating, volume and superconductivity of the cavity, or increased the strength of coherent interactions, when the coherence mechanism became more and more dominant, it was called strong coupling[38].

The strong coupling between cavity and atom has been discussed, and found the splitting phenomenon in the atomic spectrum VRS, thus demonstrating that VRS could lead to the bimodal characteristics about the transmission spectrum for the cavity. Previously, cavity electrodynamics was studied mainly in the atomic system. After a series of quantum optics experiments, the research had developed rapidly. For cavity QED, VRS could explain the quantum mechanism about the interaction between light and matter. Studies had shown that quantum nonlinearity was studied by observing the spectrum, while atomic-photon superposition of up to two photons was studied by spectral pump and probe techniques.

In a strongly coupled system, when the atom resonated with the cavity mode, there was coherent energy in the exchange between the atom and the cavity mode. In the system, atomic energy level splitting and two nondegenerate eigenmodes correspond to VRS of two sidebands, respectively.

9.2.2 Application of strong plasmon-exciton coupling interaction

In this section, a brief introduction to the applications of strong coupling interaction was given. Such applications Mainly included modifying the chemical reaction rate, processing quantum information, low-threshold laser, and promoting chemical reactions.

Modification of the chemical reactions rate

For strong coupling changing the rate of chemical reactions, studies had shown that a new mixed state, also known as Rabi splitting, will occur when the strong interaction between the two systems overcame the decoherence effect. A schematic of the Rabi splitting was shown in Fig. 9-2.

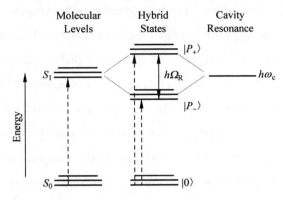

Fig. 9-2 A schematic of the Rabi splitting

Simplified energy landscape shows that there is an energy exchange between the transition (HOMO-LUMO) of molecules and the cavity, which is rapid compared to energy loss. Two hybrid states ($|p+\rangle$ and $|p-\rangle$) are formed due to strong coupling. Note that the coupled system ground-level energy $|0\rangle$ may be modified by strong coupling[54]

The prerequisite for strong coupling was that the material used must be in the optical cavity. In general, the cavity consisted of two parallel mirrors, by tuning, could resonate with the excited state. And the coupling to the vacuum field could rearrange the molecular energy levels, so strong coupling could change the reaction yield and rate. In the vacuum field, the reaction rate and thermodynamics could be corrected. Strong coupling allowed the molecule to maintain its original electronic structure during the reaction and to accelerate the reaction rate by concentrating the light.

Therefore, the cavity vacuum field could modify the chemical reaction rate and molecular properties, provided new tools for influencing useful reactions, and had an important contribution to materials science and molecular devices.

Processing of quantum information

In addition, strong coupling was also applied to quantum information processing. Information science played an increasingly important role in life.

As electronic devices such as network communication, mobile phones, and electronic computers remov the information processing restrictions, quantum information was available. It had many advantages, especially high information processing capabilities. Simultaneously, the requirements for hardware were also increasing. At present, there were program cavity-QED[55-57], ion trap[58-59] and nuclear magnetic resonance[60-61]. Among them, cavity-QED had the greatest hope. In the optical cavity, the atom was the most suitable for quantum data storage. Therefore, the cavity-QED mainly captured high-quality atoms and stored quantum data in the atomic energy state. And in the cavity, owing to the cavity field mode and atomic coupling, there was an interaction between the atoms, so it could be used to prepare the atomic entangled state, as shown in Fig. 9-3, thereby transferring atomic information, and this was for storage purposes.

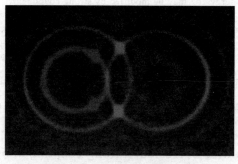

Fig. 9-3 The interaction and entangled state between atoms. Several cone pairs can be seen along the direction of the optical axis. The pairs of photon emitted along the cone intersections are entangled in polarization[62]
(For colored figure please scan the QR code on page 1)

In summary, quantum information was a novel science that encodes, processes, stores, and transmits information by using quantum mechanics in physical systems[63-64]. And through the quantum stack performance, the operating rate was quickly improved. Therefore, the development of quantum information had brought information technology to a new platform.

Low threshold laser

Compared with free space, in the microcavity, the role of spontaneous emission of atoms was different. The microcavity could suppress or enhance the spontaneous emission of atoms, so it was reversible. Thus the cavity-QED could interpret the interaction between the cavity field and the atoms. Therefore, an important application of strong coupling interactions was low threshold lasers.

In 1917, Einstein proposed two kinds of atomic excitation, stimulated radiation and spontaneous radiation. In 1946, Purcell (Fig. 9-4) proposed that spontaneously emitted atoms were affected by the environment rather than being isolated[65]. In 1960, the laser was invented. The author mainly controlled the stimulated radiation process through the optical cavity. This method became more and more dominant in the light process, resulting in various laser sources. Since the radiation was in the form of electromagnetic waves, if the atom in the excited state was located in the microcavity matching its excitation wavelength, the boundary condition did not meet the resonance requirement, and the radiation photon could not be stored, thereby suppressing the spontaneous emission[66].

Fig. 9-4 Edward Mills Purcell, American physicist, winner of the 1952 Nobel Prize in Physics for his independent discovery of in 1946 nuclear magnetic resonance (NMR) in solids and liquids

The coherence between the spontaneously emitted photons was significantly enhanced when the microcavity length was equal to the half wavelength. For a perfect microcavity, all radiated photons were coupled into a single cavity resonant mode. When the coupling coefficient of spontaneous emission was close to 1, the pump power was linearly related to the laser output power, and the threshold was gradually reduced to zero. If it was a closed cavity, the emissivity about a single atom could be described as

$$V_c = A(S+1)N, \qquad (9\text{-}16)$$

where, A was the spontaneous radiation rate, S was the number of photons, and N was the number of atoms. From Eq. (9-16) the could get the formula of the rate as follows

$$\frac{dN}{dt} = H - A(S+1)N \qquad (9\text{-}17)$$

and

$$\frac{dN}{dt} = A(S+1)N - \kappa s, \qquad (9\text{-}18)$$

where κs and H were the photon escape rate and the pumping rate, respectively. The static solution was

$$s = \frac{H}{\kappa}, \qquad (9\text{-}19)$$

$$N = \frac{1}{A} \frac{H\kappa}{(H+\kappa)}, \qquad (9\text{-}20)$$

where $H = \kappa s$, the output light intensity was proportional to the pumping rate and the laser operates without threshold. Therefore, low-threshold laser was an important application about microcavity for controlling spontaneous radiation. To achieve resonance, the author could increase the possibility about spontaneous emission at a specific wavelength by changing the microcavity size, and the process of spontaneous emission was reversible.

Promotion of chemical reactions

Surface plasmons (SPs) were produced by collective oscillations about free

CHAPTER 9 Physical Mechanism of Plasmon-Exciton Coupling Interaction   369

electrons on the conductor surface. If the SP was coupled to phonons to form SPPs, a mixed excitation was formed. SPPs could travel between the metal surface and the medium with a gradual decrease in energy[67]. SPPs were widely used in waveguides, such as SERS[68], surface plasmon resonance (SPR) sensors[69], photothermal cancer therapy[70], and cloaking[71]. At present, SPs had attracted more and more attention in chemistry.

In 2010, Sun et al. found that hot electrons were produced by plasmon decay, which was widely used in plasmon-induced surface oxidation or reduction catalytic reactions. Nevertheless, the conversion efficiency between hot electrons and SPs was weak, so increasing the conversion efficiency of the reaction became a major challenge that the authors needed to achieve, so as to accelerate the chemical reaction.

In addition, Ding et al. built a hybrid system about graphene and silver nanowires(Ag NWs) to promote chemical reactions by prolonging electron lifetime. As was demonstrated in Fig. 9-5, the advantages about the hybrid device was shown by ultrafast pump-probe transient absorption(UPPTRA) spectroscopy.

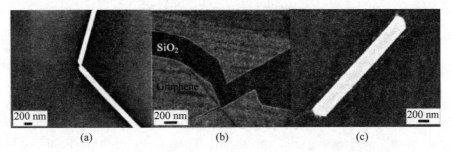

Fig. 9-5 SEM image of Ag NWs, single-layer graphene, a single NW veiled with monolayer graphene, and UPPTRA spectroscopy

(a)~(c) High-resolution SEM image of Ag NW, graphene on SiO_2/Si substrate, and a single Ag NW hybridized with monolayer graphene, respectively; (d) UPPTRA spectroscopy of Ag NW excited by laser, of which the high excitation wavelength is 400 nm and fitted dynamic curve at 532 nm; (e), (f) UPPTRA spectroscopy of both VIS and NIR of graphene are excited by 400 nm laser and the dynamics fitted at 532 nm and 1103 nm, respectively; (g), (h) UPPTRA spectroscopy of both VIS and NIR of hybrid graphene excited by 400 nm laser and the dynamics fitted at 532 nm and 1103 nm, respectively[38]

(For colored figure please scan the QR code on page 1)

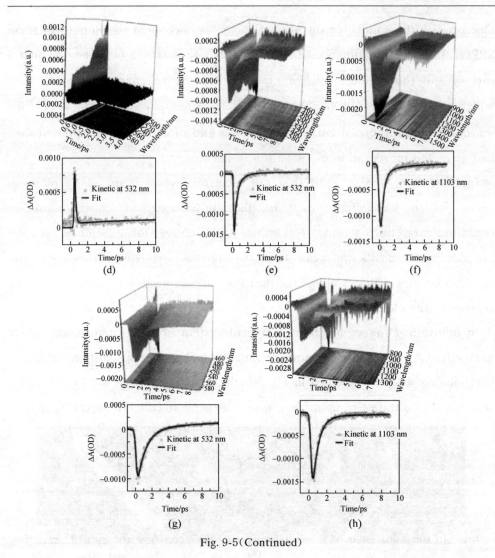

Fig. 9-5(Continued)

Studies had shown that in the visible (Vis)-near-infrared (NIR) region, UPPTRA spectroscopy was applied to study the femtosecond dissolution plasmon-extraction interaction of graphene-Ag NWs(G-Ag NWs)hybrids. It was known that Ag NWs plasmon lifetime was approximately (150 ± 7)fs(Table 9-1). For single-layer graphene, the fast dynamic process with time (275 ± 77)fs was caused by the excitation of graphene excitons. However, the interaction between graphene photons and plasmon hot electrons made the reaction process slower with a time of (1.4 ± 0.3)ps.

CHAPTER 9   Physical Mechanism of Plasmon-Exciton Coupling Interaction

Table 9-1   Fitted lifetimes of monolayer graphene, Ag NWs, and the hybrid system consist of Ag NWs and graphene detected in the Vis and NIR, respectively[38]

System	Spectrum Range of Probe light	Lifetime of Fast Process $\tau_1(f_1)$	Lifetime of Slow Process $[\tau_1(\rho_1)]$
Silver Nanowires	Vis	150 ± 7(4.4%)	
Graphene	Vis	275 ± 77(22.4%)	1.4 ± 0.3(27.9%)
Graphene	NIR	320 ± 46(14.5%)	2.5 ± 0.6(22.8%)
Silver Nanowires/Graphene	Vis	534 ± 108(20.2%)	3.2 ± 0.8(26.1%)
Silver Nanowires/Graphene	NIR	780 ± 92(11.8%)	3.9 ± 0.9(24.3%)

For the mixed system of G-Ag NWs, the time during which the hot electrons were moved to the graphene by plasmon was (534 ± 108) fs. In the Vis region, the graphene plasmon needed to be obviously enhanced by the metal plasmon for a time (3.2 ± 0.8) ps, which was used for the plasmon. Drive the chemical reaction. Studies had shown that G-Ag NWs hybridization could accumulate high-density hot electrons while improving plasmon-to-electron conversion efficiency, which was caused by strong plasmon-exciton coupling interactions.

9.2.3   Weak plasmon-exciton coupling interaction

In 1974, Fleischmann discovered Raman strength enhancement on the surface of rough silver film[72], and reported the interaction of SPs with organic molecules for the first time. This promoted the development of SERS, which could achieve a large increase in the light field in a small range, thereby achieving the aim of scattering enhancement[73]. And SPs could also induce molecular fluorescence enhancement, for example, SPs-induced absorption enhancement[74] and SP-coupled emission enhancement (SPCE)[75], which was also a weakly coupled interaction. Current SPCE studies characterized steady-state spectroscopy.

In the weak coupling mechanism, the wave function and EM mode of the plasmon and excitons were undisturbed during their interaction. This was called the SP EM field with exciton dipole coupling[76]. In previous studies, in the near-the-planar-region surface, this model could be used to research the variation of the decay emissivity of the transmitted dipole. In general, the weak coupling phenomenon mainly included the enhancement of the absorption cross section, the increase of the radiation rate, and the energy exchange between the plasmon and the excitons.

In the above, the authors had mentioned three important parameters g, γ' and κ related to the coupling, which represented the energy transfer rate between matter and light, the spontaneous emission rate about free space in the cavity and the escape of light from the cavity rate, respectively. Note that γ' was different from γ. Two important systems of cavity-QED were strong coupling and weak coupling. Strong coupling mainly had g control, while for weak coupling, the value of κ and γ' were greater than g. The weakly coupled spontaneous emission was irreversible, and it played an important role whether it was enhanced or suppressed.

Here, the authors briefly introduced the weak coupling conditions about the cavity-QED. Purcell, in 1995, pointed out that when atoms were in free space or microcavities, they produced different spontaneous emissivity, and the nuclear magnetic transition rate in the cavity gradually increased. Studies had shown that spontaneous emission was determined by the environment and transmitter, not just a factor. The ratio of free space to corrected emissivity was the Purcell factor, and it could be described as

$$F_g = \frac{\gamma_g}{\gamma_0}, \qquad (9-21)$$

where γ_g was the spontaneous emission rate about the emitter not in the free space, while γ_0 was the spontaneous emission rate about the emitter in the free space.

As mentioned above, both the environment and the emitter affected the R value, and radiance conversion was calculated by using the Fermi golden rule here,

$$\gamma = 2\pi \frac{\rho(\omega)}{\hbar^2} | \langle i | \hat{E}_{cav} \cdot \hat{\mu} | j \rangle |^2, \qquad (9-22)$$

where, $|j\rangle$ was the initial excited state and the $|i\rangle$ was the final state, μ was the electric dipole, $\rho(\omega)$ was the density of final state of the photon, $\hat{E}_{cav}$ was the operator about the vacuum field. Wherein, in vacuum, $\rho(\omega)$ could be described as

$$\rho(\omega) = \rho_0 V = \frac{\omega^2 V}{c^3 \pi^2}, \tag{9-23}$$

where, V was arbitrary volume, and ρ_0 was the local density about the state in vacuum. However, the elements of transition matrix in Eq. (9-16) were averaged out in all possible directions, so,

$$|\langle i|\hat{E}_{cav} \cdot \hat{\mu}|j\rangle|^2 = \frac{(E_{cav} \cdot \mu_{ij})^2}{3}, \tag{9-24}$$

$$\mu_{ij} = -e\langle i|r|j\rangle, \tag{9-25}$$

where, r was the operator of position, e was the electron charge. And the electric field E_{cav} could be getten from

$$\int \varepsilon_0 E_{cav}^2 dV = \frac{1}{2}\hbar\omega, \tag{9-26}$$

therefore,

$$E_{cav} = \frac{\sqrt{2}}{2} \cdot \left(\frac{\hbar\omega}{V\varepsilon_0}\right)^{-\frac{1}{2}}. \tag{9-27}$$

So, the spontaneous emission rate about the emitter in free space could be described as

$$\gamma_0 = \frac{1}{3}\frac{\mu_{ij}^2 \omega^3}{\hbar \pi \varepsilon_0 c^3}. \tag{9-28}$$

It could be seen from the above that the zero-point fluctuation about the EM field caused the excited state spontaneous emission. It was worth noting that when the emitter was in the cavity, the result was the same. Due to the different cavities, their shapes, scales, and components were different, so they were only suitable for EM-specific modes. Equation (9-23) showed that optical resonators compressed light to a small range for long periods of storage, with a significan increase in emissivity. However, in theory this was contradictory, and stricter restrictions were often accompanied by more serious losses. In recent years, improvements in emission characteristics had focused primarily on resonators.

9.2.4 Application of weak plasmon-exciton coupling interaction

The spontaneous emission about a material was not inherent to the material, but was caused by the interaction of the local EM and the material itself. If the field of the local EM was changed[77], the nature of the spontaneous emission also changed. Photonic crystals and plasmon metal NP were two core structures about nanophotonics and mainly used to change spontaneous emission[78]. SERS, surface-enhanced fluorescence, and solar cells were all weakly coupled applications. Here, the authors focused on the application of surface enhanced fluorescence and solar cells.

The interaction of phosphors and metals had attracted increasing attention. When light struck the metal nanostructures surface, it reacted with sub-wavelength metals while the fluorophores were adsorbed on the metal surface. Studies had shown that metal-generated SPs could significantly enhance the luminescence intensity about fluorescent substances. The interaction of the phosphor with the plasmon could reasonably change the luminescence rate and photostability of the material and reduce the fluorescence lifetime.

Assuming that the fluorescent molecule was attached to the metal surface, the enhancement of the fluorescence detection signal could be described as

$$F = \frac{\eta}{\eta_0} \frac{|\mu \cdot E_L|^2}{|\mu \cdot E|^2}, \tag{9-29}$$

where η was quantum efficiency, it could be described as

$$\eta = \frac{\gamma_R}{\gamma_T}, \tag{9-30}$$

where γ_T and γ_R were the total decay rate about the molecule and radiative rate, respectively. Among them,

$$\gamma_T = \gamma_R + \gamma_{NR}, \tag{9-31}$$

where γ_{NR} was the rate of the nonradiative decay.

As was demonstrated in Fig. 9-6 that when the local EM field and quantum efficiency were both amplified, the fluorescence will be significantly enhanced. In a previous study, Yang et al. built a hybrid system about Ag NPs and MoS$_2$, which modulated the metal particles diameter and produced different fluorescence enhancements as shown in Fig. 9-7.

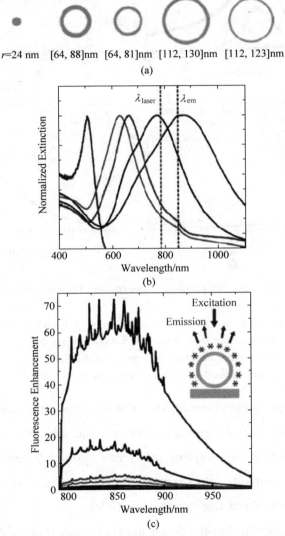

Fig. 9-6 Schematics of the relation between Raman intensity and diameter of Au NPs
(a) Schematic of Au NPs (colloidal and core-shell) as a substrate to enhance fluorescence; (b) Spectral measurements corresponding to the different metal nanoparticles in (a) as substrate; (c) Fluorescence spectra of indocyanine green (ICG) combined with substrates of different nanostructures, adjusted for surface area to be available for fluorophore conjugation. The black curve represents the fluorescence of ICG without metal nanoparticles[79]

(For colored figure please scan the QR code on page 1)

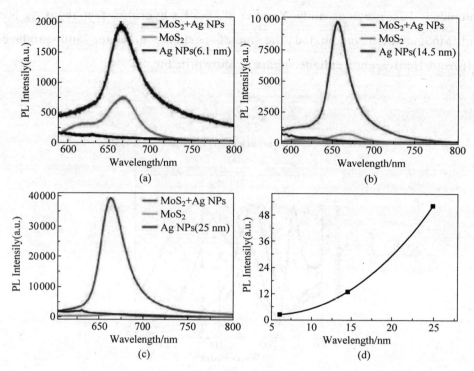

Fig. 9-7 Schematics of the relation between the fluorescence intensity and the diameter of Au NPs

(a) ~ (c) LSPR on the surface of different diameter metal particles. Different fluorescence enhancement results could be obtained, where the diameters of Ag NPs are 6 nm, 14 nm, and 25 nm, respectively; (d) Enhancement factors for different thicknesses[80]

(For colored figure please scan the QR code on page 1)

Metal nanostructures affected the fluorescence emission of molecules, adjusted the scattering of nanostructures and the rate of radiation decay of molecules by growing the coupling efficiency about fluorescence emission to the far field, and controlled the experimental process by adjusting the distance, geometric parameters and size between particles.

Next, the authors briefly introduced another important application of weak coupling, namely in solar cells. In general, photovoltaic(PV)cells could enable the conversion of energy, ie the conversion of light energy into electrical energy. PV could solve the problem of energy shortage. At present, it played an increasingly important role. Since most solar cells were produced by silicon crystals,

CHAPTER 9   Physical Mechanism of Plasmon-Exciton Coupling Interaction   377

in order to reduce the cost of producing solar cells, it was appropriate to reduce the use of silicon crystals. But this might affect the conversion efficiency.

Studies had shown that for solar cells, plasmon interconnect was a sensible choice to achieve nanoscale light localization and propagation in plasmonic field. Due to the unique properties of plasmon, it was gradually used in the field of solar cells. This could significantly increase the absorption of light by the PV device, thereby reducing the thickness of the absorber layer of the PV device. This result had led to the rise of solar cell research to a new platform.

In summary, plasmon-exciton coupling interactions have become more widely used in various fields, such as quantum information, biology, and chemistry. In the future, researchers most likely will do more research in this area.

9.2.5 Plexcitons

Plexcitons are polaritonic modes that result from coherently coupled plasmons and excitons.

Plexcitons aid direct energy flows in exciton energy transfer (EET). Plexcitons travel for 20 μm, nearly equal to the width of a human hair.

Plasmons are a quantity of collective electron oscillations.

Excitons are excited electrons bound to the hole produced by their excitation.

Plexcitons were found to emerge from an organic molecular layer (excitons) or a metallic film (plasmons).

9.3 Application

9.3.1 Plasmonic electrons-enhanced resonance Raman scattering and electrons-enhanced fluorescence spectra

Recently, the authors operated EERRS and EEF spectroscopy by adjusting the bias voltage and gate voltage[81]. Studies had shown that by tuning the bias voltage or the gate voltage, it was possible to suppress or enhance the intensity of the fluorescence and Raman spectra. And the intensity enhancement was caused by the thermal electrons generated by the decay of the plasmon, or by the plasmon EM field, rather than by the current or the bias voltage. The results showed that plasmon electrons could enhance the fluorescence resonance and Raman intensity effectively.

With the discovery of SERS and TERS spectra, it had been extensively used in ultra-sensitive spectral analysis at the nanoscale single-molecule level. When the SERS spectrum of the sample was measured, it was found that its fluorescence intensity was obviously enhanced, and with strong enhanced resonance Raman spectroscopy, while the plasmon-enhanced resonance and the resonance Raman intensity were significantly enhanced, however, the signal intensity of the resonant Raman was weak compared to the fluorescence intensity. Therefore, it was difficult to distinguish the Raman spectrum from the fluorescence spectrum. Therefore, it was a major challenge to realize simultaneous detection of the sample's fluorescence spectrum and plasmon-enhanced resonance Raman spectroscopy.

In previous study, for *in situ* simultaneous measurements of EERRS and EEF spectra on graphene-based SERS substrates, the intensity of fluorescence and Raman was measured by adjusting the bias and gate voltage. As was demonstrated in Fig. 9-8, the hybrid graphene-SERS(G-SERS) substrate used by

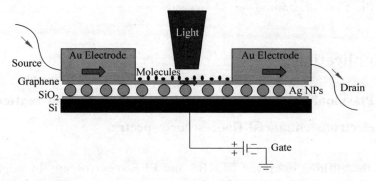

Fig. 9-8 The electrical device design for electro-optical synergic G-SERS substrate[81]
(For colored figure please scan the QR code on page 1)

the authors was successfully used for plasmon-exciton co-driven surface catalytic reactions, which were controlled by bias and gate voltages. Experiments had shown that by adjusting these two voltages, the plasmon EEF and EERRS spectra enabled simultaneous *in situ* measurements.

To investigate the effects of single-layer graphene on fluorescence and SERS, first, the authors studied fluorescence and resonance Raman spectroscopy on G-SERS substrates and SERS substrates, noting that no holes or electrons were doped. As was demonstrated in Fig. 9-9, (a), presented that the graphene was monolayer and its uniformity was shown by Raman spectroscopy at four different locations of the single-layer graphene, where $I_{2D}/(I_G \approx 2.7)$. Then, holes and electrons were doped by adjusting the bias and

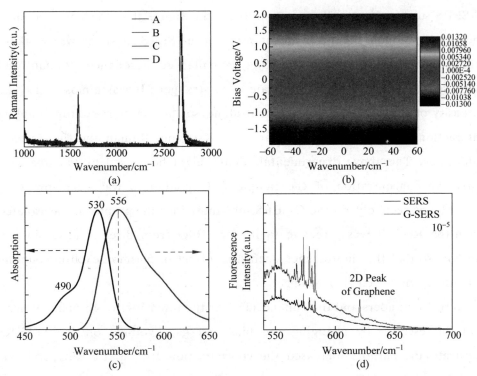

Fig. 9-9 (a) Raman spectra of graphene at four different positions; (b) The gate-and bias-voltage-dependent current in the electric device shown in Fig. 9-8; (c), (d) The measured resonance Raman and fluorescence spectra on the substrate on G-SERS[82]

(For colored figure please scan the QR code on page 1)

gate voltage using the above experimental apparatus(Fig. 9-8). Figure 9-9(b) showed the 2D gate and bias voltages related current of the device, which indicated that the device could be effectively applied to the measurement of the spectrum regulated (c) showed the fluorescence and absorption spectra of Rhodamine 6G(R6G). It was apparent that the intensity of the absorption spectrum was large at a wavelength of 530 nm, and the Raman spectrum of R6G excited by a laser with 532 nm was a resonance Raman spectrum, at the same time, the fluorescence intensity was strong and the peak value was 556 nm approximately.

Next, Fig. 9-9(d) showed the resonant Raman and fluorescence spectra of plasmon-enhanced and plasmon-exciton coupled co-enhancement on SERS and G-SERS substrates. It was apparent from the figure that the corresponding individual plasmon enhancement intensity on the SERS substrate was weaker, while the plasmon-exciton coupling co-enhanced resonance Raman and fluorescence on the G-SERS substrate were stronger. It was almost twice the intensity of the plasmon enhancement alone, so the plasmon-exciton coupling interaction was more likely to enhance the resonance Raman and fluorescence intensities. The above experimental results might be caused by two reasons. First, the concentration of the sample R6G used here was very low, only 10^{-5} M, and secondly, on the G-SERS substrate, since the graphene was isolated from the R6G, it was effective to prevent R6G from directly adsorbing on Ag NPs, so that the fluorescence quenching effect in plasmon-exciton coupling effectively enhanced fluorescence.

To further understand the enhancement of the intensities of resonant Raman and fluorescence by changing the bias and gate voltage on the G-SERS substrate, the authors increased the concentration 10^{-3} M of R6G. As was demonstrated in Fig. 9-10, resonance Raman and fluorescence spectra associated with the gate voltage when $V_{bias} = 0$ V, it could be seen from the figure that as the gate voltage increased, the fluorescence intensity gradually increased, with $V_{gate} = 0$ as a reference. When $V_{gate} = 60$ V, the fluorescence intensity was further enhanced, and when $V_{gate} = -60$ V, the intensity of fluorescence was decreased. Therefore, the intensity of the resonant Raman and fluorescence was enhanced or attenuated by adjusting the magnitude of the gate voltage. When the voltage was positive, the intensity of the resonant Raman and fluorescence increased, and conversely, when the gate voltage was negative, its intensity decreased.

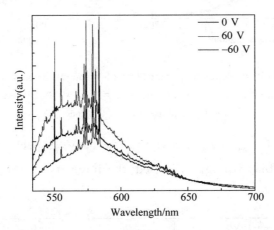

Fig. 9-10　The gate-voltage-dependent EERRS and EEF spectra[82]
(For colored figure please scan the QR code on page 1)

Next, the author further studied the effect of bias voltage on the resonance Raman and fluorescence intensity. As was demonstrated in Fig. 9-11, the intensity of resonance Raman and fluorescence slowly decreased with the increased of bias voltage when $V_{gate} = 60$ V, the result was the same when $V_{gate} = -60$ V. This was because the electrons were carriers driven by a bias voltage here, not the hot electrons generated by the decay of the plasmon, and the electrons and the phonons in the graphene were strongly scattered, thereby the intensity of plasmon resonance Raman scattering and fluorescence was reduced.

In order to directly observe the influence of the bias voltage on the resonance Raman spectrum, the authors removed the fluorescence spectrum. A resonance Raman spectrum of the collection as the bias voltage was increased

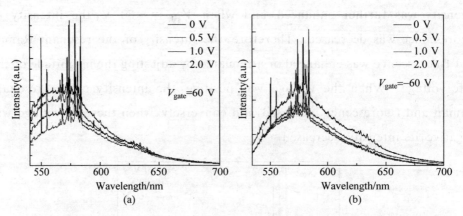

Fig. 9-11　Bias voltage dependent intensities of resonant Raman and fluorescence spectra at $V_{gate} = 60$ V(a), and $V_{gate} = -60$ V(b), respectively[82]

(For colored figure please scan the QR code on page 1)

under conditions of $V_{gate} = 60$ V and $V_{gate} = -60$ V, respectively, was shown in Fig. 9-12. As the bias voltage increased, the intensity of the resonant Raman decreased.

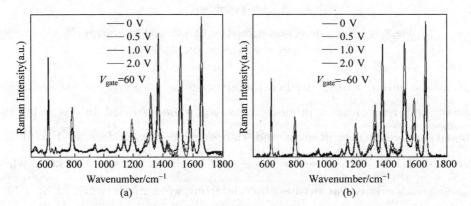

Fig. 9-12　Bias-voltage-dependent intensities of resonant Raman spectra at $V_{gate} = 60$ V(a), and $V_{gate} = -60$ V(b), respectively[82]

(For colored figure please scan the QR code on page 1)

In addition, the authors reduced the resonance excitation and fluorescence effects, measured the normal Raman spectrum excited at 633 nm, and observed the 2D peak of graphene at a concentration of 10^{-3} M. And Fig. 9-13 showed the 2D peak of graphene when $V_{gate} = -60$ V.

CHAPTER 9 Physical Mechanism of Plasmon-Exciton Coupling Interaction

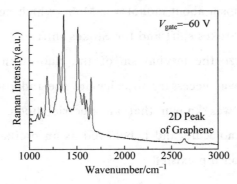

Fig. 9-13 Plasmon-enhanced normal Raman spectroscopy excited at 633 nm[82]

In this section, we have experimentally demonstrated that strong control of the resonant Raman and fluorescence spectra was related to plasmon hot electrons, independent of current.

9.3.2 Tip-enhanced photoluminescence spectroscopy

Recently, Meng et al. studied the tip-enhanced PL spectrum about molybdenum disulfide (MoS_2)[83]. The optical properties of MoS_2 were important for plasmon-enhanced spectroscopy, optoelectronic devices, sensing, and catalysts[84-89]. MoS_2 was a 2D transition-metal dichalcogenides (TMDs) and belonged to a novel semiconductor material with an indirect band gap of 1.3 eV for bulk MoS_2 and a direct band gap of 1.8 eV for monolayer MoS_2. Therefore, the band gap could be changed by adjusting the thickness of the MoS_2, and the strength of the PL was changed. Due to the unique optical properties, monolayer MoS_2 had a broader application prospect. The authors studied the fluorescence and Raman properties of monolayer MoS_2 using TERS. In the strong coupling system, the distance of the tip-MoS_2-film was optimized, the Raman enhancement factor can reach 4.5×10^8, and the fluorescence enhancement factor could reach 3.3×10^3.

In previous veferences, the TERS configuration was the perfect substrate for studying monolayer MoS_2, and the scanning probe microscopy (SPM) was able to control the metal tip precisely. The authors mainly studied the fluorescence enhancement mechanism about monolayer MoS_2, which needed to be discussed in ferms of the non-Stokes shift and the Stokes shift.

In order to analyze the mechanism of the fluorescence enhancement for monolayer MoS_2, it was necessary to calculate the radiation and nonradiative decay processes. It was known that an electric dipole source was used to simulate an excited molecule, and it behaved as an oscillating electric dipole. The following relationship existed:

$$\frac{\gamma}{\gamma_0} = \frac{P}{P_0}, \tag{9-32}$$

where γ was the rate of decay and P was the Radiated power in the TES configuration, among them, the subscript 0 indicated that it was in free space. The relationship between radiative and non-radiative decay enhancement could be described as

$$\frac{\gamma_{tot}}{\gamma_{tot,0}} = \frac{\gamma_r}{\gamma_{r,0}} + \frac{\gamma_{nr}}{\gamma_{nr,0}}, \tag{9-33}$$

where, $\gamma_r/\gamma_{r,0}$ was the radiative decay enhancement, $\gamma_{nr}/\gamma_{nr,0}$ was the non-radiative decay enhancement, and $\gamma_{tot}/\gamma_{tot,0}$ was the total decay enhancement. Fluorescence enhancement came from a combination of emission and excitation, and the local plasmon field provided tremendous excitation enhancement and radiation decay rate. It was known that in TERS configuration, the emission enhancement reflects the quantum yield variation of a monolayer MoS_2, wherein the quantum yield could be described as

$$Q = \frac{\gamma_r/\gamma_{r,0}}{\gamma_r/\gamma_{r,0} + \gamma_{nr}/\gamma_{nr,0}}, \tag{9-34}$$

where Q was the quantum yield. And the enhancement factors (EFs) could be described as

$$EF = |M(\omega_{ex})|^2 Q(\omega_{em}), \tag{9-35}$$

where, ω_{ex} was the excitation frequency, and ω_{em} was the emission frequency.

As was demonstrated in Fig. 9-14(a), the distance-dependent electric field enhancement of Ag-tip-Ag substrate in the TES configuration was demonstrated with ($D = 2$ nm) or without ($D = 1, 2$ nm) a monolayer of MoS_2. When $D = 1$ nm, a different SPR band was observed at a wavelength of around 650 nm, and it covered the resonance absorption of a monolayer MoS_2. At this time, the SPR of the TERS configuration significantly enhanced the fluorescence excitation rate. The SPR peak at approximately 650 nm belonged to the dipole coupling mode between the substrate and the tip. As the value of D increased, the electric field enhancement ($|M|^2$) value blue shifted and decreased. When placed a single layer of MoS_2 on an Ag substrate, it was found that $|M|^2$ increased significantly (as displayed by the green line in Fig. 9-14(b)), indicating fluorescence excitation rate and Raman enhancement increased in the presence about monolayer MoS_2.

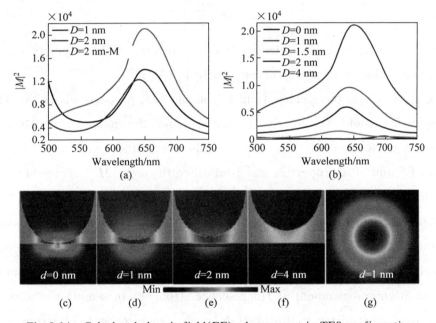

Fig. 9-14 Calculated electric field (EF) enhancement in TES configuration

(a) Comparisons of electric field enhancement ($|M|^2$) of TES configuration with ($D = 2$ nm) and without ($D = 1, 2$ nm) monolayer MoS_2; (b) Dependence of the electric field enhancement on the tip-MoS_2-film distance, plotted as a function of wavelength; (c)~(f) Side views and (g) Top view of the normalized electric field enhancement distributions with different tip-MoS_2-film distance under 660 nm laser excitation[83]

(For colored figure please scan the QR code on page 1)

In a TERS configuration, the tip-substrate distance, the tip radius, the vibrational orientation of the molecules, the angle of incidence of the excitation light,etc. ,all affeced the electric field enhancement. When the gold tip radius was between 50 nm and 80 nm on the gold film, the electric field enhancement exhibited a maximum value. Studies had shown that among these influencing factors, the tip-substrate distance had the greatest influence on the electric field enhancement. Figure 9-14(b) showed that when a single layer of MoS_2 was placed on the Ag substrate, $|M|^2$ was markedly lowered, and the SPR peak was blue-shifted when the distance of the tip-MoS_2-film was gradually increased, from 1 nm to 4 nm. And the enhancement of electric field could be defined as

$$|M|^2 = \left|\frac{E_L}{E_0}\right|^2, \tag{9-36}$$

where E_L and E_0 were local electric field and incident field, respectively. And the default value of E_0 was generally set to 1.0 V/m.

The E_L here was the electric field EF, so the SPR property about the TERS configuration could be clearly expressed from the near field. Here the definition of Raman EF was $|M|^4$. Therefore, the relationship between Raman EF and SPR properties was established through $|M|^2$. It could be seen from Fig. 9-14(b) that when the distance between the tip-MoS_2-film was 1 nm, the Raman enhancement of the single-layer MoS_2 given the maximum intensity at a wavelength of 650 nm, and the value was up to 4.5×10^8. The resonance absorption of the monolayer MoS_2 was well-matched to the SPR band, making the excitation enhancement of fluorescence stronger. In summary, studies had shown that short-distance tip-substrate made the electric field stronger and exhibited large plasmon coupling. However, when the distance between the tip-MoS_2-film was 0 nm and the distance between the substrate and tip was 1 nm, the electric field was enhanced to 0. This was due to the charge exchange between the single layer MoS_2 and the tip, they came into contact with one another, thereby reducing field enhancement. Figures 9-14(c)~(f) were the side views of a series of normalized near-field distributions. It could be seen that at a wavelength of 660 nm, the "hotspots" were located in the nanogap between substrate and tip, resulting in a strong Raman signal. Note that when the distance between the monolayer MoS_2 and the tip was 0 nm, no electric field enhancement was observed as shown in (c). And (g) showed a top view, as well as a large electric field enhancement on a monolayer of MoS_2 film, which further demonstrated the above experimental results.

To further investigate the fluorescence properties, the authors studied the fluorescence emission about a single layer of MoS_2 in the condition of a tip exist. As was demonstrated in Fig. 9-15 (a), the tip-MoS_2-film distance-dependent quantum yield, and nonradiative and radiative decay rates were displayed. The results showed that the quantum yield was very sensitive to the distance of the tip-MoS_2-film, so the fluorescence emission of the single-layer MoS_2 was strongly influenced by plasmon coupling. As the tip-MoS_2-film distance decreased, the quantum yield increased greatly. When the distance was 1nm, the quantum yield was as high as 0.16. Figure 9-15(b) showed that when the wavelength was greater than 600 nm, different SPR peaks were observed, in this region, the radiation decay rate reached a maximum, and the quantum yield was greatly increased.

Studies had shown that in TES configurations, SPR could greatly enhance the process of radiation attenuation, since the non-radiative decay rate dominated the fluorescence emission of the radiated portion of the entire spectral region, demonstrating strong non-radiative energy transfer from a single layer of MoS_2 to a TES system as displayed in Fig. 9-15(b) and (c), this results in a decrease in quantum yield. As the tip-MoS_2-film distance decreased, both decay rates increased significantly. When the distance of the tip-MoS_2-film was further reduced until 0 nm, the increase in the rate of radiation decay could not be obtained. At this time, the charge transfers between the single layer MoS_2 and the tip reduced the non-radiative and radiation attenuation channels, and the non-radiative decay rate dominated the entire fluorescence emission in the

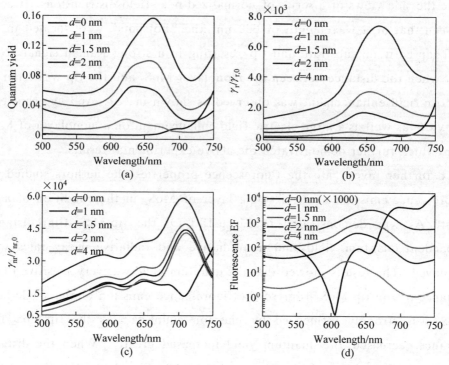

Fig. 9-15 Calculated emission enhancement and fluorescence EF of monolayer MoS_2 in TERS configuration. Tip-MoS_2-film-distancedependent (a) Quantum yield, (b) Radiative decay rate enhancement, (c) Nonradiative decay rate enhancement, and (d) Fluorescence EF of monolayer MoS_2 in TERS configuration[83]

(For colored figure please scan the QR code on page 1)

single layer MoS_2, thereby reducing the quantum yield. Moreover, when the tip-MoS_2-film distance was greater than 2 nm, the fluorescence quenching through metal substrate, and at this time, the fluorescence emission on the radiation portion was dominated by the non-radiation decay rate (Sub Fig. 9-15(b) and (c) for details).

As was demonstrated in Fig. 9-15(d), the fluorescence EF about the tip-MoS_2-film distance-dependent of the monolayer MoS_2 was displayed, and studies had shown that as the distance between the tip-MoS_2-film decreased, the fluorescence EF increased sharply. When the distance was 1 nm, the fluorescence EF reached a maximum value of 3.3×10^3. As could be seen from the figure, when the gap distance was greater than 2 nm, in the TES configuration, the fluorescence of the single layer MoS_2 was quenched by the metal substrate instead of being enhanced by the local electric field effectively. When the gap distance was 0 nm, charge transfer existed between the tip-MoS_2-film, resulting in a decrease in both quantum yield and excitation enhancement, and thus fluorescence enhancement EF was also lowered.

However, the emission wavelength was generally greater than the excitation wavelength, so the Stokes shift between the excitatione and mission wavelengths needed to be considered. In Fig. 9-16 (a) and (b) showed the measured tip-MoS_2-film distance-dependent fluorescence EF at excitation wavelengths of 630 nm and 660 nm, respectively. Studies had shown that as the tip-MoS_2-film distance decreased, from 4 nm to 1 nm, and the fluorescence EF increased significantly. When the distance was 1 nm, the fluorescence EF reached a maximum value, and when the distance decreased to 0 nm, the fluorescence EF was 0. This was consistent with the above findings. As was demonstrated in (c), the fluorescence EF about tip-MoS_2-film distance-dependent was shown when the emission wavelength was 680 nm, among them, the excitation wavelengths were set to 630 nm and 660 nm. As was apparent from the figure, when the tip-MoS_2-film distance was gradually increased, from 0 nm to 1 nm, the fluorescence EF was remarkably increased, however, as the distance continued to increase, the fluorescence EF was remarkably lowered. Fluorescence EF reached a maximum at a distance of 1 nm. Next, the best excitation wavelength needed to find. The authors studied the relationship between the fluorescence EF and the tip-MoS_2-film ((d)). At this time, the emission wavelength was set at 680 nm. The value of fluorescence EF decreased with increasing distance. When the distance was 1 nm, the fluorescence EF reached the maximum value, and the corresponding excitation wavelength was 650 nm.

In summary, in this section, both fluorescence enhancement and Raman enhancement are very sensitive to the effect of the tip-MoS_2-film distance. The tip-film distance of the SPR tip under conventional operation is typically

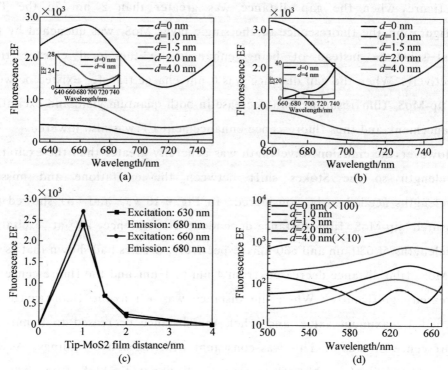

Fig. 9-16 Calculated fluorescence EF under the assumption that Stokes shift between the emission and excitation wavelength was considered. Tip-MoS_2-film-distance-dependent fluorescence EF of monolayer MoS_2 at excitation wavelength of (a) 630 nm and (b) 660 nm. (c) Fluorescence EF of MoS_2 at emission wavelength of 680 nm, plotted as a function of tip-MoS_2-film distance. (d) Tip-MoS_2-film-distance-dependent fluorescence EF of MoS_2[83]

(For colored figure please scan the QR code on page 1)

less than 1 nm, which is measured by the tunneling current and bias voltage between the tip and the film. For most TES configurations, piezoelectric ceramic tubes are used to achieve precise tip-film distances from 1 nm to micron by adjusting the voltage.

9.3.3 Femtosecond pump-probe transient absorption spectroscopy

Recently, Lin et al. studied that plasmon-exciton coupling physical mechanism by using femtosecond pump-probe transient absorption spectroscopy[90]. The physical mechanism was about the plasmon-exciton coupling interactions in a single-layer MoS_2-Ag NPs hybrid system, but it is still unclear. The local surface plasmon resonance(LSPR) induced by Ag NPs significantly enhanced the properties of excitons in a single layer of MoS_2. The authors studied the interaction of femtosecond transient absorption spectroscopy with plasmon-exciton coupling, which was a dynamic process. Their eresults shown that there were three lifetimes, i. e. electron-electron interaction, Auger scattering and electron-phonon interaction, and explained the reasons for the expansion of life in hybrid systems.

In general, femtosecond transient absorption and transmission spectra can be used to study ultrafast dynamic processes and photon-substance interactions in single or few layers of MoS_2. The plasmon-exciton coupling interaction could be reasonably controlled by LSPR, and the LSPR could be reasonably measured by detecting the size of Ag NPs. At the same time, the pump-probe femtosecond transient absorption spectrum was used to research the dynamic process about plasmon-exciton coupling interaction, especially to extend the lifetime.

This experiment used high-quality monolayer MoS_2. And the interaction between the substrate and the monolayer MoS_2 at three absorption peaks was shown in Fig. 9-17. It could be seen that there was a strong interaction near 425 nm, and the interaction of the other two weak absorption peaks was weaker, corresponding to 637. 6 nm and 595. 1 nm, respectively. Figures 9-18

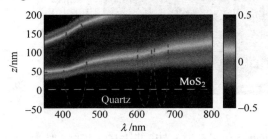

Fig. 9-17 The near electric field distribution of MoS_2 above the substrate
(For colored figure please scan the QR code on page 1)

(a) and (b) showed the photoluminescence(PL) and Raman spectra of a single layer of MoS_2, respectively. Among them, A_{1g} and E_{2g}^1 were two different vibration modes, indicating that the prepared MoS_2 was a 2-H structure.

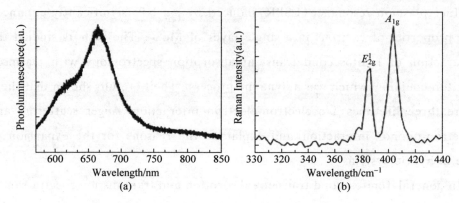

Fig. 9-18 (a) The PL spectrum of monolayer MoS_2, and (b) The Raman spectrum of monolayer $MoS^{2[90]}$

(For colored figure please scan the QR code on page 1)

In Fig. 9-19(a) and (b) showed atomic force microscopy(AFM) images of Ag NPs with different radii for femtosecond transient absorption spectroscopy measurements. while in (c), the transmission spectra of different sizes of Ag NPs were shown. It could be clearly seen that the absorption intensity increased gradually, and the SPR peak gradually red-shifted as the size of Ag NPs increased.

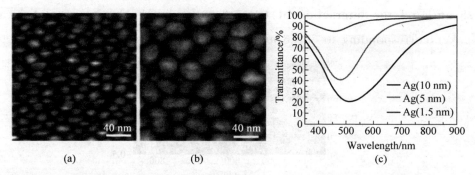

Fig. 9-19 AFM images of Ag NPs with different size: (a) 14.5 nm and (b) 21 nm, respectively; (c) The transmission spectroscopy of Ag NPs with different sizes on quartz[90]

(For colored figure please scan the QR code on page 1)

In order to understand the strong plasmon-exciton coupling interaction, Fig. 9-20(a)~(c) showed that the authors measured the UV-Vis transmission spectra about single-layer MoS_2-Ag NPs hybridized on quartz crystals. The sizes of Ag NPs used were 6.1 nm, 14.5 nm and 21 nm, respectively. Studies had shown that SPR peaks with different sizes of Ag NPs strongly influence the transmission peak about plasmon-exciton hybridization, with the peak

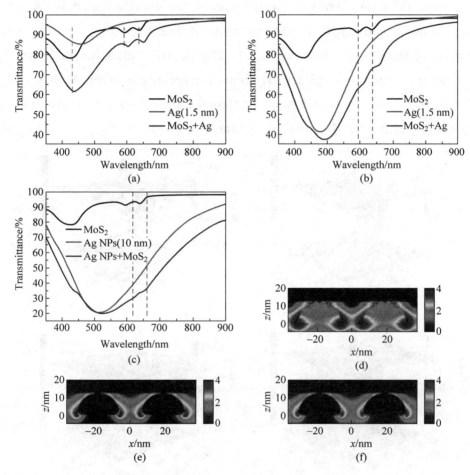

Fig. 9-20 (a)~(c) The transmission spectra of Ag NPs, monolayer MoS_2 and monolayer MoS_2-Ag NPs with different size of Ag NPs; (d)~(f) The electric field distribution of monolayer MoS_2-Ag NPs at 490 nm, 617 nm, 658 nm, where the size of Ag NPs is 21 nm[90]

(For colored figure please scan the QR code on page 1)

intensity increased and the shift changed. (d)~(f) showed the electric field distribution of a single-layer MoS$_2$-Ag NPs hybrid. The size of the Ag NPs used here was 21 nm. And the incident laser light wavelength were 490 nm, 617 nm, and 658 nm, corresponding to three strong absorption peaks, respectively. Observing this series of graphs, it was concluded that the electric field strength strongly depended on the matching between the LSPR peak of the hybrid system and the laser. When the wavelength of the laser increased and was far away from the strongest absorption peak of the hybrid system, the total intensity of the electric field decreased. The electric field strength caused by the coupling interaction was greatly reduced, especially between Ag NPs.

Next, the authors measured the femtosecond transient absorption spectra of a single-layer MoS$_2$-Ag NPs hybrid system, as shown in Fig. 9-21. The two

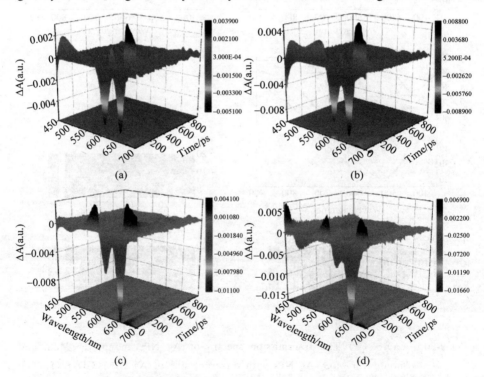

Fig. 9-21 The femtosecond transient absorption spectra
(a) MoS$_2$, (b)~(d) MoS$_2$-Ag NPs hybrid with different sizes of 6.1 nm, 14.5 nm and 21 nm, respectively[90]
(For colored figure please scan the QR code on page 1)

exciton states of MoS₂ corresponding to the absorption peaks at 637 nm and 595 nm were exciton states A and B, respectively. And, the plasmon absorption of Ag NPs having a size of 6.1 nm was far from the two exciton states. From (a) and (b), the total intensity of the exciton states A and B was greatly increased owing to the strong plasmon-exciton coupling interaction, and the relative absorption intensity was substantially unchanged. Studies had shown that as the size of Ag NPs increased, the plasmon absorption peak gradually shifted red, toward the direction of exciton states A and B, resulting in an obviously increase in the absorption intensity of exciton state A at 637 nm, from (d) the selectivity enhancement was evident. And (a) showed that PL was mainly absorbed by exciton state A. The PL of MoS₂ was significantly enhanced as the size of Ag NPs increased. Therefore, this only selectively enhanced the absorption intensity of exciton state A, so plasmon-enhanced fluorescence could be described as

$$\sigma_{PEF}(\lambda_L, \lambda, d_{av}) = |M_1(\lambda_L, d_{av})|^2 |M_2(\lambda_L, d_{av})|^2 \times \frac{\sigma_{FL}(\lambda_L, \lambda)}{|M_d(\lambda_L, d_{av})|^2}, \quad (9\text{-}37)$$

where, $|M_1|^2$ and $|M_2|^2$ were the enhancement factors related to the coupling effect, $\sigma_{EF}(\lambda_L, \lambda)$ was the scattering cross section of MoS₂ in free space, $\sigma_{PEF}(\lambda_L, \lambda, d_{av})$ was the scattering cross section of MoS₂ in the local field, λ and λ_L were the emission wavelength and the excitation wavelength, respectively, d_{av} was the average distance between the metal substrate and the fluorescent molecules, and $|M_d|^2$ represented the energy transfer factor. Studies had shown that the plasmon mainly enhanced the fluorescence and absorption intensity of exciton state A. (c) also showed that the plasmon-enhanced PL was mainly contributed by the exciton state A.

To further understand the plasmon-exciton coupling interaction of a single-layer MoS₂-Ag NPs hybrid system, a 2D femtosecond ultrafast transient absorption spectrum was shown in Fig. 9-22. As was demonstrated in (a), A single-layer MoS₂ femtosecond transient absorption spectrum, using a 400 nm pump laser. Studies had shown that the dynamic process about exciton-exciton interaction was independent of exciton interaction, but by exciton interaction. The lifetimes of exciton states A and B were severely affected by SP resonance of Ag NPs.

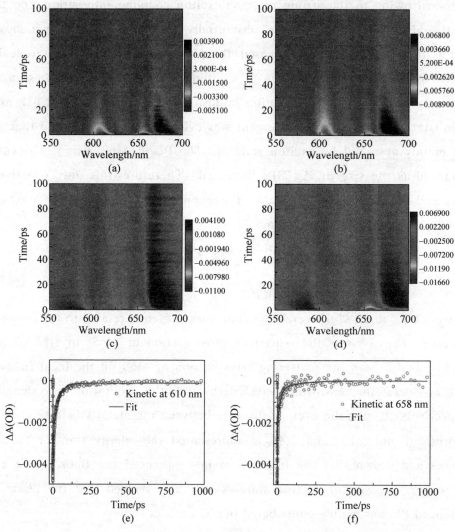

Fig. 9-22 The 2D femtosecond transient absorption spectra of (a) Monolayer MoS_2, and (b)~(d) Monolayer MoS_2-Ag NPs with different size of Ag NPs as 6.1 nm, 14.5 nm and 210 nm, respectively. The fitted transient absorption spectra of (e)~(f) Monolayer MoS_2, and (g),(h) Monolayer MoS_2-Ag NPs hybrid where the size of Ag NPs is 21 nm[90]

(For colored figure please scan the QR code on page 1)

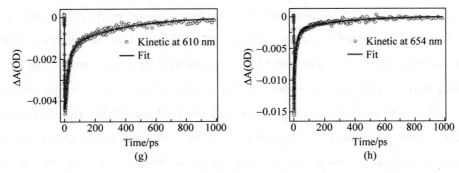

Fig. 9-22(Continued)

Figures 9-22(e)~(h) showed the lifetime of the exciton states A and B of the author-fitting's single-layer MoS$_2$ and hybrid system, which revealed the internal mechanism of the plasmon-exciton coupling interaction, and using the Ag NPs size of 21 nm, the fit data was shown in Table 9-2. Plasmon-exciton coupling interactions had great advantages in extending exciton lifetime. For example, the interaction between hot phonons and hot electrons could significantly increase the lifetime of electron-phonon interactions.

Table 9-2 The lifetime of excitonic states A and B of monolayer MoS$_2$ and monolayer MoS$_2$-Ag NPs hybrid system[90]

Materials	Peak/nm	Parameters	Lifetime/ps	Error
MoS$_2$	610	$T_0 = 0.08505$ ps		
		$A_1 = 0.0264$	$t_1 = 0.15 \pm 0.035$	23%
		$A_2 = -0.0025$	$t_2 = 2.46 \pm 0.497$	20.2%
		$A_3 = -0.00275$	$t_3 = 29.99 \pm 2.92$	9.7%
	658	$T_0 = 0.1509$ ps		
		$A_1 = 0.0104$	$t_1 = 0.24 \pm 0.0733$	30.5%
		$A_2 = -0.0066$	$t_2 = 0.8498 \pm 0.263$	30.9%
		$A_3 = -0.00241$	$t_3 = 21 \pm 3.1$	14.7%
Ag + MoS$_2$	610	$T_0 = 0.06718$ ps		
		$A_1 = 0.023$	$t_1 = 0.1522 \pm 0.0349$	22.9%
		$A_2 = -0.00299$	$t_2 = 15.31 \pm 1.52$	9.93%
		$A_3 = -0.00118$	$t_3 = 252.6 \pm 63.6$	25.2%
	654	$T_0 = 0.3845$ ps		
		$A_1 = -0.0127$	$t_1 = 0.27 \pm 0.0635$	23.5%
		$A_2 = -0.00474$	$t_2 = 6.02 \pm 1.04$	17.3%
		$A_3 = -0.00238$	$t_3 = 61.53 \pm 14.1$	22.9%

In summary, in this section, the authors focus on the principle of plasmon-exciton coupling interactions which are primarily studied by pumping probe femtosecond transient absorption and transmission spectroscopy. It is worth noting that the degree of plasmon-exciton-coupling interaction can be properly controlled here so that the experiment can proceed smoothly. In addition, the femtosecond transient absorption spectra of plasmon-exciton-coupled interactions vividly reveal the dynamic process of the MoS_2-Ag NPs hybrid system. In short, the above findings help people to further understand the nature about plasmon-exciton coupling interactions and promote its widespread application in other areas of research.

References

[1] ZAKHARKO Y, ROTHER M, GRAF A, et al. Radiative pumping and propagation of plexcitons in diffractive plasmonic crystals[J]. Nano. Lett. ,2018,18: 4927-4933.

[2] CROß H, HAMM J M, TUFARELLI T, et al. Near-field strong coupling of single quantum dots[J]. Sci. Adv. ,2018,4: 4906.

[3] ZHOU N, YUAN M, GAO Y H, et al. Silver nanoshell plasmonically controlled emission of semiconductor quantum dots in the strong coupling regime[J]. ACS Nano, 2016,10: 4154-4163.

[4] WURDACK M, LUNDT N, KLAAS M, et al. Observation of hybrid Tamm-plasmon exciton-polaritons with GaAs quantum wells and a $MoSe_2$ monolayer [J]. Nat. Commun. ,2017,8: 259.

[5] CHIKKARADDY R, DE NIJS B, BENZ F, et al. Single-molecule strong coupling at room temperature in plasmonic nanocavities[J]. Nature,2016,535: 127-130.

[6] CHEVRIER K, BENOIT J M, SYMONDS C, et al. Organic exciton in strong coupling with long-Range surface plasmons and waveguided modes[J]. ACS Photonics,2018,5: 80-84.

[7] TODISCO F, DE GIORGI M, ESPOSITO M, et al. Ultrastrong plasmon-exciton coupling by dynamic molecular aggregation[J]. ACS Photonics,2018,5: 143-150.

[8] THOMAS R, THOMAS A, PULLANCHERY S, et al. Plexcitons: the role of oscillator strengths and spectral widths in determining strong coupling[J]. ACS Nano,2018,12: 402-415.

[9] FOFANG N T, GRADY N K, FAN Z Y, et al. Plexciton dynamics: exciton − plasmon coupling in a J-Aggregate-Au nanoshell complex provides a mechanism for nonlinearity

[J]. Nano Lett. ,2011,11: 1556-1560.

[10] NAN F,ZHANG Y F,LI X G,et al. Unusual and tunable one-photon nonlinearity in gold-dye plexcitonic fano systems[J]. Nano Lett. ,2015,15: 2705-2710.

[11] SCHLATHER A E,LARGE N,URBAN A S,et al. Near-field mediated plexcitonic coupling and giant Rabi splitting in individual metallic dimers[J]. Nano Lett. ,2013, 13: 3281-3286.

[12] BALCI S,KUCUKOZ B,BALCI O,et al. Tunable plexcitonic nanoparticles: a model system for studying plasmon-exciton interaction from the weak to the ultrastrong coupling regime[J]. ACS Photonics,2016,3: 2010-2016.

[13] LEE B,LIU W J,NAYLOR C H,et al. Electrical tuning of exciton-plasmon polariton coupling in monolayer MoS_2 integrated with plasmonic nanoantenna lattice[J]. Nano Lett. ,2017,17: 4541-4547.

[14] ZHENG D,ZHANG S P, DENG Q, et al. Manipulating coherent plasmon-exciton interaction in a single silver nanorod on monolayer WSe_2 [J]. Nano Lett. ,2017,17: 3809-3814.

[15] ZAKHARKO Y,GRAF A,ZAUMSEIL J. Plasmonic crystals for strong light-matter coupling in carbon nanotubes[J]. Nano Lett. ,2016,16: 6504-6510.

[16] BELLESSA J,BONNAND C, PLENET J C,et al. Strong coupling between surface plasmons and excitons in an organic semiconductor [J]. Phys. Rev. Lett. , 2004, 93: 036404.

[17] DINTINGER J,KLEIN S,BUSTOS F,et al. Strong coupling between surface plasmon-polaritons and organic molecules in subwavelength hole arrays[J]. Phys. Rev. B, 2005,71: 035424.

[18] SCHLATHER A E,LARGE N,URBAN A S,et al. Near-field mediated plexcitonic coupling and giant Rabi splitting in individual metallic dimers[J]. Nano Lett. ,2013, 13: 3281-3286.

[19] CHENG C W, SIE E J, LIU B, et al. Surface plasmon enhanced band edge luminescence of ZnO nanorods by capping Au nanoparticles[J]. Appl. Phys. Lett. , 2010,96: 071107.

[20] MANJAVACAS A, GARCIA DE ABAJO F J, NORDLANDER P. Quantum plexcitonics: strongly interacting plasmons and excitons[J]. Nano Lett. ,2011,11: 2318-2323.

[21] FEDUTIK Y,TEMNOV V V,SCHOPS O,et al. Exciton-plasmon-photon conversion in plasmonic nanostructures[J]. Phys. Rev. Lett. ,2007,99: 136802.

[22] AKIMOV A V,MUKHERJEE A,YU C L,et al. Generation of single optical plasmons in metallic nanowires coupled to quantum dots[J]. Nature,2007,450: 402-406.

[23] CHANG D, SØRENSEN A S, HEMMER P R, et al. Quantum optics with surface plasmons[J]. Phys. Rev. Lett. ,2006,97: 053002.

[24] DITLBACHER H,HOHENAU A,WAGNER D,et al. Silver nanowires as surface plasmon resonators[J]. Phys. Rev. Lett. ,2005,95: 257403.

[25] GRAMOTNEV D K,BOZHEVOLNYI S I. Plasmonics beyond the diffraction limit [J]. Nat Photon,2010,4: 83-91.
[26] TAME M S,MCENERY K R,ÖZDEMIR Ş K,et al. Quantum plasmonics[J]. Nat. Phys. ,2013,9: 329-430.
[27] WANG L L,ZOU C L,REN X F,et al. Exciton-plasmon-photon conversion in silver nanowire: polarization dependence[J]. Appl. Phys. Lett. ,2011,99: 061103.
[28] KOLESOV R,GROTZ B,BALASUBRAMANIAN G,et al. Wave-particle duality of single SPPs[J]. Nat. Phys. ,2009,5: 470-474.
[29] DAI D,DONG Z,FAN J. Giant photoluminescence enhancement in SiC nanocrystals by resonant semiconductor exciton-metal surface plasmon coupling [J]. Nanotechnology,2013,24: 025201.
[30] GOVOROV A O,LEE J,KOTOV N A. Theory of plasmon-enhanced Förster energy transfer in optically excited semiconductor and metal nanoparticles[J]. Phys. Rev. B, 2007,76: 125308.
[31] VASA P,POMRAENKE R,SCHWIEGER S,et al. Coherent exciton-surface-plasmon-polariton interaction in hybrid metal-semiconductor nanostructures[J]. Phys. Rev. Lett. ,2008,101: 116801.
[32] SUN M T, XU H X. A novel application of plasmonics: plasmon-driven surface-catalyzed reactions[J]. Small,2012,8: 2777-2786.
[33] SUN M T, ZHANG Z L, ZHENG H R, et al. In-situ plasmon-driven chemical reactions revealed by high-vacuum tip-enhanced Raman spectroscopy[J]. Sci. Rep. , 2012,2: 647.
[34] FANG Y R, LI Y Z, XU H X, et al. Ascertaining p, p'-dimercaptoazobenzene produced from p-aminothiophenol by selective catalytic coupling reactions on silver nanoparticles[J]. Langmuir,2010,26: 7737-7746.
[35] SUN M T,LIU S,CHEN M,et al. Direct visual evidence for the chemical mechanism of surface-enhanced resonance Raman scattering via charge transfer[J]. J Raman Spectrosc,2009,40: 137-143.
[36] SUN M T,WAN S,LIU Y,et al. Chemical mechanism of surface-enhanced resonance Raman scattering via charge transfer in pyridine-Ag_2 complex[J]. J Raman Spectrosc, 2008,39: 402-408.
[37] WURTZ G A,EVANS P R,HENDREN W,et al. Molecular plasmonics with tunable exciton-plasmon coupling strength in J-aggregate hybridized Au nanorod assemblies[J]. Nano Lett. ,2007,7: 1297-1303.
[38] CAO E,LIN W H, SUN M T, et al. Exciton-plasmon coupling interactions: from principle to applications[J]. Nanophotonics,2018,7(1): 145-167.
[39] PORRAS D,CIUTI C,BAUMBERG J J,et al. Polariton dynamics and Bose-Einstein condensation in semiconductor microcavities[J]. Phys. Rev. B,2002,66: 085304.
[40] MCKEEVER J,BOCA A, BOOZER A D, et al. Experimental realization of a one-atom laser in the regime of strong coupling[J]. Nature,2003,425: 268-271.

[41] XIA K, TWAMLEY J. All-optical switching and router via the direct quantum control of coupling between cavity modes[J]. Phys. Rev. X, 2013, 3: 031013.

[42] CHANG D E, SØRENSEN A S, DEMLER E A, et al. A single-photon transistor using nanoscale surface plasmons[J]. Nat Phys., 2007, 3: 807-812.

[43] DUFFERWIEL S, SCHWARZ S, WITHERS F, et al. Exciton-polaritons in van der Waals heterostructures embedded in tunable microcavities[J]. Nat Commun, 2015, 6: 8579.

[44] TISCHLER J R, BRADLEY M S, ZHANG Q, et al. Solid state cavity QED: strong coupling in organic thin films[J]. Org. Electron., 2007, 8: 94-113.

[45] LIU W, LEE B, NAYLOR C H, et al. Strong exciton-plasmon coupling in MoS_2 coupled with plasmonic lattice[J]. Nano Lett., 2016, 16: 1262-1269.

[46] JOHN S. Strong localization of photons in certain disordered dielectric superlattices [J]. Phys. Rev. Lett., 1987, 58: 2486-2489.

[47] YOKOYAMA H, NISHI K, ANAN T, et al. Enhanced spontaneous emission from GaAs quantum wells in monolithic microcavities[J]. Appl. Phys. Lett., 1990, 57: 2814-2816.

[48] YAMAUCHI T, ARAKAWA Y, NISHIOKA M. Enhanced and inhibited spontaneous emission in GaAs/AlGaAs vertical microcavity lasers with two kinds of quantum wells [J]. Appl. Phys. Lett., 1991, 58: 2339-2341.

[49] RAIZEN M, THOMPSON R J, BRECHA R J, et al. Normal-mode splitting and linewidth averaging for two-state atoms in an optical cavity[J]. Phys. Rev. Lett., 1989, 63: 240.

[50] AGARWAL G S. Vacuum-field Rabi oscillations of atoms in a cavity[J]. J Opt. Soc. Am. B, 1985, 2: 480-485.

[51] SANCHEZ-MONDRAGON J J, NAROZHNY N B, EBERLY J H. Theory of spontaneous-emission line shape in an ideal cavity[J]. Phys. Rev. Lett., 1983, 51: 550-553.

[52] ZHU Y, GAUTHIER D J, MORIN S E, et al. Vacuum Rabi splitting as a feature of linear-dispersion theory: analysis and experimental observations [J]. Phys. Rev. Lett., 1990, 64: 2499-2502.

[53] WEISBUCH C, NISHIOKA M, ISHIKAWA A, et al. Observation of the coupled exciton-photon mode splitting in a semiconductor quantum microcavity[J]. Phys. Rev. Lett., 1992, 69: 3314-3317.

[54] HUTCHISON J A, SCHWARTZ T, GENET C, et al. Modifying chemical landscapes by coupling to vacuum fields[J]. Angew. Chem. Int. Ed. Engl., 2012, 51: 1592-1596.

[55] ZHENG S B, GUO G C. Efficient scheme for two-atom entanglement and quantum information processing in cavity QED[J]. Phys. Rev. Lett., 2000, 85: 2392-2395.

[56] PELLIZZARI T, GARDINER S A, CIRAC J I, et al. Decoherence, continuous observation, and quantum computing: a cavity QED model[J]. Phys. Rev. Lett., 1995, 75: 3788-3791.

[57] TURCHETTE Q A, HOOD C J, LANGE W, et al. Measurement of conditional phase-shifts for quantum logic[J]. Phys. Rev. Lett. ,1995,75: 4710-4713.

[58] STEANE A. The ion trap quantum information processor[J]. Appl. Phys. B Lasers. Opt. ,1997,64: 623-642.

[59] CIRAC J I, ZOLLER P. Quantum computations with cold trapped ions[J]. Phys. Rev. Lett. ,1995,74: 4091-4904.

[60] GERSHENFELD N A, CHUANG I L. Bulk spin-resonance quantum computation[J]. Science,1997,275: 350-356.

[61] JONES J A, MOSCA M, HANSEN R H. Implementation of a quantum search algorithm on a quantum computer[J]. Nature,1998,393: 344-346.

[62] MONROE C. Quantum information processing with atoms and photons[J]. Nature, 2002,416: 238-246.

[63] BENNETT C H, DIVINCENZO D P. Quantum information and computation[J]. Nature,2000,404: 247-255.

[64] NIELSEN M A, CHUANG I. Quantum computation and quantum information[J]. AAPT,2002,70: 558-559.

[65] KELLY K L, CORONADO E, ZHAO L L, et al. The optical properties of metal nanoparticles: the influence of size, shape, and dielectric environment[J]. J. Phys. Chem. B,2003,107: 668-677.

[66] BJöRK G, MACHIDA S, YAMAMOTO Y, et al. Modification of spontaneous emission rate in planar dielectric microcavity structures[J]. Phys. Rev. A,1991,44: 669-681.

[67] KNOLL W. Interfaces and thin films as seen by bound electromagnetic waves[J]. Annu. Rev. Phys. Chem. ,1998,49: 569-638.

[68] MOSKOVITS M. Surface-enhanced Raman spectroscopy: a brief retrospective[J]. J Raman Spectrosc,2005,36: 485-496.

[69] ZHANG S, BAO K, HALAS N J, et al. Substrate induced Fano resonances of a plasmonic nanocube: a route to increased-sensitivity localized surface plasmon resonance sensors revealed[J]. Nano Lett. ,2011,11: 1657-1663.

[70] LAL S, CLARE S E, HALAS N J. Nanoshell-enabled photothermal cancer therapy: impending clinical impact[J]. Acc. Chem. Res. ,2008,41: 1842-1851.

[71] ERGIN T, STENGER N, BRENNER P, et al. Three-dimensional invisibility cloak at optical wavelengths[J]. Science,2010,328: 337-339.

[72] FLEISCHMANN M, HENDRA P J, MCQUILLAN A J. Raman spectra of pyridine adsorbed at a silver electrode[J]. Chem. Phys. Lett. ,1974,26: 163-166.

[73] CAMPION A, KAMBHAMPATI P. Surface-enhanced Raman scattering[J]. Chem. Soc. Rev. ,1998,27: 241-250.

[74] SCHAADT DM, FENG B, YU ET. Enhanced semiconductor optical absorption via surface plasmon excitation in metal nanoparticles[J]. Appl. Phys. Lett. , 2005, 86: 063106.

[75] YEH D M, SHI Y, CHEN M. Localized surface plasmon-induced emission

enhancement of a green light-emitting diode[J]. Nanotechnology,2008,19: 345201.

[76] GOVOROV A O,BRYANT G W,ZHANG W,et al. Exciton-plasmon interaction and hybrid excitons in semiconductor-metal nanoparticle assemblies[J]. Nano Lett.,2006, 6: 984-994.

[77] RIKKEN G,KESSENER Y. Local field effects and electric and magnetic dipole transitions in dielectrics[J]. Phys. Rev. Lett.,1995,74: 880-883.

[78] TAM F,GOODRICH G P,JOHNSON B R,et al. Plasmonic enhancement of molecular fluorescence[J]. Nano Lett.,2007,7: 496-501.

[79] LI J F, HUANG Y F, DING Y, et al. Shell-isolated nanoparticleenhanced Raman spectroscopy[J]. Nature,2010,464: 392-395.

[80] YANG X,YU H, GUO X, et al. Plasmon-exciton coupling of monolayer MoS_2-Ag nanoparticles hybrids for surface catalytic reactions[J]. Materials Today Energy,2017, 5: 72-78.

[81] CAO E,GUO X,ZHANG L,et al. Electrooptical synergy on plasmon-exciton-codriven surface reduction reactions[J]. Advanced Materials Interfaces,2017,4: 1700869.

[82] WANG X X,CAO E,ZONG H,et al. Plasmonic electrons enhanced resonance Raman scattering(EERRS) and electrons enhanced fluorescence(EEF) spectra[J]. Applied Materials Today,2018,13: 298-303.

[83] MENG L Y,SUN M T. Tip-enhanced photoluminescence spectroscopy of monolayer MoS_2[J]. Photonics Research,2017,5: 745-749.

[84] DU J,WANG Q,JIANG G,et al. Ytterbium-doped fiber laser passively mode locked by few-layer molybdenum disulfide (MoS_2) saturable absorber functioned with evanescent field interaction[J]. Sci. Rep.,2014,4: 6346.

[85] SMOLYANITSKY A, YAKOBSON B I, WASSENAAR, T A, et al. a MoS_2-based capacitive displacement sensor for DNA sequencing [J]. ACS Nano.,2016, 10: 9009-9016.

[86] PONRAJ J S,XU Z,DHANABALAN S C,et al. Photonics and optoelectronics of two-dimensional materials beyond graphene[J]. Nanotechnology,2016,27: 462001.

[87] HE R,HUA J, ZHANG A, et al. Molybdenum disulfide-black phosphorus hybrid nanosheets as a superior catalyst for electrochemical hydrogen evolution[J]. Nano Lett.,2017,17: 4311-4316.

[88] YANG X,YU H, GUO X, et al. Plasmon-exciton coupling of monolayer MoS_2-Ag nanoparticles hybrids for surface catalytic reactions[J]. Materials Today Energy,2017, 5: 72-78.

[89] WANG H,ZHANG C,CHAN W,et al. Ultrafast response of monolayer molybdenum disulfide photodetectors[J]. Nat. Commun.,2015,6: 8831.

[90] LIN W,SHI Y, YANG X, et al. Physical mechanism on exciton-plasmon coupling revealed by femtosecond pump-probe transient absorption spectroscopy[J]. Mater. Today Phys.,2017,3: 33-40.

CHAPTER 10

Plasmon-Exciton-Co-Driven Surface Catalysis Reactions

In previous studies, the authors studied plasmon-driven degradation reactions of various molecules, such as hydrogen[1], nitrogen[2], water[3-4], and carbon dioxide[5], thereby deepening the comprehension about plasmon-driven catalytic reactions[6]. Moreover, it had been reported that plasmon-exciton co-driven surface catalytic reaction had more advantages than only plasmon-driven catalytic reaction[1, 7-13]. Plasmon-exciton coupling interactions in graphene-Ag NPs hybrid systems were widely used in surface catalytic reactions[14-15]. In addition to being controlled by plasmon-exciton coupling interactions, surface catalytic reactions could also be regulated by bias voltage and gate voltage[1]. In this section, we introduce the mechanism about plasmon-exciton co-driven surface reduction and oxidation catalytic reactions[14-16]. At the same time, a method for uniform treatment of oxidation and reduction is provided[17]. Plasmon-driven and Plasmon-exciton co-driven surface catalytic reactions have important applications in the fields of chemistry, physics, environment and energy development[18-45].

10.1 Plasmon-Exciton-Co-Driven Surface Oxidation Catalysis Reactions

Recently, Lin et al. studied an electro-optical device based on graphene-Ag NPs (G-Ag NPs) hybrid, which was fabricated as a substrate for graphene (mediated)

SERS(G-SERS),and reasonably adjusted by the gate and bias voltags. Studies had shown that the gate voltage could reasonably adjust the densities of states of holes and electrons on graphene,and the current or bias voltage controlled the kinetic energy of electrons and holes, thereby promoting the plasmon-exciton co-driving surface catalytic reaction. The results showed that the contribution of the plasmon-exciton co-driven oxidation reaction mainly came from the hot holes on the graphene,and the contribution of the hot electrons was almost negligible. The electro-optical device proposed by the author could reasonably control the plasmon-exciton to drive the oxidation or reduction reaction by adjusting the gate voltage and the bias voltage. Next,explain in detail.

It was known that the lifetime about hot electrons generated by plasmon decay was short,only a few hundred femtoseconds,and its lifetime was greatly prolonged for a plasmon-exciton interaction hybrid system. The surface-catalyzed reaction driven by the plasmon-exciton was better performed due to the increase in the lifetime of the system compared to the catalytic reaction driven only by the plasmon. Moreover,the G-Ag NPs hybrid system was used here,and its Fermi level was bigger than 0.4 eV,compared with the coarse Ag substrate. The authors attempted to understand the physical mechanism about the plasmon-exciton driving the surface catalytic reaction. Experiments had shown that the electro-optic device based on the G-Ag NPs hybrid system oxidizes p-aminothiophenol(PATP) to p,p'-dimercaptoazobisbenzene(DMAB). Moreover, the properties about electrons and holes on the graphene could be reasonably adjusted by the gate voltage,and the bias voltage could reasonably control the kinetic energy of the charge. The results of the study revealed that the surface oxidation reaction was primarily driven by hot holes.

Figure 10-1 showed the electrical properties about electro-optical devices based on G-Ag NPs hybridization, with different gate voltages and bias voltages causing differences in current. To further explain the effect of Ag NPs on the device, it was necessary to study the electrical properties of graphene overlying the SiO_2 substrate without Ag NPs. In Fig. 10-1,(a) and (b) showed the electrical properties about the devices with and without Ag

NPs coverage, respectively. (c) and (d) showed the linear relationship of the conductance of devices with and without Ag NPs, respectively. As could be seen that, when Ag NPs were between the SiO_2/Si substrate and graphene, the current intensity was reduced by about 1/10.

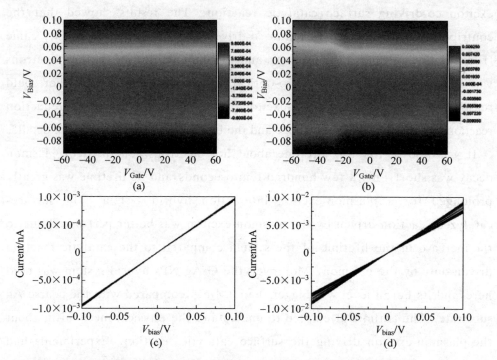

Fig. 10-1 (a),(b) The electrical properties of electro-optical devices with and without Ag NPs; (c),(d) The gate-voltage-dependent $I - V$ curves of electro-optical devices with and without Ag NPs, where the gate voltage is from -60 V to 60 V[14]
(For colored figure please scan the QR code on page 1)

Before investigating the plasmon-exciton co-driven oxidation reaction, it was necessary to understand the Raman spectra about DMAB, PATP and graphene, as shown in Fig. 10-2, and the graphene here was single-layered. In the sub figure (a), G and 2G represent two Raman peaks of graphene. It could be seen that at the wavenumber of 1590 cm^{-1}, the Raman peak about PATP and graphene overlapped, while the DMAB did not exhibit a Raman peak at a wavenumber of 1590 cm^{-1}, and a Raman peak appeared at 1572 cm^{-1}. In previous studies, it was difficult to achieve the direct oxidation of PATP to

DMAB on graphene. As was demonstrated in (b), the authors further investigated the laser-power-dependent plasmon-exciton driving on the G-SERS substrate to drive the oxidation about PATP to DMAB. And the intensities of the two Raman peaks at wave numbers 1148 cm^{-1} and 1440 cm^{-1} increased as the laser power increased(marked by the green line in the figure).

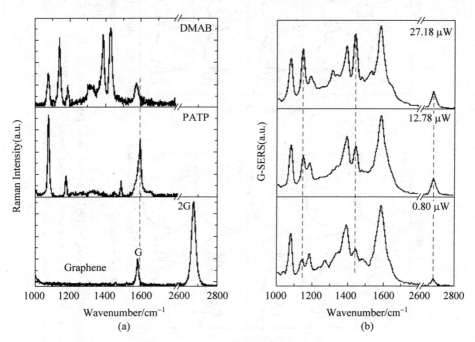

Fig. 10-2 (a) The Raman spectra of PAPT, DMAB and graphene, and (b) The laser-power-dependent G-SERS of PATP oxidation to DMAB[14]

(For colored figure please scan the QR code on page 1)

Next, the authors studied the effect of gate-voltage regulation on the plasmon-exciton co-driven oxidation reaction, and measured the gate-voltage-dependent G-SERS spectrum when the bias voltage was fixed, and the laser power used here was 0.8 μW, (as shown in Fig. 10-3) Among them, subfigure (a) showed the G-SERS spectra of the gate voltages of −80 V, −40 V, 0 V, 40 V, 80 V when V_{bias} = 0 V. When the voltage was increased in the negative direction, from 0 to −40 V to −80 V, the probability about oxidation reaction increasingly increased, and the intensities of the two Raman peaks about DMAB

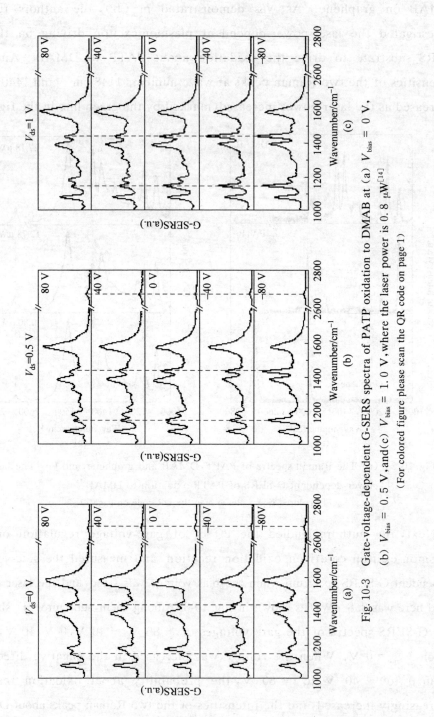

Fig. 10-3 Gate-voltage-dependent G-SERS spectra of PATP oxidation to DMAB at (a) V_{bias} = 0 V, (b) V_{bias} = 0.5 V, and (c) V_{bias} = 1.0 V, where the laser power is 0.8 μW[14]

(For colored figure please scan the QR code on page 1)

CHAPTER 10 Plasmon-Exciton-Co-Driven Surface Catalysis Reactions   409

appearing at wave numbers of 1148 cm^{-1} and 1440 cm^{-1} were significantly increased. When the voltage increased in positive direction, that is, from 0 V to 40 V to 80 V, the probability of the oxidation reaction and the change in the intensities of the two Raman peaks of the DMAB were negligible. In addition, (b) showed that the oxidation reaction probability slightly changed when V_{bias} = 0.5 V, and when the bias voltage was further increased to 1.0 V, as shown in (c), compared with (a), the gate voltage increased in the negative direction, from 0 V to −40 V, the probability of oxidation reaction increased greatly. However, when the gate voltage was positive, the probability of the oxidation reaction was almost constant as the bias and gate voltages increased.

The physical mechanism about the plasmon-exciton co-driven oxidation reaction was not explained too much here. Studies had shown that the contribution about such a reaction mainly came from hot holes, and the contribution of hot electrons was very small. On the contrary, the main contribution to this kind of reaction was provided by hot electrons.

Next, the authors studied the effect of bias voltage on this reaction. As was demonstrated in Fig. 10-4, bias-voltage-dependent G-SERS spectroscopy for such reactions was shown, and the power of laser was 0.8 μW. The fixed gate and bias voltage values were continuously adjusted to plot the G-SERS spectrum. When V_{gate} = 0 V, as shown in subfigure (a), the value of the bias voltage slowly increased from 0 V to 0.5 V to 1.0 V. As could be seen that the intensities of the DMAB Raman peaks at wave numbers of 1148 cm^{-1} and 1440 cm^{-1} sustainedgly increased. This was because the increased bias voltage controlled the current vising so that hot electrons or hot holes obtained greater kinetic energy on the graphene, thereby promoting the oxidation reaction. Comparing Fig. 10-4 (a) with (b), when V_{bias} = 1.0 V, the gate voltage was increased from 0 V to −40 V, and the oxidation reaction was further promoted due to the presence of holes in the valence band of graphene. When the gate voltage was further increased, V_{gate} = −80 V, when comparing Fig. 10-4 (b) with (c) at V_{bias} = 0.5 V, it was apparent that the probability of the oxidation reaction was greatly increased. Therefore, both the gate voltage and the bias voltage could appropriately promote the hot-hole-driven oxidation reaction on the graphene. The results strongly supported the promotion of thermal holes on graphene driven by gate voltage and bias voltage in such a reaction. Subfigares (d) and (e)

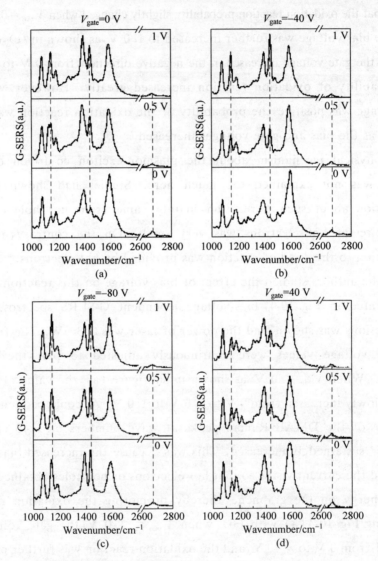

Fig. 10-4 The bias-voltage-dependent plasmon-exciton co-driven surface oxidation reactions at different gate voltages 0 V(a), −40 V(b), −80 V(c), 40 V (d), and 80 V(e), where the laser power is 0.8 μW[14]

(For colored figure please scan the QR code on page 1)

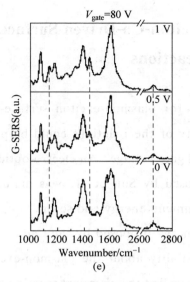

Fig. 10-4(Continued)

showed the G-SERS spectra of the oxidation reaction at gate voltages of 40 V and 80 V, respectively. At this time, the physical mechanism about the plasmon-exciton driving the oxidation reaction was mainly caused by the hot electrons on the graphene. As could be seen from the figure, the probability of the oxidation reaction was slowly increased. Therefore, it was concluded that when the gate voltage was positive, the probability about oxidation reaction mainly driven by hot electrons on graphene was much smaller than that of oxidation reaction mainly driven by hot holes when the gate voltage was negative.

In summary, in this section, we introduced electro-optical devices fabricated based on G-Ag NPs hybridization, and studied plasmon-exciton co-driven oxidation reactions. We could conclude that the probability of such reaction was mainly caused by hot holes on the graphene, and in the reduction reaction, the contribution of promoting the reaction mainly came from the hot electrons on the graphene. In the next chapter, we will explain in detail.

10.2 Plasmon-Exciton-Co-Driven Surface Reduction Catalysis Reactions

In the previous section, for plasmon-exciton surface-catalyzed reactions, the efficiency and probability of the reactions could be promoted and suppressed by changing the bias and gate voltage. An electro-optical device based on G-Ag NPs hybridization proposed by Sun et al. was mainly used in the fields of catalysis, sensors, environment, energy and so on.

The electro-optical synergy proposed by the Cao et al. effectively improved the efficiency and probability about the plasmon-exciton co-driven chemical reaction, and further controlled the chemical reaction by adjusting the gate and bias voltage. The gate voltage controlled the Fermi level about the G-Ag NPs hybrid system, thereby further increasing the density of state (DOS) about the hot electrons, and the current controlled by the bias voltage causes the plasmon hot electrons to obtain greater kinetic energy. Under these conditions, the aim of further increasing the probability and efficiency about the plasmon-exciton co-driven surface catalytic reaction could be achieved.

As was demonstrated in Fig. 10-5, (a) Raman spectroscopy, showed that graphene was a single layer, and the ratio of the 2D mode to the G mode was $I_{2D}/I_G \approx 3.5$. (b) A comparison between the transmission spectra of graphene, Ag NPs, and G-Ag NPs hybrid systems, marked with red, blue, and green lines, respectively. This revealed the coupling interaction between single-layer graphene and Ag NPs. In the transmission spectrum, the strongest absorption of Ag NPs was about 500 nm, and the plasmon-exciton strongest coupling interaction was about 500 nm.

Electrical measurement of the device, at room temperature, and in a vacuum environment. In Fig. 10-6, (a) and (b) showed the 2D gate voltage dependent and bias voltage dependent currents for devices with and without molecules,

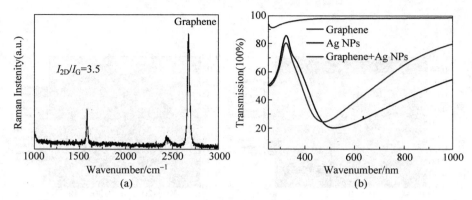

Fig. 10-5 (a) Raman spectrum of graphene, (b) Transmission spectra of monolayer graphene, Ag NPs and the hybrid system of monolayer G-Ag NPs[15]

(For colored figure please scan the QR code on page 1)

respectively. And (c) and (d) further illustrated a linear relationship of the current of the device as a function about bias voltage. In addition, (e) and (f) showed graphs of conductance as a function of gate voltage for devices without and with molecules, respectively. Studies had shown that graphene dominated electron transport, and electrons could be transferred from Ag NPs to single-layer graphene due to the coupling interaction of graphene-Ag NPs. Its physical mechanism will not be explained too much here.

In order to further study, the electro-optical synergy effect of plasmon-exciton driving surface catalytic reaction, Cao et al. studied the kinetics of plasmon-exciton co-driven surface catalytic reaction without voltage control. As was demonstrated in Fig. 10-7, The Raman spectra of the DMAB powder was at the top, and 4NBT powder was at the bottom of the figure, respectively. A series of G-SERS spectra in the middle showed the entire kinetics as the laser intensity decays. As could be seen from the figure, when the laser power was 5.36 μW, the initial stage of the reaction was observed at this time, which was equal to the Raman spectrum of the reactant 4NBT powder. It was worth noting that due to the interaction between the substrate and 4NBT, the $-NO_2$ vibration mode of 4NBT shifts on the SERS substrate. Studies had shown that as the laser intensity increased, the Raman peaks about 4NBT at wavenumbers

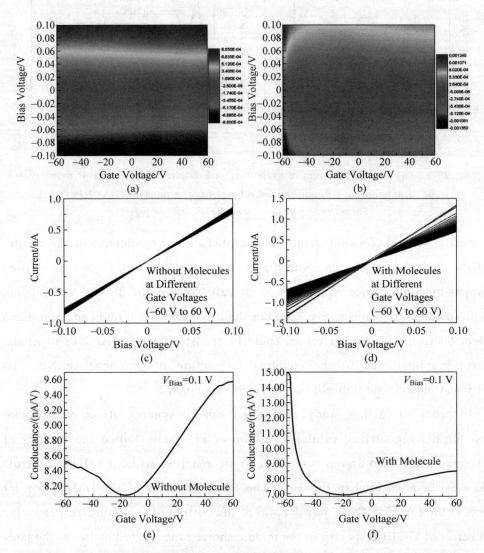

Fig. 10-6 (a) and (b) Gate-and bias-voltage-dependent electrical current for without and with molecules; (c) and (d) The current varies with bias voltage at different gate voltages, and (e) and (f) Gate-voltage-dependent conductance for device without and with molecules, where $V_{bias} = 0.1$ V[15]

(For colored figure please scan the QR code on page 1)

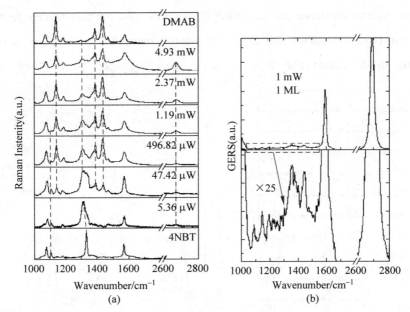

Fig. 10-7 (a) Raman spectra of 4NBT and DMAB powders, the laser-intensity-dependent SERS spectra of 4NBT to DMAB, and the laser powers for the measurements of G-SERS (b) Laser-power-dependent Raman spectra of 4NBT adsorbed on the graphene without Ag NPs, where the molecular concentration is 1 mol/L[15]

(For colored figure please scan the QR code on page 1)

$1109~cm^{-1}$ and $1306~cm^{-1}$ gradually decreased until it disappeared. On the contrary, the DMAB gradually increased at the wavenumbers $1390~cm^{-1}$ and $1438~cm^{-1}$. The SERS spectra about DMAB and DMAB powders were consistent, indicating that the plasmon-exciton co-driven surface-catalyzed reaction was complete. Figure 10-7(b) showed the laser power-dependent Raman spectrum about 4NBT adsorbed on graphene without Ag NPs. The results revealed that when the laser power was higher, the molecular concentration was higher, if there was no Ag NPs on the graphene. Then there would be no chemical reaction, and its Raman peak could not be observed because its strength was too weak.

Next, the authors studied that bias-dependent plasmon-exciton co-driven the surface catalytic reaction when the gate voltage was constant. As shown in Fig. 10-8, (a) showed that when $V_{gate} = 0$ V, the bias voltage depended on the plasmon-exciton co-drive surface-catalytic reaction. As the bias voltage increased, the Raman peak intensity about DMAB gradually increased at 1438 cm^{-1}, and the vibration mode about the Raman peak here belonged to the —N = N—

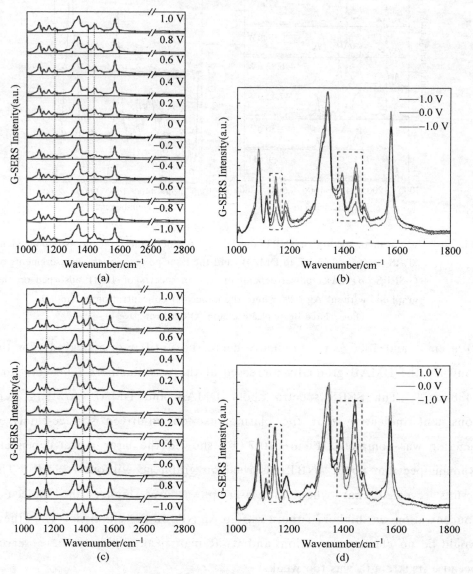

Fig. 10-8 Bias-voltage-dependent plasmon-exciton-co-driven surface-catalytic reaction at different gate voltages, where (a) and (b) $V_{gate} = 0$ V, (c) and (d) $V_{gate} = 40$ V, and (e) $V_{gate} = -40$ V[15]

(For colored figure please scan the QR code on page 1)

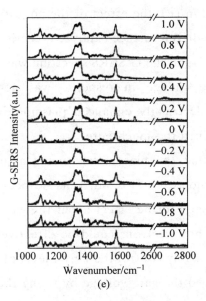

Fig. 10-8(Continued)

vibration mode. It indicated that the efficiency and probability of the catalytic reaction were gradually increasing. (b) showed the G-SERS spectrum with gate, voltages of -1.0 V, 0 V, and 1.0 V, respectively. The value of the gate voltage was 0 V, and the value of the bias voltage was 1.0 V or -1.0 V, the intensity of the Raman peak of DMAB was significantly increased.

The authors speculated that as the gate voltage increased, the probability and efficiency of the catalytic reaction will increase. Subfigures (c) and (d) showed the Raman peak of DMAB with —N═N— vibration mode at a wavenumber of 1438 cm^{-1} when the gate voltage was large and the value was 40 V. The intensity was gradually increased, and the G-SERS spectrum could be directly compared with (d) when the bias voltage was -1.0 V, 0 V, 1.0 V, respectively. Comparing Fig. 10-8(b) and (d), it shown that when the gate voltage was 40 V, the ratio of I_{1432}/I_{1340} was significantly larger than the ratio of the gate voltage value to 0 V. It was concluded that when the gate voltage was large, the bias voltage could greatly increase the plasmon-exciton co-driven surface-catalytic reaction.

As was demonstrated in (e), the effect of the polarity of the bias voltage on the palsmon-exciton co-driven surface-catalyzed reaction was studied, when the gate voltage was -40 V, and the G-SERS spectrum was drawn. Even if the gate voltage was large, the change in response was negligible. The physical mechanism by which electrically enhanced plasmon-exciton co-drive surface catalysis was not explained too much here.

Studies had shown that the bias voltage could adjust the probability and efficiency about plasmon-exciton co-driven surface-catalytic reaction. In Fig. 10-9, (a) and (c) showed the effect about the gate voltage on the catalytic reaction. The DOS of hot electrons could be effectively controlled since the gate voltage could reasonably control the Fermi level of Ag NPs and graphene. Here the authors changed the gate voltage by fixing the bias voltage to detect the effect of the gate voltage on the catalytic reaction. (a) showed also that as the gate voltage increased, the probability about catalytic reaction slowly increased, which was mainly contributed by the DOS of the increased hot electrons in the graphene conduction band. When $V_{bias} = 1$ V, the hot electrons had higher kinetic energy, which was caused by the current, so that the catalytic reaction was more efficient as shown in (b). In addition, (c) showed that when $V_{bias} = -1$ V, the effect about the polarity of the bias voltage on the reaction was investigated, and a comparison between (b) and (c) was made, the effect of the direction of the current on the reaction was negligible.

In summary, in this section, we mainly introduce the plasmon-exciton co-driven surface catalytic reduction reaction studied on the G-SERS substrate regulated by gate voltage and bias voltage. The probability and efficiency of the reaction can be appropriately increased. The physical mechanism about plasmon-exciton co-driven reduction and oxidation reactions is different, so how to deal with these two reactions in a unified mechanism will be explained in detail in the next chapter.

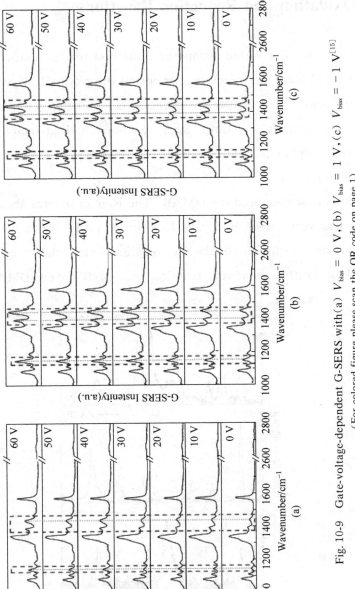

Fig. 10-9 Gate-voltage-dependent G-SERS with (a) $V_{bias} = 0$ V, (b) $V_{bias} = 1$ V, (c) $V_{bias} = -1$ V[15]
(For colored figure please scan the QR code on page 1)

10.3 Unified Treatment for Plasmon-Exciton-Co-Driven Oxidation and Reduction Reactions

In this section, the authors proposed a method for the uniform treatment of oxidation and reduction reactions, supported the physical mechanism of the reaction, and promoted a deeper comprehension about the plasmon-exciton co-driven surface-catalytic reaction. Such reactions had attracted widespread attention in many fields, such as physics, chemistry, energy, materials, and environmental science.

The authors first measured the DMAB. The Raman spectra about PATP and 4NBT powders were shown in Fig. 10-10, because PATP was a reactant of the plasmon-exciton co-driven oxidation reaction, and the target molecule corresponding to the reduction reaction was 4NBT, and DMAB was the

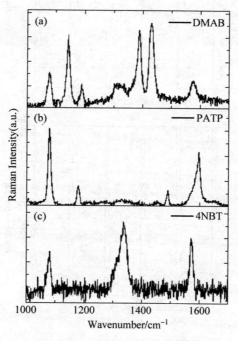

Fig. 10-10 Raman spectra of (a) DMAB, (b) PATP, and (c) 4NBT powder[17]

reaction product. Therefore, there were two ways to generate DMAB from PATP and DMAB according.

Next, the authors measured the plasmon-driven potential-dependent reduction about 4NBT on a rough Ag electrode in a liquid environment as shown in Fig. 10-11(a). When the potential was 0 V and −0.1 V, the SERS curve of 4NBT was equal to the Raman spectrum about the 4BNT powder shown in Fig. 10-11(b). Studies had shown that when the potential gradually increased from the negative direction of −0.2 V, the reaction gradually completed. When the potential was −0.2 V, the reaction DMAB could be observed. When the potential was raised to −1.2 V, the reaction proceeded to

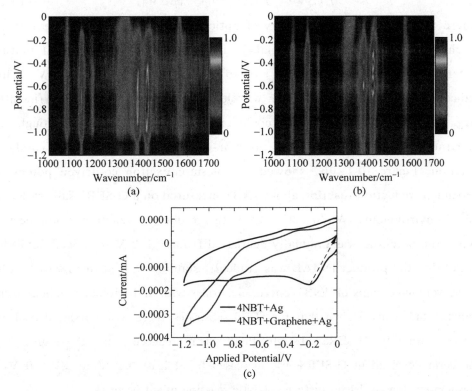

Fig. 10-11 (a) Plasmon-driven reduction reactions; (b) Plasmon-exciton co-driven reduction reactions; (c) Cyclic voltammograms for the SERS and G-SERS substrates[17]

(For colored figure please scan the QR code on page 1)

completion, and 4NBT was converted to DMAB completely. When the potential was gradually decreased from -1.2 V to 0 V, it was observed that the reaction product DMAB was stabilized. In general, as the negative potential increased, a higher Fermi level occurred on the rough Ag substrate, thereby lowering the barrier of the reaction. And the plasmon-exciton co-driven reduction reaction was mainly driven by hot electrons created by plasmon decay. The authors assumed that, when the potential was increased to the Fermi level about the Ag substrate, and when the energy level of the hot electron was increased to 0.2 eV, the relationship between the electrode potential shift and the light absorption energy could be described as

$$E(\Delta V) = E(\Delta V = 0) - e\beta \Delta V, \qquad (10\text{-}1)$$

where, the $E(\Delta V)$ was the energy of optical absorption, ΔV was the amount of change in the cathode potential V, in general, $\beta \leqslant 1$. Note that in this experiment, $E(\Delta V = 0) = 2.33$ eV, so $E(\Delta V = -0.2) = 2.53$ eV. This indicated that the light absorption wavelength was 490 nm, also, a sufficiently large light absorption frequency could excite the LSPR about the rough Ag substrate, thereby promoting the plasmon-exciton co-driven surface catalytic reaction. Figure 10-11(b) showed the plasmon-exciton co-driven potential-dependent reduction reaction about 4NBT measured on a G-SERS substrate in a liquid environment. When $\Delta V = 0$ V, this reduction reaction could occur. When the potential was gradually decreased from -1.2 V to 0 V, it could be found that the product DMAB was gradually stabilized. And in Fig. 10-11(c), cyclic voltammograms of 4NBT on Ag electrodes with and without graphene were shown. Studies had shown that for SERS substrates, there was a reduced peak of -0.2 V, while for G-SERS substrates there was no reduced peak. Therefore, when the Fermi level about G-SERS was increased by at least 0.2 V, at $\Delta V = 0$ V, a plasmon-exciton co-driven surface-catalytic reaction might occur.

Next, the authors measured the plasmon-driven oxidation reaction as displayed in Fig. 10-12(a), and the such oxidation reaction was displayed in Fig. 10-12(b). When the potential was raised to 0.7 V, the plasmon-driven oxidation reaction about PATP to DMAB occurred, and as the potential gradually increased until 0.9 V, the oxidation reaction proceeded completely, and PATP was converted into DMAB completely. When the potential was gradually lowered and returned to 0 V, the reaction product DMAB was gradually stabilized. When the potential was 0.2 V, it could be observed that the plasmon-exciton co-driven PATP to DMAB oxidation reaction occurred, which indicated that compared with the plasmon-driven oxidation reaction, the plasmon-exciton co-driven oxidation reaction required a low potential and could effectively reduce the barrier of PATP to DMAB oxidation. When the potential continued to increase to 0.9 V, the reaction proceeded to completion, and when the potential was lowered from 0.9 V to 0 V, the product DMAB was stable. Figure 10-12(c) showed a cyclic voltammogram of the G-SERS and

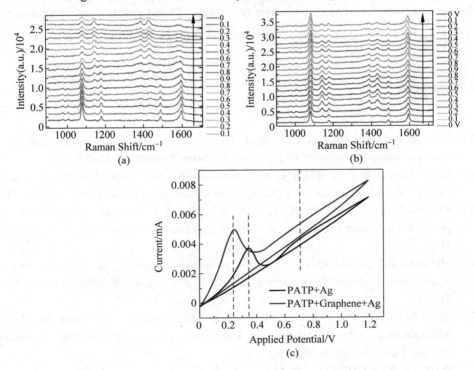

Fig. 10-12 (a) Plasmon-driven oxidation reactions; (b) Plasmon-exciton co-driven oxidation reactions; (c) Cyclic voltammograms of SERS and GSERS substrates[17]

(For colored figure please scan the QR code on page 1)

SERS substrates, showing that the reactant PATP was adsorbed on the rough Ag electrode, and there were two oxidation peaks at potentials of 0.3 V and 0.7 V, respectively. The oxidation peak at 0.3 V was caused by Ag itself, and the oxidation peak at 0.7 V was caused by the oxidation reaction about PATP to DMAB. For PATP adsorption on the G-SERS substrate, only one oxidation peak appeared at a potential of 0.2 V, which corresponded to the plasmon-exciton co-driven oxidation reaction. Therefore, the plasmon-exciton coupling interaction could effectively reduce the reaction barrier.

In summary, in this section, we described the nature of the plasmon-exciton co-driven surface catalytic reaction in electrochemical SERS. It is also concluded that hot electrons and electrons are mainly used for plasmon-exciton co-driven reduction reactions, while hot holes and holes are mainly used for plasmon-exciton co-driven oxidation reactions. These reaction mechanisms are widely used in the fields of spectral analysis, energy, and environment.

References

[1] MUKHERJEE S, LIBISCH F, LARGE N, et al. Hot electrons do the impossible: plasmon-induced dissociation of H_2 on Au[J]. Nano Lett. ,2013,13: 240-247.

[2] MARTIREZ J M P, CARTER E A. Prediction of a low-temperature N_2 dissociation catalyst exploiting near-IR-to-visible light nanoplasmonics[J]. Sci. Adv. ,2017,3: 4710.

[3] LIU Z, HOU W, PAVASKAR P, et al. Plasmon resonant enhancement of photocatalytic water splitting under visible illumination[J]. Nano Lett. ,2011,11: 1111-1116.

[4] MUBEEN S, LEE J, SINGH N, et al. An autonomous photosynthetic device in which all charge carriers derive from surface plasmons [J]. Nature Nanotechnol. , 2013, 8: 247-251.

[5] HOU W, HUNG W H, PAVASKAR P, et al. Photocatalytic conversion of CO_2 to hydrocarbon fuels via plasmon-enhanced absorption and metallic interband transitions [J]. ACS Catal. ,2011,1: 929-936.

[6] WANG X X, WANG J G, SUN M T, et al. Plasmon-driven molecular photodissociations [J]. Applied Materials Today,2019,15: 212-235.

[7] DING Q Q, SHI Y, CHEN M D, et al. Ultrafast dynamics of plasmon-exciton interaction of Ag nanowire-graphene hybrids for surface catalytic reactions[J]. Sci.

Rep. ,2016,6: 32724.
[8] KUMAR D,LEE A,LEE T,et al. Ultrafast and efficient transport of hot plasmonic electrons by graphene for Pt free, highly efficient visible-light responsive photocatalyst [J]. Nano Lett. ,2016,16: 1760-1767.
[9] WANG Y J,CHEN H L,SUN M T,et al. Ultrafast carrier transfer evidencing graphene electromagnetically enhanced ultrasensitive SERS in graphene/Ag-nanoparticles hybrid [J]. Carbon,2017,122: 98-105.
[10] PARK J Y, KIM S M, LEE H, et al. Nedrygailov, hot-electron-mediated surface chemistry: toward electronic control of catalytic activity[J]. Acc. Chem. Res. ,2015, 48: 2475-2483.
[11] BRONGERSMA M L,HALAS N J,NORDLANDER P. Plasmon-induced hot carrier science and technology[J]. Nat. Nanotechnol. ,2015,10: 25-34.
[12] WANG P J,LIU W, LIN W H, et al. Plasmon-exciton co-driven surface catalytic reaction in electrochemical G-SERS[J]. J. Raman Spectrosc. ,2017,48: 1144-1147.
[13] LIU W,LIN W, ZHAO H, et al. The nature of plasmon-exciton codriven surface catalytic reactions[J]. J Raman Spectrosc. ,2018,49: 383-387.
[14] LIN W,CAO E,ZHANG L,et al. Electrically enhanced hot hole driven oxidation catalysis at the interface of a plasmon-exciton hybrid [J]. Nanoscale, 2018, 10: 5482-5488.
[15] CAO E,GUO X,ZHANG L,et al. Electrooptical synergy on plasmon-exciton-codriven surface reduction reactions[J]. Advanced Materials Interfaces,2017,4: 1700869.
[16] SUN M T,CAO E, LIN W H. Photoelectric synergistic surface plasmon-exciton catalytic reactions device and preparation method. China Pat: 201710609332.3[P]. 2017.
[17] LIN W,CAO Y, WANG P, et al. Unified treatment for plasmon-exciton co-driven reduction and oxidation reactions[J]. Langmuir,2017,33: 12102-12107.
[18] FANG Y R,LI Y Z, XU H X, et al. Ascertaining p, p'-dimercaptoazobenzene produced from p-aminothiophenol by selective catalytic coupling reactions on silver nanoparticles[J]. Langmuir,2010,26: 7737-7746.
[19] HUANG Y F,ZHU H P, LIU G K,et al. When the signal is not from the original molecule to be detected: chemical transformation of para-aminothiophenol on Ag during the SERS measurement[J]. J. Am. Chem. Soc. ,2010,132: 9244-9246.
[20] DONG B,FANG Y R,CHEN X W,et al. Substrate-,wavelength-,and time-dependent plasmon-assisted surface catalysis reactions of 4-nitrobenzenethiol dimerizing to p,p'-dimercaptoazobenzene on Au,Ag,and Cu films[J]. Langmuir,2011,27: 10677-10682.
[21] SUN M T, XU H X. A novel application of plasmonics: plasmon-driven surface-catalyzed reactions[J]. Small,2012,8: 2777-2786.
[22] UENO K,MISAWA H. Surface plasmon-enhanced photochemical reactions[J]. J. Photochem. Photobiol. C,2013,15: 31-52.
[23] YAMAMOTO Y S, OZAKI Y, ITOH T. Recent progress and frontiers in the

electromagnetic mechanism of surface-enhanced Raman scattering[J]. J. Photochem. Photobiol. C,2014,21: 81-104.

[24] ZHANG Z,XU P,YANG X,et al. Surface plasmon-driven photocatalysis in ambient, aqueous and high-vacuum monitored by SERS and TERS [J]. Journal of Photochemistry and Photobiology C: Photochemistry Reviews,2016,27: 100-112.

[25] ZHANG Z,FANG Y,WANG W,et al. Propagating SPPs: towards applications for remote-excitation surface catalytic reactions[J]. Adv. Sci. ,2016,3: 1500215.

[26] LEE J,MUBEEN S,JI X,et al. Plasmonic photoanodes for solar water splitting with visible light[J]. Nano Lett. ,2012,12: 5014-5019.

[27] HUANG Y F,WU D Y,ZHU H P,et al. Surface-enhanced Raman spectroscopic study of paminothiophenol[J]. Phys. Chem. Chem. Phys. ,2012,14: 8485-8497.

[28] XU P,KANG L L,MACK N H,et al. Mechanistic understanding of surface plasmon assisted catalysis on a single particle: cyclic redox of 4-aminothiophenol[J]. Sci. Rep. ,2013,3: 2997.

[29] CHEN X J,CABELLO G, WU D Y, et al. Surface-enhanced Raman spectroscopy toward application in plasmonic photocatalysis on metal nanostructures [J]. J. Photochem. Photobiol. C,2014,21: 54-80.

[30] HUANG Y F,ZHANG M,ZHAO L B,et al. Activation of oxygen on gold and silver nanoparticles assisted by surface plasmon resonances[J]. Angew. Chem. Int. Ed. , 2014,53: 2353-2357.

[31] SAMBUR J B,CHEN P. Approaches to single-nanoparticle catalysis[J]. Annu. Rev. Phys. Chem. ,2014,65: 395-422.

[32] WANG F,LI C H,CHEN H J,et al. Plasmonic harvesting of light energy for suzuki coupling reactions[J]. J. Am. Chem. Soc. ,2013,135: 5588-5601.

[33] KANG L L,HAN X J,CHU J Y,et al. In situ surface-enhanced Raman spectroscopy study of plasmon-driven catalytic reactions of 4-nitrothiophenol under a controlled atmosphere[J]. Chem Cat Chem. ,2015,7: 1004-1010.

[34] INGRAM D B,LINIC S. Water splitting on composite plasmonic-metal/semiconductor photoelectrodes: evidence for selective plasmon-induced formation of charge carriers near the semiconductor surface[J]. J. Am. Chem. Soc. ,2011,133: 5202-5205.

[35] WANG J L, ANDO R A, CAMARGO P H C. Controlling the selectivity of the surfaceplasmon resonance mediated oxidation of p-aminothiophenol on aunanoparticles by charge transfer from UV-excited TiO_2[J]. Angew. Chem. ,2015,127: 7013-7016.

[36] SUN M T, ZHANG Z L, ZHENG H R, et al. In-situ plasmon-driven chemical reactions revealed by high-vacuum tip-enhanced Raman spectroscopy[J]. Sci. Rep. , 2012,2: 647.

[37] VAN SCHROJENSTEIN L E M,DE PEINDER P,MANK A J G,et al. Separation of time-resolved phenomena in surface-enhanced Raman scattering of the photocatalytic reduction of p-nitrothiophenol[J]. ChemPhysChem,2015,16: 547-554.

[38] ZHANG Z L,CHEN L,SUN M T,et al. Insights into the nature of plasmon-driven

catalytic reactions revealed by HV-TERS[J]. Nanoscale,2013,5: 3249-3252.

[39] MERLEN A,CHAIGNEAU M,COUSSAN S. Vibrational modes of aminothiophenol: a TERS and DFT study[J]. Phys. Chem. Chem. Phys. ,2015,17: 19134-19138.

[40] KUMAR N, STEPHANIDIS B, ZENOBI R, et al. Nanoscale mapping of catalytic activity using tip-enhanced Raman spectroscopy[J]. Nanoscale,2015,7: 7133-7137.

[41] ZHANG Z L,SHENG S X,WANG R M,et al. Tip-enhanced Raman spectroscopy[J]. Anal. Chem. ,2016,88: 9328-9346.

[42] FANG Y R, ZHANG Z L, SUN M T, et al. High vacuum tip-enhanced Raman spectroscope based on a scanning tunneling microscope[J]. Rev. Sci. Instrum. ,2016, 87: 033104.

[43] HARVEY C E,WECKHUYSEN B M. Surface-and tip-enhanced Raman epectroscopy as operando probes for monitoring and understanding heterogeneous catalysis [J]. Catal. Lett. ,2015,145: 40-57.

[44] HARTMAN T, WONDERGEM C S, KUMAR N, et al. Surface-and tip-enhanced Raman spectroscopy in catalysis[J]. J. Phys. Chem. Lett. ,2016,7: 1570-1584.

[45] KIM H,KOSUDA K M,VAN DUYNE R P,et al. Resonance Raman and surface-and tip-enhanced Raman spectroscopy methods to study solid catalysts and heterogeneous catalytic reactions[J]. Chem. Soc. Rev. ,2010,39: 4820-4844.

CHAPTER 11

Nonlinear Optical Microscopies of CARS, TPEF, SHG, SFG and SRS

In 1965, Terhune and Maker in the Science Laboratory at Ford Motor Company reported the coherent anti-Stokes Raman scattering (CARS) phenomenon for the first time[1]. They used a pulsed ruby laser to study the third-order response of several materials. First, a Raman laser beam with a frequency of ω was transmitted using a Raman shifter, a second beam was generated at a frequency of $\omega - \omega_v$, and then the two beams were simultaneously irradiated to the sample. They pointed out that when the pulses of the two beams overlapped in time and space, a blue-shifted CARS signal was produced at frequency $\omega + \omega_v$, and studies had shown that when the frequency of the sample matches the frequency difference between the two beams, The CARS signal was significantly increased. Their research was called "three wave mixing experiments." In 1974, Begley et al. proposed the name of the CARS spectrum at Stanford University[2]. Since then, this nonlinear optical technology had been called CARS technology.

In the past few years, nonlinear optical microscopes had been widely used in various fields such as biology, medicine, and pharmacy[3-7]. Compared with general radiation imaging, optical imaging methods had the advantages of accuracy, practicability, sensitivity and non-invasiveness for cell development and detection and diagnosis of viral molecules. In optical imaging, nonlinear optical microscopes had the advantages of high sensitivity, ultra-high resolution and deep penetration. Studies had shown that nonlinear optical microscopy could accurately locate and completely remove tumors, analyze living cells, and directly track different proteins and chemical bonds, such as CARS[8-9], two-photon excited fluorescence(TPEF)[10], second harmonic generation(SHG)[11], sum frequency generation (SFG)[12], and stimulated Raman scattering (SRS)[13-14].

11.1 Principles of Nonlinear Optical Microscopies

Brief introduction

In the fields of medicine and biology, since optical microscopy was limited by phototoxicity, in order to avoid the effects of phototoxicity, the minimum average energy of the laser was required to be measured. In general, the use of NIR lasers for nonlinear excitation microscopy could effectively reduce phototoxicity, scattering of tissue and living cells, thereby increasing the penetration of nonlinear microscopes. Nonlinear optical microscopy was an important tool for medical drug diagnosis[15].

CARS spectroscopy was broadly used in the fields of physics, chemistry, biology and materials science[1,2,16-18]. Its energy level diagram was shown in Fig. 11-1. The CARS spectrum used multiple photons to study Raman active molecular vibrations, resulting in the coherent Raman signal. CARS belonged to a third-order nonlinear optical process, it mainly included three laser beams, namely a pump laser, a probe laser and a Stokes laser, wherein their corresponding frequencies were ω_p, ω_{pr} and ω_s. These three lasers interacted with the sample at the anti-Stokes frequency to produce a coherent optical signal, where the anti-Stokes frequency was $\omega_{pr} + \omega_p - \omega_s$. When $\omega_{pr} = \omega_p$, the anti-Stokes frequency could be expressed as

$$\omega_{\text{anti-Stokes}} = 2\omega_{\text{pump}} - \omega_{\text{Stokes}}. \tag{11-1}$$

The CARS microscope had a 3D sectioning abilities that scanned the sample

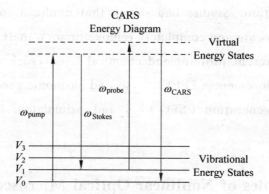

Fig. 11-1 The energy diagrams of CARS[24]
(For colored figure please scan the QR code on page 1)

to produce a CARS image when the Stokes and pump laser beam were focused onto the sample[7, 19-23]. The optical path diagram of the CARS instrumentation and microscope was displayed in Fig. 11-2. Here, a femtosecond laser was used as the incident light, which was then split into two beams, one as the probe light and the pump light, and the other as the Stokes beam, which became a collinear probe, pump and Stoke light beam through a dichroic mirror. The beam was focused on the sample through the objective lens and the CARS signal was detected by a photomultiplier(PMT)using a bandpass filter[24].

SRS was a third-order nonlinear optical process that was a type of Raman scattering that directly imaged the vibrating fingerprint information of living cells and tissues using picosecond or femtosecond lasers. SRS technology had many advantages, such as high sensitivity, ultrahigh resolution, and fast speed. It was capable of freely labeling molecules, enabling imaging of specific molecules[25], detecting intracellular extracts[26-29], glucose metabolism[2], and tumor cells[14], and imaging multiple proteins in situ[13, 30-33]. According to the CARS microscope, an electro-or acousto-optic modulator was added to the Stokes beam, and a lock-in amplifier was added to the photomultiplier to generate an SRS signal. A schematic of this was shown in Fig. 11-3.

For a second-order nonlinear optical process, when $\omega_{pr} = \omega_p$, it was defined as SHG, and when $\omega_{pr} \neq \omega_p$, it was defined as SFG. The energy level diagrams of

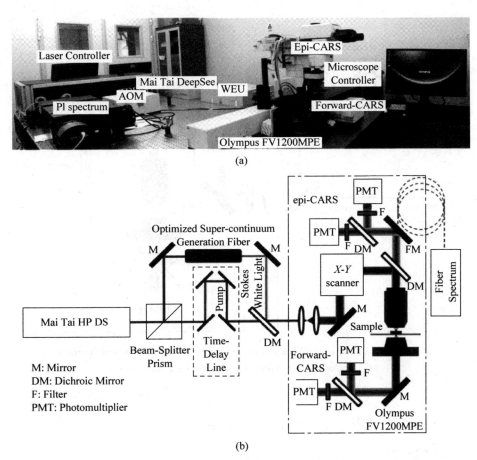

Fig. 11-2 (a) CARS instrument, and (b) CARS optical path[24]
(For colored figure please scan the QR code on page 1)

SHG and SFG were shown in Fig. 11-4. In 1961, Franken first discovered SHG[32]. It was widely used in biological molecular detection, medical diagnosis, drug testing and other fields. For the detection of collagen in living cells and tissues, the diagnosis of muscles and microtubules, SHG microscopy played an important role. The contractile integrity of muscles was measured by sarcomeric myosin imaging using SHG microscopy. It was worth noting that SHG only corresponded to materials that had no inversion symmetry.

TPEF was a second-order nonlinear optical process with a penetration depth of 1 mm in living cells and tissues. The technique had high sensitivity and

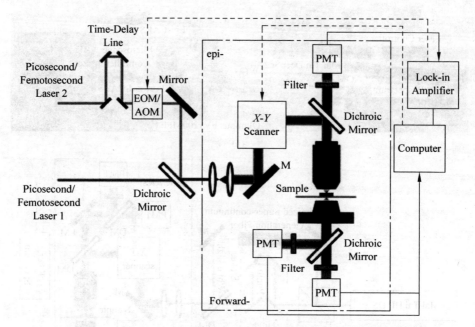

Fig. 11-3 Diagram of an SRS microscope with epi-and forward-channels[7]
(For colored figure please scan the QR code on page 1)

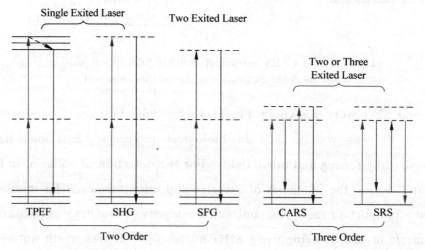

Fig. 11-4 Energy-level diagram for nonlinear optical processes of TPEF,
SHG, SFG, CARS, and SRS process[7]
(For colored figure please scan the QR code on page 1)

resolution. TPEF microscopy enabled the use of exogenous markers to successfully diagnose and detect cancer cell metabolic activity, phenotypic changes, and protein expression[33]. The schematic of the TPEF or SHG microscope were shown in Fig. 11-5.

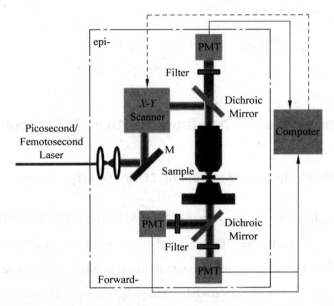

Fig. 11-5 Diagram of a TPEF or SHG microscope with epi and forward channels[7]
(For colored figure please scan the QR code on page 1)

The energy level diagrams for the nonlinear optical processes of SHG, TPEF, SFG, CARS and SRS were displayed in Fig. 11-4. For linear optics, the relationship between electric field strength and polarization could be described as

$$P(t) = \chi^{(1)} E(t), \tag{11-2}$$

where, $\chi^{(1)}$ was the linear susceptibility, and the $P(t)$ was the polarization. The author accurately described a nonlinear optical microscope in which the relationship between electric field strength and polarization could be expressed by[34],

$$P(t) = \varepsilon_0 [\chi^{(1)} E(t) + \chi^{(2)} E^2(t) + \chi^{(3)} E^3(t) + \cdots], \tag{11-3}$$

where, ε_0 was the free space permittivity, $\chi^{(2)}$ was the second-order susceptibility, and $\chi^{(3)}$ was the third-order susceptibility. Wherein, the polarization of the second-order nonlinear optical device could be expressed as

$$P(\omega_{pr} + \omega_p) = 2\varepsilon_0 \chi^{(2)} E_1 E_2, \quad (11\text{-}4)$$

when $\omega_{pr} = \omega_p$, it belonged to SHG, otherwise, it was SFG. For third-order nonlinear optical processes, such as SRS and CARS, their polarization could be expressed as

$$P(2\omega_{pr} - \omega_p) = 3\varepsilon_0 \chi^{(3)} E_1^2 E_2^*, \quad (11\text{-}5)$$

$$P(\omega_{pr} + \omega_p - \omega_s) = 6\varepsilon_0 \chi^{(3)} E_1 E_2 E_3^*. \quad (11\text{-}6)$$

11.2 Applications of Nonlinear Optical Microscopies

11.2.1 Optical characterizations of 2D materials

In previous studies, the authors used CARS, SHG, and TPEF nonlinear optical microscopy to characterize 2D materials, not only to visibly observe their surface topography, but also to understand their quality. These nonlinear optical microscopes played a key role in studying the properties of 2D materials.

First, the authors measured the Raman spectra of graphene grown on polished and unpolished Cu films[35] (Fig. 11-6). Obviously, the graphene on the unpolished Cu film was multiple layers, corresponding to $I_{2D}/I_G \approx 1$, and a strong Raman peak appeared at 1331 cm^{-1} owing to defects in graphene. In addition, it could be seen from Fig. 11-6(b) that the graphene produced on the polished Cu film was a single layer corresponding to $I_{2D}/I_G \approx 2.5$.

A single and mixed image of CAES, TPEF and SHG of the multilayer graphene produced on the unpolished Cu film was shown in Fig. 11-7, and a strong Raman peak appeared at 1331 cm^{-1}. The clear streaks on the surface of the graphene could be seen from the CARS image, and its quality was revealed, thereby visually reflecting the properties of graphene produced on the unpolished Cu film. A mixed image of TPEF and SHG of graphene was shown in Fig. 11-7(b), and the strong fluorescence produced was caused by

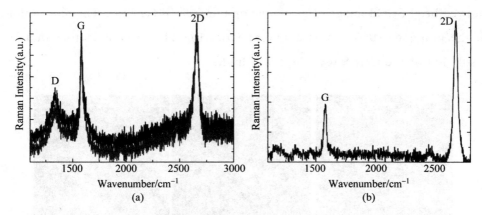

Fig. 11-6 (a) The Raman spectrum of graphene grown on unpolished Cu foils; (b) Raman spectrum of monolayer graphene[35]

(For colored figure please scan the QR code on page 1)

graphene defects. It was worth noting that the TPEF measurement area had a wavelength range of 400 nm to 480 nm, and the SHG signal could also be detected in the TPEF. To further investigate whether the above results were correct, the authors measured the SHG image separately (Fig. 11-7(c)), and could see the streaks on the graphene surface, but the sharpness was far less than the CARS image. This also showed that the CARS technology had a high vibration resolution.

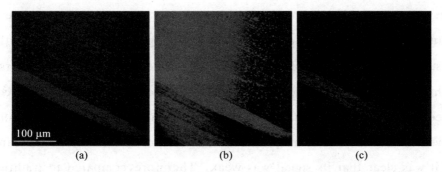

Fig. 11-7 (a) The CARS; (b) The TPEF+SHG; (c) SHG image of graphene at 1360 cm^{-1} [35]

(For colored figure please scan the QR code on page 1)

Subsequently, the authors measured the CARS image of graphene at 1584 cm^{-1} (Fig. 11-8(b)) and the TPEF+SHG image (Fig. 11-8(c)), while Fig. 11-8(a) showed bright field image of graphene on unpolished Cu film. The morphology of the graphene surface was apparent from Fig. 11-8(b), indicating that the resolution of the CARS technique was high.

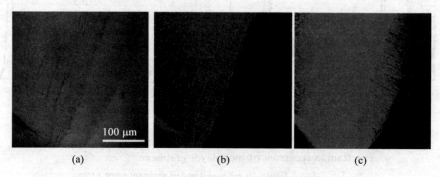

Fig. 11-8 (a) Bright field of graphene; (b) CARS image of graphene; (c) TPEF + SHG image of graphene[35]
(For colored figure please scan the QR code on page 1)

To further demonstrate the above results, the authors measured the SHG image of multilayer graphene. Studies had shown that the large defects and zigzag edges of graphene could further enhance the fluorescence intensity, thereby increasing the SHG signal. Since the defects of the single-layer graphene were relatively small, the intensity of the D peak was too weak to be observed as a single-layer graphene (Fig. 11-9). To further investigate the advantages of CARS technology in characterizing 2D materials, the authors measured the CARS image of single-layer graphene at 1331 cm^{-1}. First, a bright-field image of a single-layer graphene was shown(Fig. 11-9(a)). As could be seen from Fig. 11-9(b), the CARS image had high quality. Figure 11-9(c) showed the SHG + TPEF image measured over the 400 nm to 480 nm range, and it was clear that its signal was weak. Therefore, compared to traditional Raman spectroscopy, CARS technology had high resolution. Figure 11-9(d) showed a combined image of bright field, CARS, SHG and TPEF. So as to improve the CARS resolution, a smaller scale bar (10 μm) CARS image and SHG + TPEF image were required.

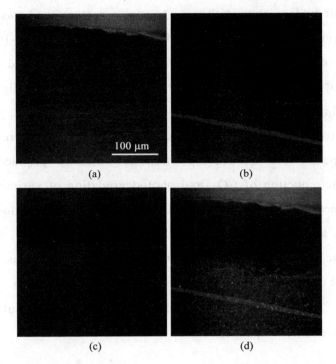

Fig. 11-9 (a) Bright field of monolayer graphene; (b) CARS image of graphene; (c) TPEF + SHG image of graphene; (d) The merged image of bright field + CARS + TPEF + SHG at Raman mode 1360 cm^{-1} [35]

(For colored figure please scan the QR code on page 1)

In summary, for the characterization of 2D materials, nonlinear optical microscopy could not only clearly observed the surface morphology of graphene, but also fully revealed its quality. In particular, CARS technology, with high resolution, played an significant role in the study of the surface of graphene.

11.2.2 Highly efficient photocatalysis of g-C_3N_4

In recent years, people had been paying more and more attention to the development of renewable energy, especially for the use of solar energy, such as visible-light-photocatalysis technology, which had attracted wide interest of researchers. Kong et al. researched the physical mechanism of g-C_3N_4[36], this was a graphite hydrocarbon with tris-s-triazine structures and was a polymeric metal-free semiconductor having a medium bandgap. Figure 11-10(a) showed the structure of g-C_3N_4. When a vacancy defect was produced at the equal position about the N atom produced, the two C atoms gradually approached each other, Fig. 11-10(b) displayed the bond length was 1.52Å. g-C_3N_4 had a 2D layered structure and exhibited unique properties, which were widely used in many aspects, such as decomposition of water to produce hydrogen, photocatalytic reduction of CO_2, and photodegradation of various pollutants. However, since the bulk g-C_3N_4 had fewer active sites, a lower specific surface area, and a faster electron-hole binding rate, thereby inhibiting its photocatalytic reaction[37-38], the authors had adopted various methods to improve the photocatalytic properties of g-C_3N_4, such as surface modification[39-40], doping foreign elements[41-43], and manufacturing of heterostructure nanocomposites[44-45].

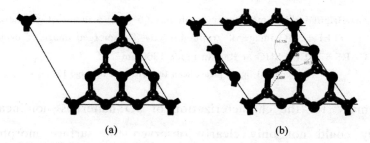

Fig. 11-10 (a) and (b) Cell structure of intrinsic and N defecting g-C_3N_4[38]
(For colored figure please scan the QR code on page 1)

The authors used an acid treatment method to prepare porous 2D g-C_3N_4 nanosheets with high-quality and good-photocatalytic activity. SHG, CARS and TPEF nonlinear optical microscopes could be used to characterize the defects and optical properties of g-C_3N_4.

First, the authors measured bright field images about g-C_3N_4 and ECNV-2.5, Fig. 11-11, subfigures (a) and (b). For the optical properties of both, (c) and (d) showed the TPEF images of both, (e) and (f) showed CARS microscope

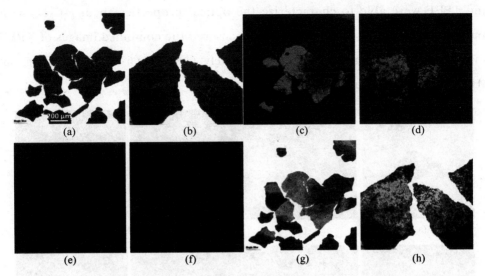

Fig. 11-11 (a),(b) Bright field of g-C_3N_4 and ECNV-2.5; (c),(d) TPEF of g-C_3N_4 and ECNV-2.5; (e),(f) CARS of g-C_3N_4 and ECNV-2.5; (g),(h) Merged CARS and TPEF of g-C_3N_4 and ECNV-2.5[36]

(For colored figure please scan the QR code on page 1)

images of both. g-C_3N_4 had a strong photoluminescence (PL), and the PL strength was significantly reduced due to the defect of ECNV-2.5. Since the PL of g-C_3N_4 was very strong, its CARS signal strength was very weak. As the PL of ECNV-2.5 decreased significantly, the CARS signal gradually increased. Hence, nonlinear optical microscopy could reasonably characterize the optical properties about g-C_3N_4 with and without defects. (g) and (h) showed images of the combined TPEF and CARS of the two, indicating that the TPEF and CARS techniques were capable of characterizing ECNV-2.5 and g-C_3N_4.

The authors directly demonstrated the optical properties of ECNV-2.5 and g-C_3N_4 in Fig. 11-12, (a) and (b) showed the PL of both, (c) and (d) showed the TPEF images of both, and Fig. 11-12(e) and (f) shown the SHG images of both. Since g-C_3N_4 had strong PLs, when the ECNV-2.5 defect increased continuously, the PLs were significantly reduced. Due to the increase of defects, the roughness of the g-C_3N_4 surface was increased and the aperture was larger, thereby significantly enhancing the SHG signal intensity. Thus, SHG and TPEF were able to characterize the optical properties about g-C_3N_4 with and without defects. The authors then measured the combined images of TPEF and SHG for g-C_3N_4 and ECNV-2.5, respectively, ((g) and (h)), demonstrating that SHG and PL could characterize them.

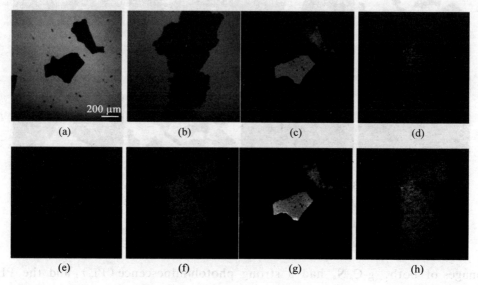

Fig. 11-12 (a),(b) Bright field of g-C_3N_4 and ECNV-2.5; (c),(d) TPEs of g-C_3N_4 and ECNV-2.5; (e),(f) SHGs of g-C_3N_4 and ECNV-2.5; (g),(h) Merged SHG and TPEF of g-C_3N_4 and ECNV-2.5[36]

(For colored figure please scan the QR code on page 1)

In summary, the authors revealed the physical mechanism of high-efficiency photocatalysis. Studies had shown that nonlinear optical microscopy could reasonably characterize 2D photocatalytic materials.

11.2.3 Optical characterizations of 3D materials

In previous studies, CARS microscopy was able to characterize porous carbon materials, especially 3D CARS images, which clearly demonstrated the structure and properties of the interior and surface of these materials. SHG demonstrated that the reverse symmetry of porous carbon materials was destroyed and they were not centrally symmetric. The TPEF image indicated that the porous carbon material had weak auto-fluorescence characteristics, which directly proved that photon momentum could be transferred to the electronic system. Therefore, nonlinear optical microscopy could be used for optical characterization of micro-nanoscale materials.

The authors first measured the Raman and CARS spectra about porous carbon materials. In Fig. 11-13, (a) showed a bright-field image of the kind of material without a nickel skeleton. Studied had shown that porous carbon could be synthesized on the nickel skeleton, but it was necessary to remove the

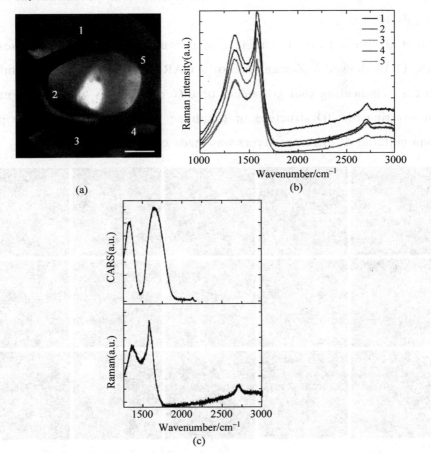

Fig. 11-13 (a) The bright field of porous carbon without the nickel skeleton; (b) The Raman spectra of sample at five different points; (c) The comparison between CARS and Raman spectra of porous carbon materials[22]

(For colored figure please scan the QR code on page 1)

skeleton during optical measurement. (b) depicted the Raman spectrum for the five positions in (a), ranging from 1000 cm^{-1} to 3000 cm^{-1}, indicating that the spectra produced by the sample in different regions were uniform. For the activated carbon material, 2D and G band strong Raman peaks appeared at wavenumbers 1360 cm^{-1} and 1587 cm^{-1}, and the intensity of the G band Raman peak was larger than the 2D band, thereby demonstrating the graphene with a multilayer structure. Subsequently, the authors made the comparison between the Raman and CARS spectra about porous carbon materials, see (c).

Sun et al. studied the 3D structure of porous carbon by using a Z-scan[46]. Figure 11-14 showed a Z-scan slice of a CARS image with a wavenumber of 1360 cm^{-1}, indicating that graphenes of different depths were different, thus demonstrating the 3D structure of graphene. And the inside of the porous carbon material of different layers was made of carbon.

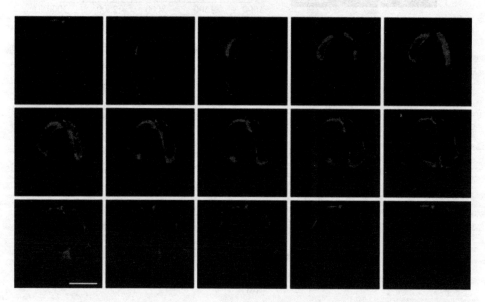

Fig. 11-14 Z-scan of CARS, where Raman modes are 1360 cm^{-1}, where the scale bar is 100 μm[22]

(For colored figure please scan the QR code on page 1)

The authors then studied the TPEF spectrum about porous carbon to produce auto-fluorescence. In Fig. 11-15, (a) showed a bright field image of porous carbon at 1360 cm^{-1} in Raman mode, while (b) depicted its CARS image. (c) and (d) showed the images of TPEF and TPEF + CARS, respectively, demonstrating that the porous carbon material had weak auto-fluorescence.

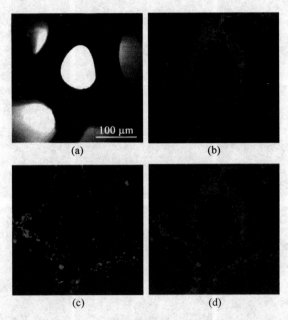

Fig. 11-15 (a) Bright field of porous carbon, (b) The CARS, (c) The TPEF and (d) The 3D CARS + TPEF images for Raman modes at 1360 cm^{-1}[22]
(For colored figure please scan the QR code on page 1)

To further investigate the 3D structure about porous carbon materials, the authors described the rotation of CARS images at different angles(Fig. 11-16), rotating from 0° to 90° and then to 180°, that is, from the top view to the side view and then to the bottom view of the structure, this series of images was a good visualization of the 3D structure. The first picture was the CARS image without rotation (the last one is also), the second was rotated 90°, and the second to last series of images was scanned at different angles of 90° to 180°. This enabled a clear observation of the back and internal structure of the porous carbon material. So the penultimate plot was a CARS image of a 180°

porous carbon material with the last corresponding rotation angle of 0°, corresponding to the back and front of the structure. It could be concluded that the front and back sides of the carbon material were made of carbon.

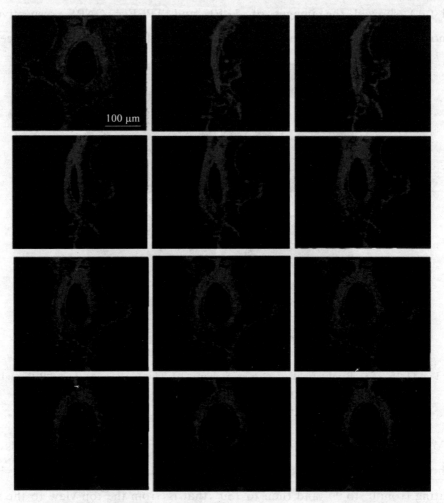

Fig. 11-16　Insight into inner shell of porous carbon materials[24]
(For colored figure please scan the QR code on page 1)

Subsequently, the authors used a nonlinear optical microscope to characterize the porous carbon material, In Fig. 11-17(a) showed its 3D bright field image, and (b) showed its SHG image. It was shown in (c) and (d) for CARS images at wavenumbers 1587 cm^{-1} and 1360 cm^{-1}, respectively. Figure 11-17(e) showed the TPEF image of the structure, it clearly observed the TPEF signal and directly demonstrated that photon momentum could be transferred to the electronic system. The authors then measured the merged image of bright field, SHG, CARS and TPEF of porous carbon materials, (f).

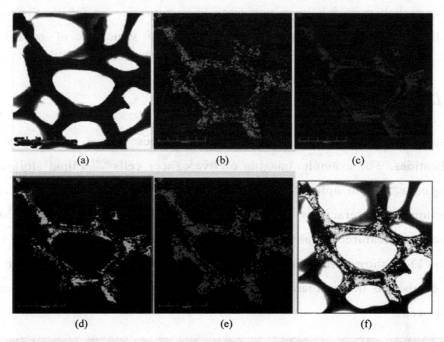

Fig. 11-17 (a) Bight field of porous carbon materials, (b) The SHG, (c) and (d) The CARS at 1587 cm^{-1} and 1360 cm^{-1}, respectively, (e) The TPEF, and (f) The merged image of bight, SHG, CARS and TPEF. The scale bar is 240 μm[22]
(For colored figure please scan the QR code on page 1)

In addition, the authors continued to study the nonlinear optical properties about thin-layer porous carbon materials. Studies had shown that as this kind of material became thinner, its resolution and signal intensity gradually decreased. Compared with the thick porous carbon material, the reverse symmetry breaking degree of the thin porous carbon material was small. For very thin porous carbon materials, photon momentum could be efficiently transferred to electronic systems. When its number of layers was greater than three, the photon momentum could still be transferred to the electronic system.

In summary, nonlinear optical microscopy could characterize the properties and structure about 3D materials, such as porous carbon materials. And the 3D CARS microscope belonged to the micro-scale level. SHG demonstrated that porous carbon materials were not centrally symmetric. So CARS, SHG, and TPEF could characterize materials in other 3D structures of micron or nanoscale.

11.2.4 Advances of biophotonics

In the field of biophotonics, CARS microscopes had a wide range of applications. For example, imaging of live cancer cells[47-48], lipid storage in living cells[49], lipid uptake of living stem cells during differentiation[50], and detection of interactions between plasmas and living cells[51]. CARS microscopes featured ultrahigh resolution, label-free, real-time and ease of use. Figure 11-18 showed the imaging of lipids with CARS and measured their size and number to detect their effects on hormone-treated prostate and breast cancer cells.

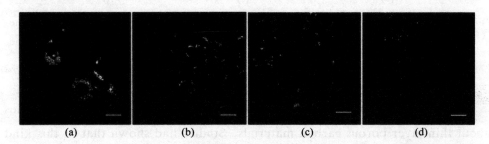

Fig. 11-18 CARS images of treated versus vehicle control living breast and prostate cancer cells

(a) Medroxyprogesterone acetate treated and (b) Untreated breast cells, (c) Synthetic androgen R1881treated, and (d) Untreated prostate cells, scale bar 30 μm[50]

Figure 11-19 showed TPEF and SHG imaging of the distribution of alpha-actinin and fluorescent drugs and sarcomere structures in live embryonic cardiomyocytes, effectively revealing long-term structural changes in live fibrofibrosis formation of individual cardiomyocytes.

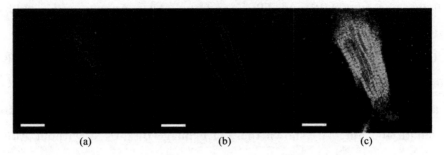

Fig. 11-19 TPEF and SHG images of living cultured neonatal cardiomyocyte
(a) TPEF image, (b) SHG image of sarcomeric structure, and (c) Merged of images
(scale bars: 10 μm)[52]
(For colored figure please scan the QR code on page 1)

In addition, the authors also studied the CARS images of lipids in early embryos and the different stages of live mouse oocyte division[4]. Collagen and apatite in living cells were imaged using SHG and CARS microscopes[53]. In summary, the nonlinear optical microscope had an ultra-high resolution and sensitivity for detecting molecules such as proteins in living cells and tissues, and played an important role in the field of biophotonics.

11.2.5 MSPR-enhanced nonlinear optical microscopy

Surface plasmons(SPs) were produced by collective coherent oscillations of free electrons, which meaningfully enhanced coherent and incoherent nonlinear processes, including SHG[54-55], TPEF[56], and plasmon-enhanced CARS[57]. In previous studies, a plasmon resonance frequency excited was used for near-field enhancement, and its enhancement factor was small for normal plasmonic nanoparticles. Thus, the search for unique nanostructured materials was the key to the problem, it could significantly enhance the double and fundamental frequency nonlinear optical signals, that is, multiple surface plasmon resonance (MSPR). The authors proved the MSPR using Au@Ag nanorods at double and fundamental frequencies of 400 nm and 800 nm, respectively[58]. Studies had shown that MSPR could significantly enhance the nonlinear optical microscopy about two-photon CARS and TPEF, and that Au@Ag nanorods could also significantly enhance the two-photon CARS about 2D materials.

Figure 11-20 showed a plasmon-enhanced TPEF image of Au@Ag nanorods covered the surface of g-C_3N_4. Subfigure (b) showed the PL(red) and absorption(black) spectra of g-C_3N_4 with strong extinction peaks around wavelength 400 nm. (c) shown the TPEF image of g-C_3N_4 without Au@Ag nanorods, while the plasmon-enhanced TPEF image of g-C_3N_4 covering Au@Ag nanorods was shown in (d). The clarity of the latter was significantly higher than the former, demonstrating that the Au@Ag nanorod-covered g-C_3N_4 could significantly enhance the TPEF signal.

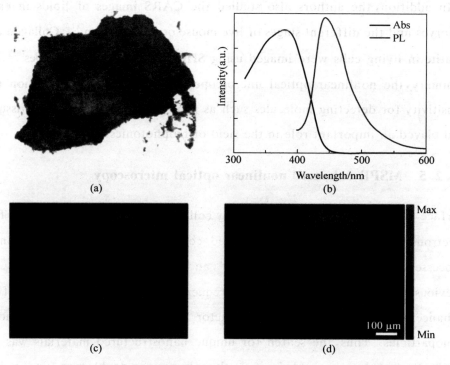

Fig. 11-20 Optical characterization of g-C_3N_4
(a) The bright field optical image of g-C_3N_4; (b) The absorption and PL spectra of g-C_3N_4; (c) The TPEF and (d) Plasmon-enhanced TPEF of g-C_3N_4 without and with the Au@Ag nanorods, respectively[58]

(For colored figure please scan the QR code on page 1)

To enhance Raman efficiency while reducing PL intensity, the authors measured plasmon enhanced CARS using the defect g-C_3N_4 at the N position on the tris-s-triazine. In Fig. 11-21, (a) ~ (d) showed the CARS and TPEF images of the defect g-C_3N_4 without Au@Ag nanorods coverage, while the corresponding CARS and TPEF patterns with Au@Ag nanorods were shown in (e)~(f). The TPEF image shown in (b) displayed that its PL intensity was significantly reduced, while its two-photon CARS signal was too weak to be detected, see (c). Au@Ag nanorods were covered on the defective g-C_3N_4, see (e), which significantly enhanced the resolution and intensity of CARS and TPEF. Studies had shown that Au@Ag nanorods significantly enhanced their CARS signal due to the relatively strong nonlinear electromagnetic field enhancement, see(g). (h) showed a combined image of the TPEF and CARS of the defective g-C_3N_4, revealing that the CARS signal was significantly enhanced.

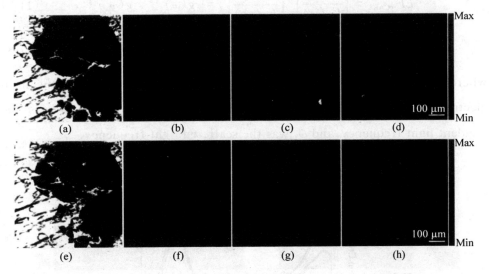

Fig. 11-21 Optical characterization of plasmon-enhanced g-C_3N_4
(a) The bright field optical image of defective g-C_3N_4. (b) The TPEF. (c) CARS and (d) Merged image of g-C_3N_4. (e) The bright field optical image of the defective g-C_3N_4 with the Au@Ag nanorods. (f) The plasmon-enhanced TPEF, (g) CARS and (h) Merged image of defected g-C_3N_4[58]
(For colored figure please scan the QR code on page 1)

To further demonstrate the above findings, the authors performed simulated electrical analysis of TPEF and CARS to control the MSPR model by adjusting the angle of incident light. Figure 11-22 showed the incident-angle-dependent extinction spectrum and the angle of incidence increased from 0° to 90°. It could be seen that the strong SPR peaks appearing at wavelengths of 400 nm and 800 nm corresponded to excitation along the short axis(90°) and the long axis (0°), respectively. A strong electromagnetic field enhancement was exhibited at the Au@Ag nanorods terminals when the incident angle was 0°, and the corresponding longitudinal mode was as shown in Fig. 11-23(a). When the incident angle was increased to 90°, the electric field on the two short sides of Au@Ag nanorods was obviously enhanced corresponding to the horizontal mode(Fig. 11-23(b)). The electric field enhancement factors of two-photon CARS and TPEF could be expressed as

$$EF_{tpCARS} = \left|\frac{E(\omega)}{E_0(\omega)}\right|^8 \left|\frac{E(\omega_s)}{E_0(\omega_s)}\right|^4 = |g(\omega)|^8 |g(\omega_s)|^4, \quad (11\text{-}7)$$

$$EF_{TPEF} = \left|\frac{E(\omega)}{E_0(\omega)}\right|^2 \left|\frac{E(\omega_s)}{E_0(\omega_s)}\right|^1 = |g(\omega)|^2 |g(\omega_s)|^1, \quad (11\text{-}8)$$

where, EF was the enhancement factor of electric field, $|g|$ was the local electromagnetic field enhancement about the probe molecule, ω was the incident light frequency, and ω_s was the scattered light frequency.

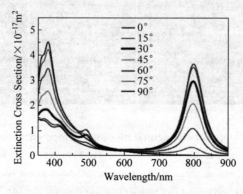

Fig. 11-22 Incident-angle-dependent extinction spectra from 0° to 90° with an increment of 15°[58]

(For colored figure please scan the QR code on page 1)

Studies had shown that when the angle of incidence was 30°, the strongest electromagnetic field enhancement was exhibited. The EFs of TPEF(Fig. 11-23 (c)) and CARS(Fig. 11-23(d)) of Au@Ag nanorods at 800 nm and 400 nm could be as high as 10^4 and 10^{16}, respectively.

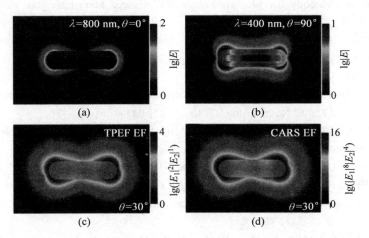

Fig. 11-23 (a),(b) Distributions of the electric field intensity of the Au@Ag nanorods with excitations of 800 nm and 400 nm, respectively; (c),(d) The respective distributions of the enhanced TPEF and CARS[58]

(For colored figure please scan the QR code on page 1)

In summary, MSPR revealed a plasmon-enhanced nonlinear optical microscope for two-photon CARS and TPEF. Studies had shown that at double excitation frequencies 800 nm and 400 nm, the optical signal of g-C_3N_4 was significantly enhanced. Therefore, MSPR could enhance the nonlinear optical signal strength, thereby improving the resolution of imaging.

References

[1] MAKER P D,TERHUNE R W. Study of optical effects due to an induced polarization third order in the electric field strength[J]. Phys. Rev. ,1965,137: A801.
[2] BEGLEY R F,HARVEY A B,BYER R L. Coherent anti-Stokes Raman spectroscopy [J]. Appl. Phys. Lett. ,1974,25: 387.
[3] OPILIK L,SCHMID T,ZENOBI R. Modern Raman imaging: vibrational spectroscopy on the micrometer and nanometer scales[J]. Annu. Rev. Anal. Chem. ,2013,6(1):

379-398.

[4] HONG W L, KARANJA C W, ABUTALEB N S, et al. Antibiotic susceptibility determination within one cell cycle at single-bacterium level by stimulated Raman metabolic imaging[J]. Anal. Chem. ,2018,90: 3737-3743.

[5] KAST R E, TUCKER S C, KEVIN K, et al. Emerging technology: applications of Raman spectroscopy for prostate cancer[J]. Cancer Metastasis Rev. , 2014, 33(2): 673-693.

[6] BRADLEY J, POPE L, MASIA F, et al. Quantitative imaging of lipids in live mouse oocytes and early embryos using CARS microscopy[J]. Development, 2016, 143(12): 2238-2247.

[7] LI R, WANG X X, ZHOU Y, et al. Advances in nonlinear optical microscopy for biophotonics[J]. J. Nanophoton. ,2018,12(3): 033007.

[8] EVANS C L, POTMA E O, PUORIS'HAAG M, et al. Chemical imaging of tissue in vivo with video-rate coherent anti-Stokes Raman scattering microscopy[J]. Proc. Natl. Acad. Sci. U. S. A. ,2005,102(46): 16807-16812.

[9] ROMEIKE B F M, MEYER T, REICHART R, et al. Coherent anti-Stokes Raman scattering and two-photon-excited fluorescence for neurosurgery[J]. Clin. Neurol. Neurosurg. ,2015,131: 42-46.

[10] LI X S, JIANG M J, LAM J W Y, et al. Mitochondrial imaging with combined fluorescence and stimulated Raman scattering microscopy using a probe of the aggregation-induced emission characteristic[J]. J. Am. Chem. Soc. ,2017,139(47): 17022-17030.

[11] AWASTHI S, LZU L T, MAO Z L, et al. Multimodal SHG-2PF imaging of microdomain Ca^{2+}-contraction coupling in live cardiac myocytes[J]. Circ. Res. ,2016, 118(2): e19-e28.

[12] KIM H, KIM D Y, JOO K, et al. Coherent Raman imaging of live muscle sarcomeres assisted by SFG microscopy[J]. Sci. Rep. ,2017,7(1): 9211.

[13] HU F H, LAMPRECHT M R, WEI L, et al. Bioorthogonal chemical imaging of metabolic activities in live mammalian hippocampal tissues with stimulated Raman scattering[J]. Sci. Rep. ,2016,6: 39660.

[14] JI M B, ORRINGER D A, FREUDIGER C W, et al. Rapid, label-free detection of brain tumors with stimulated Raman scattering microscopy[J]. Sci. Transl. Med. , 2013,5(201): 201-119.

[15] MANTULIN W W, MASTERS B R, SO P T C. Handbook of biomedical nonlinear optical microscopy[J]. J. Biomed. Opt. ,2009,14(1): 019901.

[16] CHENG J X, KEVIN JIA Y, ZHENG G F, et al. Laser-scanning coherent anti-Stokes Raman scattering microscopy and applications to cell biology[J]. Biophys. J. ,2002, 83: 502-509.

[17] SCHIE I W, WEEKS T, MCNERNEY G P, et al. Simultaneous forward and epi-CARS microscopy with a single detector by time-correlated single photon counting[J]. Opt.

Express,2008,16: 2168-2175.
[18] EVANS C L,POTMA E O,PUORIS'HAAG M,et al. Chemical imaging of tissue in vivo with video-rate coherent anti-stokes Raman scattering microscopy[J]. Proc. Natl. Acad. Sci. U. S. A. ,2005,102(46): 16807-16812.
[19] DUNCAN M D, REINTJES J, MANUCCIA T J. Effective reflectance of oceanic whitecaps[J]. Opt. Lett. ,1982,7: 350-352.
[20] ZUMBUSCH A,HOLTOM G R,XIE X S. Three-dimensional vibrational imaging by coherent anti-Stokes Raman scattering[J]. Phys. Rev. Lett. ,1999,82: 4142.
[21] CHENG X,XIE X S. Coherent anti-Stokes Raman scattering microscopy: instrumentation,theory,and applications[J]. J. Phys. Chem. B,2004,108: 827-840.
[22] CHENG X J. Coherent anti-Stokes Raman scattering microscopy [J]. Appl. Spectrosc. ,2007,61: 197A.
[23] MEDYUKHINA A, VOLGLER N, LATKA I,et al. 3D CARS image reconstruction and pattern recognition on SHG images[J]. Proc. SPIE,2012,8427H.
[24] LI R,WANG L,MU X J,et al. Nonlinear optical characterization of porous carbon materials by CARS,SHG and TPEF[J]. Spectrochimica Acta Part A: Molecular and Biomolecular Spectroscopy,2019,214: 58-66.
[25] LONG R,ZHANG L Y,SHI L Y, et al. Two-color vibrational imaging of glucose metabolism using stimulated Raman scattering[J]. Chem. Commun. ,2018,54(2): 152-155.
[26] SAAR B G,FREUDIGER C W,REICHMAN J,et al. Video-rate molecular imaging in vivo with stimulated Raman scattering[J]. Science,2010,330(6009): 1368-1370.
[27] FREUDIGER C W, MIN W, SAAR B G, et al. Label-free biomedical imaging with high sensitivity by stimulated Raman scattering microscopy[J]. Science, 2008, 322 (5909): 1857-1861.
[28] TIPPING W J, LEE M, SERRELS A, et al. Imaging drug uptake by bioorthogonal stimulated Raman scattering microscopy[J]. Chem. Sci. ,2017,8(8): 5606-5615.
[29] ITO T,OBARA Y, MISAWA K. Single-beam phase-modulated stimulated Raman scattering microscopy with spectrally focused detection[J]. J. Opt. Soc. Am. B,2017, 34(5): 1004-1015.
[30] ZHANG X,ROEFFAERS M B J,BASU S,et al. Label-free live-cell imaging of nucleic acids using stimulated Raman scattering microscopy[J]. Chem Phys Chem, 2012, 13(4): 1054-1059.
[31] YANG W L, LI A, SUO Y Z, et al. Simultaneous two-color stimulated Raman scattering microscopy by adding a fiber amplifier to a 2 ps OPO-based SRS microscope [J]. Opt. Lett. ,2017,42(3): 523-526.
[32] FRANKEN P A,HILL A E,PETERS C W,et al. Generation of optical harmonics[J]. Phys. Rev. Lett. ,1961,7(4): 118-119.
[33] LIU Z R,TANG Y H,XU A,et al. A new fluorescent probe with a large turn-on signal for imaging nitroreductase in tumor cells and tissues by two-photon microscopy

[J]. Biosens. Bioelectron. ,2017,89: 853-858.
[34] BOYD R W. The nonlinear optical susceptibility[M]. Burlington: Academic Press, 2008: 1-67.
[35] LI R,ZHANG Y J, XU X F, et al. Optical characterizations of two-dimensional materials using nonlinear optical microscopies of CARS, TPEF, and SHG [J]. Nanophotonics,2018,7(5): 873-881.
[36] KONG L R, MU X J, FAN X X, et al. Site-selected N vacancy of g-C_3N_4 for photocatalysis and physical mechanism[J]. Appl. Mater. Today,2018,13: 329-338.
[37] CAO S W,LOW J X, YU J G, et al. Polymeric photocatalysts based on graphitic carbon nitride[J]. Adv. Mater. ,2015,27(13): 2150-2176.
[38] CHANG S Q, XIE A Y, CHEN S, et al. Enhanced photoelectrocatalytic oxidation of small organic molecules by gold nanoparticles supported on carbon nitride[J]. J. Electroanal. Chem. ,2014,719(4): 86-91.
[39] WANG X L,FANG W Q, YAO Y F, et al. Switching the photocatalytic activity of g-C_3N_4 by homogenous surface chemical modification with nitrogen residues and vacancies[J]. RSC Adv. ,2015,5(27): 21430-21433.
[40] MENG Y L,SHEN J, CHEN D, et al. Photodegradation performance of methylene blue aqueous solution on Ag/g-C_3N_4 catalyst[J]. Rare Metals,2011,30: 276-279.
[41] YAN S C,LI Z S, ZOU Z G. Photodegradation of rhodamine B and methyl orange over boron-doped g-C_3N_4 under visible light irradiation[J]. Langmuir, 2010, 26(6): 3894-3901.
[42] ZHOU Y J,ZHANG L X, LIU J J, et al. Brand-new P-doped g-C_3N_4: enhanced photocatalytic activity for H_2 evolution and rhodamine B degradation under visible light[J]. J. Mater. Chem. A,2015,3(7): 3862-3867.
[43] BU Y Y,CHEN Z Y. Effect of oxygen-doped C_3N_4 on the separation capability of the photoinduced electron-hole pairs generated by O-C_3N_4 @ TiO_2 with quasi-shell-core nanostructure[J]. Electrochim. Acta,2014,144: 42-49.
[44] LI Q,ZHANG N, YANG Y, et al. High-efficiency photocatalysis for pollutant degradation with MoS_2/C_3N_4 heterostructures [J]. Langmuir, 2014, 30 (29): 8965-8972.
[45] WANG X J,YANG W Y,LI F T,et al. In situ microwave-assisted synthesis of porous N-TiO_2/g-C_3N_4 heterojunctions with enhanced visible-light photocatalytic properties [J]. Ind. Eng. Chem. Res. ,2013,52(48): 17140-17150.
[46] VAN STRYLAND E W, VANHERZEELE H, WOODALL M A, et al. Two-photon absorption,nonlinear refraction and optical limiting in semiconductors[J]. Opt. Eng. , 1985,24: 613.
[47] POTCOAVA M C,FUTIA G L,AUGHENBAUGH J,et al. Raman and coherent anti-Stokes Raman scattering microscopy studies of changes in lipid content and composition in hormone-treated breast and prostate cancer cells[J]. J. Biomed. Opt. , 2014, 19(11): 111605.

[48] ISHITSUKA K, KOIDE M, YOSHIDA M, et al. Identification of intracellular squalene in living algae, aurantiochytrium mangrovei with hyper-spectral coherent anti-Stokes Raman microscopy using a sub-nanosecond supercontinuum laser source[J]. J. Raman Spectroscopy, 2016, 48(1): 8-15.

[49] DAEMEN S, VAN ZANDVOORT M A M J, PAREKH S H, et al. Microscopy tools for the investigation of intracellular lipid storage and dynamics[J]. Mol. Metab., 2016, 5(3): 153-163.

[50] DI NAPOLI C, POPE L, MASIA F, et al. Quantitative spatiotemporal chemical profiling of individual lipid droplets by hyperspectral CARS microscopy in living human adipose-derived stem cells[J]. Anal. Chem., 2016, 88(7): 3677-3685.

[51] FURUTA R, KURAKE N, ISHIKAWA K, et al. Intracellular-molecular changes in plasma-irradiated budding yeast cells studied using multiplex coherent anti-Stokes Raman scattering microscopy [J]. Phys. Chem. Chem. Phys., 2017, 19 (21): 13438-13442.

[52] LIU H H, QIN W, MA Z, et al. Myofibrillogenesis in live neonatal cardiomyocytes observed with hybrid two-photon excitation fluorescence-second harmonic generation microscopy[J]. J. Biomed. Opt., 2011, 16(12): 126012.

[53] HOFEMEIER A D, HACHMEISTER H, PILGER C, et al. Label-free nonlinear optical microscopy detects early markers for osteogenic differentiation of human stem cells[J]. Sci. Rep., 2016, 6: 26716.

[54] CAMPAGNOLA P J, CLARK H A, MOHLER W A, et al. Second-harmonic imaging microscopy of living cells[J]. J Biomed Opt., 2001, 6: 277-286.

[55] CHEN C K, CASTRO A R B, SHEN Y R. Surface-enhanced second-harmonic generation[J]. Phys. Rev. Lett., 1981, 46: 145-148.

[56] SANCHEZ E J, NOVOTNY L, XIE X. Near-field fluorescence microscopy based on two-photon excitation with metal tip[J]. Phys. Rev. Lett., 1999, 82: 4014-4017.

[57] ICHIMURA T, HAYAZAWA N, HASHIMOTO M, et al. Tip-enhanced coherent anti-Stokes Raman scattering for vibrational nanoimaging[J]. Phys. Rev. Lett., 2004, 92: 220801.

[58] MI X H, WANG Y Y, LI R, et al. Multiple surface plasmon resonances enhanced nonlinear optical microscopy[J]. Nanophotonics, 2019, 8(3): 487-493.

致谢

本教材获得北京科技大学研究生教材建设项目资助。